HOW THE
WORLD BREAKS

HOW THE WORLD BREAKS

Life in Catastrophe's Path, from the Caribbean to Siberia

Stan Cox and Paul Cox

THE NEW PRESS

NEW YORK
LONDON

Requests for permission to reproduce selections from this book should be mailed to: Permissions Department, The New Press, 120 Wall Street, 31st floor, New York, NY 10005.

Published in the United States by The New Press, New York, 2016
Distributed by Perseus Distribution

ISBN 978-1-62097-012-6 (hc)
ISBN 978-1-62097-013-3 (pbk)
CIP data is available

The New Press publishes books that promote and enrich public discussion and understanding of the issues vital to our democracy and to a more equitable world. These books are made possible by the enthusiasm of our readers; the support of a committed group of donors, large and small; the collaboration of our many partners in the independent media and the not-for-profit sector; booksellers, who often hand-sell New Press books; librarians; and above all by our authors.

www.thenewpress.com

All illustrations by Priti Gulati Cox
Composition by dix!
This book was set in Minion

Printed in the United States of America

10 9 8 7 6 5 4 3 2 1

To the memory of Lucille Brewer Cox
and all the others who lost their lives in Gainesville, Georgia,
and Tupelo, Mississippi, in the tornadoes of April 6, 1936

CONTENTS

Acknowledgments   ix

Introduction   1

1. Fire Regimes: Australia and Siberia   15

2. Leave It Up to Batman: The Philippines   38

3. Neighbors to the Sky: New York City   64

4. Every Silver Lining . . .   88

5. Gray Goo: East Java, Indonesia   122

6. How to Booby-Trap a Planet   143

7. Foreshock, Shock, Aftershock: L'Aquila, Italy   160

8. Atlantis of the Americas: Miami, Florida   175

9. Engineer, Defend, Insure, Absorb, Leave   198

10. The Absorbers: Mumbai, India, and Kampala, Uganda   224

11. Vulnerability Seeps in Everywhere   238

12. Keeping the Lights On: Montserrat, West Indies   257

13. "We Do Things Big Here": Greensburg, Kansas,
        and Joplin, Missouri   278

14. When Mountains Fall: Uttarakhand State, India   298

    Epilogue: Rainbow of Chaos   317

Notes   325

Index   387

ACKNOWLEDGMENTS

For the invaluable help and insight they provided us, we want to thank John Araneta, Francisco Artigas, Florian Boer, David Bowman, Henry Briceño, Richard Davies, Celeste de Palma, Gudi Devi and Kunwar Singh, Mauro Dolce, Philip Drake, Colin Foord, Vineet Gahalaut, Charlie and John Giordano, Goldi Guerra, Nicole Hernandez Hammer, Peter Harlem, Ilya Jalal, Michael Jarvis, Leigh Johnson, Sachin Kadam, Phil and Christie Le Breton, Sharon Louden and Vinson Valega, Lilah Mejia, Arnoud Molenaar, Marilys Nepomeche, Delacey Verne Peter, Amy Peterson, Seetharam, Sherwin, Nadja Tchebakova, Harold Wanless, Judith Weis, Zelma White, and Andrew Whitehead. Our warmest gratitude for exceptional guidance as we worked to understand and find our way around the world of disaster goes to Stacy Barnes, Judith Buhay, Warren Cassell, Chris Cook, Bob Dixson, Elena Kukavskaya, Ted Lefroy, Janine Lester, Warner Marzocchi, Gean Moreno, Bruce Mowry, Tony Mukiibi, Pheonah Nabukalu, Debra Parkinson, Kerry Paul, Evgeniy Ponomarev, Dwinanto Presetyo, Hardi Prasetyo, Gray Reid, Donaldson Romeo, Phil Stoddard, Mark Tingay, and Adarsh Tribal. We are deeply, deeply thankful to Jed Bickman and Sarah Fan at The New Press, who, in the bleak aftermath of Sandy, made this book possible, and then two and a half years later made it a much better book. This would be a different and much poorer book without the unequaled talent and countless hours that Priti Gulati Cox invested in the making of maps that are both beauteous and terrifying. And as always, our warmest, deepest thanks—for love and support through the writing of this book and the many years before that—go to our families: Vinita and Cmde. Sahi, Santosh Gulati, Maurya and Bob Farah, Brenda and Tom Cox, Marylaverne and Jerry Bramel, Paula Bramel and John Peacock, Sheila Cox, and most of all the respective loves of our lives, Priti Gulati Cox and Amanda Farah.

HOW THE
WORLD BREAKS

INTRODUCTION

Ramala Khumriyal ran a small eatery and shop for pilgrims next to Kedarnath Temple, one of the holiest shrines to Shiva in all of India. He would spend his summers working here with his children, high in the Himalaya ten miles from his hometown, Guptkashi. Then in June 2013, after a period of extraordinarily intense rain, a natural glacial dam melted and collapsed above Kedarnath, emptying a swollen lake and bringing 15 million cubic feet of freezing sludge down upon the settlement. Khumriyal's shop was just high enough on the mountain slope above the temple for him to see catastrophe coming. Turning his back on the muddy, rocky tide swallowing up thousands of people below him, he had time to dash upslope into the forest, pulling his six children with him.

Up on the mountainside, it was still raining, and the chill was intense. Bodies were strewn everywhere. Those still alive seemed to be in an altered state, moving as if drunk, some collapsing right in front of Khumriyal. Many survivors spoke of a poison gas that they believed came out of the earth, choking some people and killing others; Khumriyal told us that he and his children felt severe shortness of breath.[1] Having escaped the flood only to plunge himself and his children into the unknowns of the forest, he began wondering if they might have been better off dying down below. The only way back to the nearest town, Gaurikund, was to walk, or stumble, cross-country through some of the world's steepest and most rugged landscapes. What in normal times would have been a ten-mile trek took, by his guess, twenty-five miles of meandering to avoid cliffs, landslides, and flash floods.

When he and his children reached what was left of Gaurikund, it was already thronged with survivors, and all were trapped. The road down the mountain was gone. The disaster was much larger than Khumriyal knew: the extreme burst of rain and floods had scoured villages and collapsed mountainsides far, far down the Mandakini River and the rivers it joined, all the way into the plains, where it swelled the mighty Ganges. In the stranded town there was little food other than a few cookies, packets of which were selling for four times the usual price. Tens of thousands of tourists and pilgrims waited, hoping for rescue helicopters to arrive. Others, including Khumriyal and his family, became impatient and set out walking again.

When they reached the swollen Vasukiganga River, a tributary of the Mandakini, the bridge was completely gone. A tall tree, however, had caught on the rocks in such a way that its trunk just reached the other bank. Though half submerged in the torrent, it was the only hope for getting across. Evacuees were lining up to make the crossing. Out of each group of half a dozen or so people who attempted to cross on the tree, Khumriyal estimates that two were taken by the flood. He saw the crossing attempt as the last of many tests that he and his family had been given by Lord Shiva in the past days; the presiding deity of Kedarnath, he believed, was separating the pure from the dead. Khumriyal's whole family passed the test, and they made their way safely into the town of Sonprayag, which offered some food and shelter despite being largely buried, and eventually found a way back to their home in Guptkashi. The journey home had taken six days.

Seven months later, Khumriyal had become a partner in a small soap-making business in Gaurikund with assistance from the organization iVolunteer, which fosters small-scale development. Asked if he or others who had previously served the tourist industry in Kedarnath would return for the next season, he answered, "People who had a child or family member die there surely don't want to return. They know they can make money there, but is it worth the lives of their family?" He was hopeful his new venture would work out so that he would never have to go back to Lord Shiva's abode. "I think about it every day. I have nightmares about it. Even if I went back up the mountain with you, I would not go beyond Sonprayag."[2]

Khumriyal's escape was a transformative journey, a pilgrimage in reverse. His family braved it without technology, assistance, or any defense other than courage and devotion. They survived Lord Shiva's trial, and it remade them. For Khumriyal, the transformation was a profoundly personal and spiritual one. This was his understanding. But the task of interpreting disasters never stops with the survivors themselves.

Others count Khumriyal's family as part of a global trend: in the Internal Displacement Monitoring Centre's annual *Global Estimates* report they appear unnamed, in aggregate, as some of the 22 million people displaced by natural hazards worldwide in 2013.[3] The vast majority of these displacements were the result of cyclonic storms and floods that, like the devastating Himalayan deluge, are more and more being supercharged by human-induced climate change. Other commentators, within the region and without, point to the contributions made to the disaster by the increasingly infrastructure-heavy pilgrimage and tourism business, the dam-building hydroelectric sector, and other activities to extract capital from the fragile mountains. In this interpretation, climate change was only one way in which Khumriyal's family were victims of a voracious global economic system. (See Chapter 14 for more on the Himalayan disaster and these factors.)

Yet other interpreters refuse to see Khumriyal and his children as victims at all. The journey through the charnel grounds, they say, led from a place of vulnerability to a place of resilience. Now Khumriyal has a new life and a new livelihood, high on a ridge out of the flood zone. To believers in resilience, his remarkable story of survival reveals not victimhood but a heroism and adaptability we should all learn from as we venture into an uncertain future. This is empowering, but it's a very particular sort of empowerment, one that offers little guidance out of the mountains' enchanted chaos.

READING THE RUINS

There once were events called natural disasters. Unpredictable, blind, destructive, they sprang from the planet itself and taught us humility. The word "natural" served to turn attention away from the *why* of these disasters, keeping it on the *how* of earth science. Then, in the late twentieth century, the social sciences began to chip away at the received wisdom on the subject. An alternative view left purely natural explanations behind and started to consider disasters as equal parts earthly force and human choice.[4] Nature is full of hazards, researchers said, but only some hazards wreak disaster. It is human-built structures, not the shaking ground, that kill when an earthquake strikes; large populations live and work in low-lying areas where flooding is almost a certainty; a lack of preparedness and safe shelter amplifies the impact of tornadoes and tropical cyclones; well-intentioned forest management fuels bigger fires; poverty puts people in even greater danger; and either devastated communities get help to survive and recover or they don't. Attention shifted to the vulnerability of people to hazards. In the words of the anthropologists Anthony Oliver-Smith and Susanna Hoffman, "A disaster is made inevitable by the historically produced pattern of vulnerability, evidenced in the location, infrastructure, sociopolitical structure, production patterns, and ideology, that characterizes a society," and these social, political, and economic conditions have far more influence on the scale and severity of the disaster than the physical force of the hazard itself.[5]

These and many more attacks on the concept of the natural disaster create difficulty in naming the subject of this book. The category we are exploring is one we no longer believe in; we're writing about the disasters formerly known as natural. The "all natural" label may have been a misdirection from the start, but peeling it off is tricky, for one reason in particular: everyone knows what a natural disaster is. The words carry the weight of common sense, neatly defining catastrophes that seem to arise unbidden from the Earth itself. In practice, too, most attention and effort are still directed not at the human root causes but at erecting barriers against physical hazards, rescuing victims, providing emergency relief, and rebuilding.

While we wait for a good alternative to come along, most people still talk about natural disasters, and experts do their best to talk around the issue. The United Nations, through its Office for Disaster Risk Reduction (UNISDR), now adheres to a definition of disaster that is far too open-ended to settle the matter: "A serious disruption of the functioning of a community or a society involving widespread human, material, economic or environmental losses and impacts, which exceeds the ability of the af-fected community or society to cope using its own resources." Taken at face value, this could describe anything from a tsunami to a bombing campaign to a burst economic bubble. In practice the UNISDR has a clear mandate over certain risks because of its place in the UN system; other agencies cover conflict (the Security Council and peacekeeping missions), disease (World Health Organization), crop loss (Food and Agriculture Organization), and so on.

Not being part of the UN system, we need to provide an inventory of the types of disasters that will be the subject of this book. From a broad spectrum, we have chosen sudden, dramatic events precipitated by haz-ards of geology, climate, or physical geography—let's call them geoclimatic hazards—that have the power to shift earth, water, atmosphere, and com-bustible fuel.[6] In the ten stories that follow, we primarily cover wildfires, tornadoes, tropical cyclones, floods, landslides, earthquakes, and volcanic eruptions. In intervening chapters we also mention tsunamis, avalanches, winter storms, dust storms, heat waves, droughts, and a few others.[7] In de-voting less space to this second set of hazards, we are not denying their importance. There has been much excellent research and writing done on all of them—too much for one book, it turns out.

We have not included strictly industrial disasters in this book, but we do devote significant attention to hazards that fall into an opaque area some-where between industry and nature, resulting from the collision of geocli-matic and economic forces. Disputes over assignment of blame for such hybrid hazards can have a devastating impact on response and recovery and can hamper attempts to prevent future disasters. As we will see, the label of natural disaster is frequently all about blame—or, rather, about deflecting blame. Unfortunately for the guilty, scientists are learning more every year about how much human activity can contribute to "natural" hazards, too.[8]

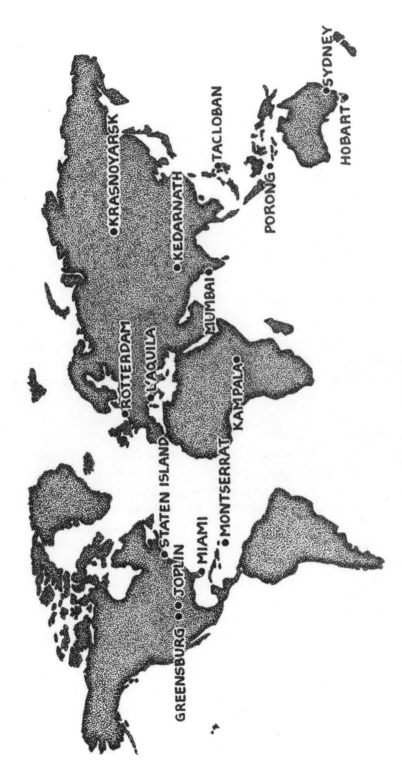

Figure 1. Sites of some of the geoclimatic disasters discussed in this book.

This didn't start with climate change, but it may end there. Climate disruption has collapsed the last walls between the human and the natural—and the storms are growing.

Even many comparatively recent books on disaster come with disclaimers and equivocations regarding the contribution of anthropogenic climate change to the frequency and intensity of extreme climatic and meteorological events. We wish to offer no such concessions to climate change denial. The scientific evidence is easy to find, and it is strong enough that we think the time has come to take a stormier future as a given and ask what comes next.[9] In fact, we believe climate change denial has entered its sunset years, and its "merchants of doubt"—to use the memorable title bestowed by historians Naomi Oreskes and Erik Conway—are as ready as we are to move on.[10] Our greater concern is what strategy the status quo will marshal next.

MERCHANTS OF OPTIMISM

The controversial political scientist Roger Pielke Jr. tends to get labeled as a climate change denier by his critics, but he is really just an optimist. In a typically contrarian piece on the data journalism site FiveThirtyEight, Pielke challenged the idea that storms were getting worse because of climate change—or even that they were getting worse at all:

> When you read that the cost of disasters is increasing, it's tempting to think that it must be because more storms are happening. They're not. All the apocalyptic "climate porn" in your Facebook feed is solely a function of perception. In reality, the numbers reflect more damage from catastrophes because the world is getting wealthier. We're seeing ever-larger losses simply because we have more to lose—when an earthquake or flood occurs, more stuff gets damaged. . . .
>
> There is some good news to be found in the ever-mounting toll of disaster losses. As countries become richer, they are better able to deal with disasters—meaning more people are protected and fewer lose their lives. Increased property losses, it turns out, are a price worth paying.[11]

Climate scientists lined up to attack the piece for its highly selective use of data, and the fight got so ugly that FiveThirtyEight dropped Pielke as a contributor. Nevertheless, his gospel of protection through wealth is a popular one among fellow optimists. It would be wonderful if the world really could buy its way out of disaster without even having to make big investments. Simply carrying on with the business of growth, in Pielke's view, could tip the balance of loss away from human lives toward the loss of humans' stuff. Disaster could be domesticated, soaked up by the economy, so we the people could all experience the event as something distant and manageable, canceled out on future balance sheets by its silver linings.

As an approach to protecting us against the threats we face, this comes with three implicit conditions. First, to the extent that generating this insulating prosperity depends on burning a lot more carbon, Pielke is asking us to ignore or minimize the connection between CO_2 in the atmosphere and more, bigger disasters. Second, the wealth solution is predicated on greater equality—that the rising tide will lift all boats and swamp none—even though growing income and wealth gaps are our reality, and the poorest still live in the most hazardous spots. Third, if the future really is bringing lower death counts and higher costs from disasters, it will become more and more necessary to ask who pays those costs.[12]

This type of optimism is, we believe, what we will have to worry about when we don't have to worry about climate change denial anymore. Economists in the United States and other wealthy countries have concluded that any geoclimatic disaster can be absorbed without long-term negative effects on the growth of their own economies.[13] But the fact that societies can take a blow and then quickly return to their growth trajectory is not a good measure of strength. From the point of view of those with vast wealth at stake, the cure for climate catastrophe—deep, ongoing restraint in production and consumption to limit greenhouse gas emissions—would be far more devastating than the worst earthquake, flood, or hurricane. Meanwhile, attempts by nations of the North to address disasters in the South almost always turn out to be extensions of the same development strategies that are already generating crisis after crisis.[14]

Not all economists are convinced by models that provide for the full domestication of disaster.[15] Nicholas Stern, lead author of the well-known 2007 Stern Report on the economics of climate change, has since characterized as inadequate all past attempts at weighing the costs of climate change mitigation versus the costs of doing nothing.[16] To Stern, the latter are badly underestimated. One of the most obvious of many problems is that the economic models project the impacts of greenhouse warming against a background of uninterrupted economic growth. That assumption of growth, which is believed to provide a buffer against disaster, is based entirely on circular reasoning: the argument that reduction in economic growth, regarded as the most undesirable outcome of climate change, can be avoided only if steady growth can be maintained indefinitely. Unending growth would require infinite physical resources and is clearly impossible. But we, along with Stern, believe that even before resources are exhausted growth will falter badly. He recommends that economic modelers "embrace a real possibility of creating an environment so hostile that physical, social, and organizational capital are destroyed, production processes are radically disrupted, future generations will be much poorer, and hundreds of millions will have to move." [17] Such prospects, Stern believes, smash a hole in all current models of the economic impacts of climate disruption. Neither the fact that progress has been made in reducing mortality from disasters nor the fact that annual economic damages have so far been held to an acceptable fraction of GDP will bring any solace to the disaster victim of the present or future who is wading through sewage-laced floodwaters, has lost everything to a landslide, has had to flee to a refugee camp far from home, is living under martial law, has descended into perpetual debt, or has been sucked into a black hole of poverty.

Books by writers across a broad spectrum of disciplines and political orientations—including Nicholas Georgescu-Roegen, Herman Daly, Wes Jackson, John Bellamy Foster, Joel Kovel, Naomi Klein, James Gustav Speth, Richard Heinberg, Ted Trainer, Tim Jackson, and others [18]—have demonstrated that the necessity for economic growth that lies at the heart of capitalism poses a grave threat to the planet and cannot be sustained. Humanity must deeply reduce the world economy's throughput—that is, the quantity of material flowing in as resources and flowing out as wastes—and the degradation of the planet's ecosystems. Doing that will, in turn, inevitably hobble or even reverse economic growth. This is why international attempts to stop greenhouse warming have been totally hamstrung. The big, wealthy nations and corporations that have created the climate crisis claim that they will do whatever is required to resolve it, as long as they are not deterred from their pursuit of wealth—which is the trigger of the crisis itself. Therefore, they cannot be expected to do anything effective. What Naomi Klein has written of professional climate change deniers is true of virtually all of the major powers' government and business leaders, with their expressions of boundless faith in ingenuity and opportunity: "their deep fear is that if the free market system really has set in motion physical and chemical processes that, if allowed to continue unchecked, threaten large parts of humanity at an existential level, then their entire crusade to morally redeem capitalism has been for naught." [19] Thus, whether or not they publicly acknowledge that the global ecological crisis is real, they can do nothing effective to prevent catastrophe without undermining the economic system that is the source of their power.

That said, the deep need is for more than simply curbing growth—it is for climate justice, which would encompass not only reduced emissions from the North but also, in the words of the Institute for Social Ecology's Brian Tokar, "reparations for the ecological and climate debts owed by the richest countries to those most affected by resource extraction and climate-related disasters."[20] Climate justice could also provide a foundation for ensuring protection and restoration to people threatened by disasters of all kinds.[21] The chapters that follow, which tell the stories of recent geoclimatic catastrophes and the people who have struggled with the consequences, document some moments in the blossoming of a disaster justice movement in both the South and the North.[22]

FROM DISASTER TO RESILIENCE AND BACK

It is right and just that cost-benefit analyses take a back seat to loss of life, which is always the first measure of any disaster. This tradition affirms that every life has high, and equal, value. But it falls short, and stands to fall even shorter, if we define a life only in the binary, as a death avoided. What is also lost is the share of life that has to be invested in repair. First there has to be survival; then mutual aid; then reconstructing home, body, psyche, and community; then paying off the debt of lost livelihood incurred in the years of recovery. It's difficult to calculate this sort of toll and even harder to quantify human misery, which is why survivors' accounts are another tradition in the reporting of disaster. The headline gives the death toll, but tales from within the disaster complicate the binary of life and death.

Without such stories, another form of faith may keep us too comfortable on our descent into a catastrophic future: the faith in resilience. In a short span of years, resilience has been deployed everywhere and defined too seldom. Its ring of familiarity (partly due to an older, not quite related usage in psychology) makes it sound like a self-evident virtue that has been around forever.[23] However, its use in the neoliberal governance of disaster draws on a particular definition that emerged from the study of ecosystems in the 1970s (which we'll explore in Chapter 1).

Its current shape can be seen in the UN's Sendai Framework for Disaster Risk Reduction 2015–2030, the preeminent international action plan signed by 186 governments. This employs the UN definition of resilience as "the ability of a system, community or society exposed to hazards to resist, absorb, accommodate to and recover from the effects of a hazard in a timely and efficient manner, including through the preservation and restoration of its essential basic structures and functions."[24] In this definition we can still make out the language of ecology, as human resilience mirrors the natural ability of a forest or wetland to return to balance after a disruption. Resilience's utility for the policymaker lies in what it leaves out, the two factors that are absent in natural ecosystems because they are unique to human systems: labor and political power. Political power gives those who wield it the ability to evacuate to the high ground of setting societal priorities, while others are left behind to be resilient, to do the hard and dirty work of survival.

The UN's promotion of a universal language of disaster resilience began with a 2002 review of success stories titled *Living with Risk*. In it, the UN Office for Disaster Risk Reduction described the labor that will be required of the hazard-exposed. They should create "a healthy and diverse economy that adapts to change and recognizes social and ecological limits" and captures "opportunities for social change during the 'window of opportunity' following disasters, for example by utilizing the skills of women and men equally during reconstruction." Most important, "it is crucial for people to understand that they have a responsibility towards their own survival and not simply wait for governments to find and provide solutions." Furthermore,

> this requires understanding and accepting the values of changed behavior, having access to the necessary technical and material resources, and accepting personal responsibilities to carry through the efforts involved. Communities are frequently inattentive to the hazards they face, underestimate those they identify, and overestimate their ability to cope with a crisis. They also tend not to put much trust in disaster reduction strategies and rely heavily on emergency assistance when the need arises. These viewpoints underline the need for tools to create a culture of prevention against all forms of hazards within communities.[25]

To address these communities' putative shortcomings, says the disaster establishment, resilience thinking must be instilled as a cultural value at all levels. But resilience doctrine may have finally ventured too far, stepping into a minefield of critical analysis. Theorists Brad Evans and Julian Reid, in their book *Resilient Life: The Art of Living Dangerously*, see the current demotion of human beings to the status of "resilient subjects" as part of the dark, even nihilistic value system of a future that has abandoned all pretense of higher goals:

> To be resilient, the subject must disavow any belief in the possibility to secure itself and accept instead an understanding of life as a permanent process of continual adaptation to threats and dangers which are said to be outside its control. It is also about "thriving" in times of unending chaos without losing the faculty of neoliberal reason. As such the resilient subject must permanently struggle to accommodate itself to the world. The resilient subject is not a political subject who on its own terms conceives of changing the world, its structure and conditions of possibility. The resilient subject is required to accept the dangerousness of the world it lives in as a condition for partaking of that world and accept the necessity of the injunction to change itself in correspondence with threats now presupposed as endemic and unavoidable.[26]

Geographer Kevin Grove has explored a real-life localized disasterscape of this kind in Jamaica, where he participated in a series of foreign-funded community resilience initiatives with the country's Office of Disaster Preparedness and Emergency Management (ODPEM). He describes it as "an immaterial, therapeutic disaster management strategy that requires people to use what is around them—environmental conditions, social relations, and so forth—in new and creative ways." The strategy calls on local shopkeepers to provide food and water on credit in time of emergency, on residents who own chainsaws to help clear roads, and on people with medical training to assist with rescue. Through these and other mechanisms, writes Grove, ODPEM seeks to "set in motion an 'adaptation machine,' the resilient community that automatically responds to environmental insecurities without external intervention."[27]

Everyone is incorporated into such an adaptation machine. Resilience turns the debate of past decades, which primarily asked why some people are more vulnerable than others, on its head, transforming the powerless into the powerful by pure fiat.[28] For five years running, the United Nations devoted its International Day for Disaster Reduction, held on October 13, to various wellsprings of resilience: children and young people (2011), women and girls (2012), people living with disabilities (2013), older people (2014), and indigenous peoples (2015). Rather than simply focusing on the inordinate structural vulnerabilities to disaster imposed on these populations in nearly every society, the proceedings put rhetorical emphasis on their

possible contributions. Children and young people "can and should be encouraged to participate in disaster risk reduction and decision making,"
women and girls "are powerful agents of change," the disabled are over-
looked in "their unique contribution to helping communities prepare for
and respond to disasters," the elderly play a "critical role . . . in resilience-
building through their experience and knowledge," and indigenous peoples
are stewards of traditions that "complement modern science and add to an
individual's and societies' resilience." [29]

   This empowering positivity is a hallmark of UN prescriptions, but
here it reveals much more. In Grove's line of work, "adaptive capacity is no
longer something limited by structural constraints such as race, class, or
gender inequalities, it now depends on individuals' psychological disposi-
tions and the wider cultural belief systems that affect their perceptions of
self-efficacy." [30] The dominant cultural belief system is one of inversions;
one in which crisis brings out the power of the powerless; one in which
every storm has a silver lining; one enamored of Friedrich Nietzsche's spurious claim that "what does not kill me makes me stronger." [31] Resilience, the
tough love dispensed by adversity, has become the brightest of silver linings.

   In the United States, for instance, these inversions are being used as
therapy for a broken prison system. As the Forest Service cuts back on its
workforce and pushes the work of fighting wildfires onto states and local
jurisdictions, prison crews have taken on much of the work. California pays
inmates a dollar an hour to work in fires and floods, saving the state $80 mil-
lion per year, while Arizona offers an hourly wage of 50¢ for firefighting.
When the *New York Times* reported on the trend in 2013, the reporter's interest was not in the hard work or the risk (five inmates and a correctional
officer died in a 1990 fire in Arizona) but in the benefits to prisoners.

> While inmates' pay is low, there are other rewards. Arizona in-
> mates work outdoors much of the year, and if they are out fight-
> ing fires, their status as inmates is not easy to discern. In California,
> inmates wear orange fire-retardant jumpsuits and sleep in separate
> camps when they are out on fire lines. But firefighting inmates in
> Arizona wear the same clothes as other wilderness firefighters. . . .
> Grant Lovato, 46, who worked two seasons with the Lewis crew
> while serving time in prison for credit card fraud and identity theft,
> is now a firefighter for the United States Fish and Wildlife Service
> in Gulfport, Miss. In a telephone interview, he said, "The idea that I
> could be in prison and still get out into the woods and get to enjoy
> the nature was a transformative experience to me." [32]

California deployed approximately four thousand prisoners to help battle the epic wildfires that gripped the state in 2015. It's a long-standing tradition in a state with big fires and big, overcrowded prisons. All of those deployed were volunteers, and despite the jumpsuits, a reporter from the BBC also found inmates who were glad for the experience. One was Henry Cruz, an inmate at San Quentin prison, who said, "It gets scary sometimes, but at the same time, it makes me feel good. Being a firefighter is a privilege—it makes you feel like you are in civilization. . . . I like saving nature, and sometimes people. It makes me feel like a hero."[33]

The old ideas of social leveling are at work here, turning subalterns and outcasts into heroes. Disaster transforms. This potential is rooted in the human imagination, but it even cross-fertilizes with the ecological concepts out of which resilience ideas arose. See this passage from a 2001 introductory book on the subdiscipline of disturbance ecology, cheerfully titled *The Silver Lining: The Benefits of Natural Disasters*:

> Reconsider the Yellowstone fires. Walt Disney had it right in *Bambi*. The environment was ablaze. It was the fire that made the animals run. In the film, we saw predators and prey all mixed together, fleeing before the flames. Foxes and rabbits ran away from the heat, together. When the great fire sweeps through the forest, no one stops to eat; all notions of competition and predation are gone. Survival is the order of the day. Disturbances change all the rules. After the fires cooled, the entire community of Yellowstone was restructured.[34]

One can't help but think there's a lot of cultural projection going on here—and not just because *Bambi* is the author's case study. Predators, like prisoners, cease to be themselves in the fire. Orange jumpsuits and orange fur fade away in the glow of the inferno. The only difference is that the foxes aren't heroes, just survivors.

CONSERVATION OF FRAGILITY

Stories that celebrate resilience, like the stories that naturalize it, are key to the operation of adaptation machines. Labor in the disasterscape is at least a matter of survival, and at most a matter of universal humanity—and in neither case does the concept of a fair wage hold much water. So if we're asking vulnerable communities to be the source of resilience, this is what we're asking of them: to work constantly toward the capacity to absorb shocks and changes so that the rest of us don't have to worry about those shocks and changes, and we can keep generating more of them. It's the expendability of their time and enjoyment of life that makes all of global capitalism resilient.[35] This link between fragility and strength may seem paradoxical, but it's exactly how resilience works—what theorists call its modularity. As explained by four researchers with the international Resilience Alliance network, including Nobel Prize–winning economist Elinor Ostrom, "Transformation at one scale in a system, which may be related to an inherent fragility in a system module, is a necessary part of maintaining resilience at other scales in the system."[36] When individual "modules" are toughened up, the system as a whole becomes insensitive to shocks, setting itself up for collapse when the big one hits.

This principle, which has been called the "conservation of fragility," was identified in biological systems and has now been applied to the social and economic realms.[37] It is a reminder that resilience can be described only within a specific place and time. Resilient parts don't add up to a resilient whole. Although the word "resilience" is at times used interchangeably with "sustainability," we see that it's very different. A sustainable household makes a city more sustainable, which makes a nation more sustainable, which makes our species' whole tenure on Earth more sustainable. Resilience isn't like that. A society can be highly resilient for a century but destroy its resource base in the process, ruining its next century. A community can bounce back from a catastrophe by virtue of outside support—say, from federal flood insurance— but enough such events can sink the insurance program. Conversely, a vulnerable community's destruction can create a buffer zone, or it can prompt more widespread action to confront the hazard. It's necessary to examine resilience at a specific scale but, in a world where cross-scale subsidies and feedbacks are what really matter, it's also insufficient.[38]

This image of how the world breaks—not all at once, but in millions of cataclysms small and large, which protect and power the whole global system—is something that we found through the stories that make up most of this book. Of the people who are spending their lives patching over these tears in the fabric of the world, we think it's an injustice to argue over whether they are heroes or victims. They are both, and choosing one designation over the other is a very deliberate act of reductive representation. Heroes don't blame others when calamity strikes; victims don't manage to find a way out of their vulnerability. Those who suffer unnatural disasters do both, as required, and that is the way the world will heal.

1

FIRE REGIMES

Australia and Siberia

In the beginning, a long time ago, the creatures of the world did not have their present shape. One day Ngundid the snake, Mulili the catfish, Galaba the kangaroo and all other beings gathered in the country of the Rembarrnga people. From Maningrida way came Nagorgo, the Father, and his son Mulnanjini. They looked at all the creatures and said, "You are not proper people and not proper animals. We must change this." Then they made a ceremony which is still performed today. They lit a fire with their fire sticks which quickly spread until it engulfed all the beings and scorched the earth and rocks. When the fire subsided all the creatures found that animals and humans had lost their strange features and looked as they do now.

—Jennifer Isaacs, *Australian Dreaming*, 1980[1]

In October 2013 the Blue Mountains burned. In one afternoon, fires destroyed or seriously damaged 473 homes on this stunning spur of Australia's Great Dividing Range edging the basin of Greater Sydney. Firefighters worked to contain three huge blazes in rugged bush for ten days before cooler weather slowed the advance. It was a month before the Rural Fire Service declared the fires extinguished. In all, 250 square miles of forest went up in flames. Although, against all odds, there was no loss of human life, the greatest destruction of homes in the Blue Mountains' history made a mark on local communities and shocked all of Sydney below. So did the blackening of a World Heritage–listed natural landscape.

Senior constable Mary-Lou Keating of the New South Wales Police loved the people and the nature of the Blue Mountains, a place where weekend retreats and commuter towns commingle with the forest. As she went about her police work in the days and weeks after the fires, she began taking photographs. Eventually a small group of locals collaborated to publish some of Keating's four thousand photos in a book, *As the Smoke Clears*.[2] The book was printed in the Blue Mountains and sold in area shops to raise money for the Mayor's Bushfire Relief Fund.[3]

Locally produced books and videos are precious fragments of the social memory of disaster. They can be traced from broadsheets printed after the 1755 Lisbon earthquake to commemorative volumes published by local newspapers, tornado and flood videos sold in midwestern gas stations, and lovingly edited tributes left on Facebook walls.[4] But among these, *As the Smoke Clears* is unique and beautiful. As the introduction explains, "The focus was always on looking beyond the devastation to the hope for the future. You will not see photos of fire trucks and flames—those things were well documented at the time, and still cause distress to some people, particularly children, affected by the fires."

Instead, most of Keating's photographs focus on the bush itself. Shocking green *Xanthorrhoea* grass trees spring out of the gray forest floor. The charred seedpods of eucalypts, banksias, and hakeas burst open, spilling seeds from a golden red interior. Dormant epicormic buds thrust out of every trunk, turning trees fuzzy with life. Animals return, recorded with the detail of a field guide: "one of the first birds sighted after the fires, a scruffy ash-coloured kookaburra," "grubs of the Variole Paropsine Beetle feasting on eucalyptus leaves," a weevil, a lizard, a dazed swamp wallaby. In this tender biota, mountain residents saw their own capacity for regeneration from fire: "the resilience of the Blue Mountains bush and its people."

"Fire is an intrinsic phenomenon in most Australian landscapes," wrote the authors of a 2011 national assessment of forest vulnerability.[5] "Native forest systems have evolved correspondingly, increasingly becoming resilient to fire. . . . For instance, a large group of woody species have the capacity to resprout after fire, including the majority of the native *Eucalyptus* species. Other species have developed reproductive strategies that are dependent on fire, such as many species within the *Banksia* genus." Sclerophylly in plants, characterized by a hard outer layer of leaf tissue that prevents loss of water, is an evolutionary adaptation to dry environments. Tough-skinned sclerophyll plants, led by the evergreen genus *Eucalyptus*, spread over Australia millions of years ago when the once-lush continent dried out. By armoring their inner tissues against sunlight and heat, and evolving weedy abilities of quick colonization and regeneration, eucalypts inherited a whole continent. Those same qualities made them serendipitously ready when the continent began to burn.

The ancient tale is told by environmental historian Stephen Pyne in his fire history of Australia, *Burning Bush*: "Since the Pleistocene it has generally been the case that, where biotas have changed, they have moved toward a state of more fire, not less. Some species were swept aside, while some accommodated, adapted, and learned to tolerate fire. Others thrived."[6]

Eucalypts and their fellow thrivers come back from fires with all the col-
orful diversity of adaptations on display in Keating's photographs. Thick
bark insulates living trunks, so only the surface is charred. Eucalypts generally shed their lower branches to keep fires from climbing up, but when the
treetops do burn, countless green buds burst out from beneath the bark to
replace them. Bulbous underground lignotubers safely store the nutrients
a tree needs to recover, even if it burns to a stump. And seed rains down
everywhere, ready to sprout in the nutrient-enriched ashbed. Some species
flower only after a burn, and Pyne describes the scene after Victoria's legendary Ash Wednesday fires of 1983, when whole hillsides were suddenly
carpeted with rare orchids.[7]

Perhaps the most important fact of any Australian landscape is its fire
regime. The term is shorthand for the pattern of recurring fires in terms
of frequency, season, type, severity, and extent, which is often essential to
maintain the reproductive cycles and growing conditions of some plants
while keeping other species from taking over.[8] Only a few places in Australia, remnants of rain forests, have no fire regime. The regime of the dry
gum forests of the Blue Mountains is well known. They experience—in fact,
need—patchy but fierce surface fires that return to the same area at intervals
between seven and thirty years.[9] Like many Australians, the creators of As
the Smoke Clears understood how natural these cycles are. They also understood better than most how painful they are. Local writer Arna Radovich
expressed the feeling in a poem titled "Mosaic of Loss":

> . . . Recovery
> when it comes
> is slow
> and fragile
>
> tender green shoots
> arising
> from the blackened
> wasteland
>
> From the scattered
> mosaic of loss
> they begin
> a new life
> piece by piece[10]

The poem, like the rest of the book, uses the language of ecology to describe the community's experience. Patchy fires create ecosystem "mosaics" with different species and ages of vegetation. According to the ecological theory of patch dynamics, mosaics created by disturbances such as fire tend to make for high biodiversity, many species niches, and an overall healthy ecosystem. But for Radovich's community, the bush's mosaic of life is also a mosaic of loss.

THE BIRTH OF RESILIENCE

Resilience, the guiding paradigm for confronting disaster in the twenty-first century, is another concept on loan from ecology. The Canadian ecologist Crawford Stanley Holling is credited with defining and popularizing the word within his discipline, beginning with a 1973 paper, "Resilience and Stability of Ecological Systems."[11] At the time, resource managers who were looking to keep forests and fisheries in stable condition focused on models of self-regulating equilibrium informed by the technologies and tactics of the Cold War. In this tradition, processes of feedback within ecosystems were supposed to balance out shocks and naturally maintain a certain equilibrium state—in everyday language, a balance of nature.

There was just one problem with stability: in the real world, it was nowhere to be found. Both managed and wild environments tended suddenly to career out of control: fish stocks collapsed, fires burned. Seeing this, Holling, a resource manager himself, argued that stability was a transient mirage. In his paper he observed how the caprices of fire demanded instability instead. Wildfires illuminated resilience from its first articulations. Holling wrote, "As an example, the random perturbation caused by fires in Wisconsin forests has resulted in a sequence of transient changes that move forest communities from one domain of attraction to another. The apparent instability of this forest community is best viewed not as an unstable condition alone, but as one that produces a highly resilient system capable of repeating itself and persisting over time until a disturbance restarts the sequence."

The balance of nature was actually a dance. Holling called this dance resilience: "Resilience determines the persistence of relationships within a system and is a measure of the ability of these systems to absorb changes of state variables, driving variables, and parameters, and still persist." Disturbances such as fire can buffet ecosystems repeatedly and fragment them into mosaics across the landscape. That doesn't make the ecosystems less resilient, however, and it might even make them more so. Resilience was

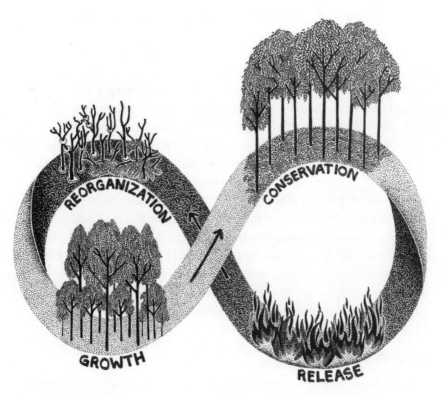

Figure 2. The adaptive cycle of resilient systems, as depicted in C.S. Holling's "figure-eight" diagram, illustrated by the stages of disturbance in an Australian eucalypt forest. See Lance Gunderson and C.S. Holling, *Panarchy: Understanding Transformations in Human and Natural Systems* (Washington, DC: Island Press, 2002).

measured only by the survival of the system, what Holling called "questions of existence or not."

While Holling's conceptual leap was replacing the mirage of stability in systems ecology, the carefully managed equilibrium of the Cold War had begun to look like a mirage, too. The balance of power, like the balance of nature, did not hold. Soon after the fall of the USSR, new risks and disasters brought resilience out of the ecology journals and into disaster studies, finance, security, urban planning, international development, and nearly every other field that contends with risk and vulnerability.

The concept's mobility was partly due to the promise of Holling's 1973 insight that the resilience approach "does not require a precise capacity to predict the future, but only a qualitative capacity to devise systems that can absorb and accommodate future events in whatever unexpected form they may take." The migration of ecological resilience out of its home discipline was also a result of Holling's own efforts. Starting in the 1990s, he has led a movement to integrate the social, economic, and natural into a general theory of resilience in social-ecological systems. In his early work Holling had written about particular ecosystems, such as the Wisconsin forest, facing particular natural or human disturbances. But in the new general theory of resilience, *every* human and ecological system can be observed as following a similar dance of adaptation through crisis, visualized in Holling's papers as a figure eight or sometimes as a Möbius strip. The dance's four stages are rapid successional growth; conservation, or equilibrium; release, or collapse; and a spontaneous reorganization for regrowth.[12]

HAVING A GO

The ridgetop town of Winmalee was visited by one of the Blue Mountains fires on October 17, 2013. Winmalee is in the lower mountains, a bedroom community an hour's drive from central Sydney, but it's very much of the bush. Houses along the western side of the ridge, like those on Buena Vista Road, faced the worst of that spring's destruction. The ridge has been the site of human settlement for hundreds of centuries and has seen countless cycles of fire. A 23,000-acre conflagration in the spring of 1968 burned more than seventy houses to the ground. The 1960s were also the decade when this town and many others on Sydney's outskirts really began to grow, prompted by new electric train services; Buena Vista Road was mapped in 1963. Winmalee grew from a population of 2,860 in 1966 to about 6,000 in 1986, despite major fires sweeping the mountains in 1968, 1976, 1977, and

1979. In the fire of 2013, forty homes were destroyed on Buena Vista Road alone.[13]

Phil Le Breton helped his parents build their home in the early 1980s, when he was a teenager. Later he and his wife, Christie, bought the house from his parents as a place to raise their own children. "My mom jokes that we didn't do a very good job of looking after it for them," Phil laughs. They were away at work, the kids at school, when the fire came through in 2013. They returned to find their house had burned so fiercely that the steel roof beams were twisted, as he put it, "like licorice sticks." The family had no choice but to rent a house down in the Sydney suburbs while their whole neighborhood entered a postapocalyptic dream. As plants in the bush were dropping their seeds into fertile ash, the yards of Winmalee were regrowing from kitchen compost of the past; in a domestic parody of the bush, a burst of cherry tomatoes and pumpkins overtook the ruins of Buena Vista Road.[14]

Within the next year, most of the residents had returned. Fourteen months later, the street was ringing with construction work and some families were preparing to celebrate Christmas at home again. The Le Bretons had managed to rent an unburned house across the street while rebuilding their own home. The task of rebuilding was all-consuming, Christie said, but there was a cycle to it as well: "In the school holidays Connor, our son, is going to go up there and he's going to help labor for the builder, so that's a new generation of rebuilding. It's both lovely and really, really hard, when you think about it." The Le Bretons knew this was where they belonged, and they were relieved to get out of Sydney and back to the familiar sounds of the bush, the returning birds and nighttime owls. Even the racket of a neighborhood under construction took on a new sort of music, Christie said. "When we first moved back in, the kids were standing on this front veranda and Bridie, our youngest, said, 'Oh, I love that sound.' We asked, 'What sound?' 'The sound of building.' 'Why?' 'Because it means they're coming back.'"

The residents of Winmalee are part of a long tradition. For the tens of thousands of years people have been in Australia, they have been coming back. Stephen Pyne tells the story of the Aboriginal settlement of the continent and how, "with uncanny mimicry, the genus *Homo* recapitulated the experience of the genus *Eucalyptus* . . . discovering cultural equivalents to sclerophylly."[15] The fire continent became home to a fire people, who practiced "firestick farming" and managed whole landscapes to meet their needs, creating "the biggest estate on Earth," to borrow the title of another recent history.[16] When Europeans took over the continent, they picked up the firestick and used it to clear farmland and pasture, transforming the land into something they thought they recognized. The sunburnt culture of the bush, the Australian "let's have a go" ethos, is a product of an extreme environment and the extreme measures taken to live there.[17] And the bush-fire was always the most extreme spectacle of all.

In the late twentieth century a new type of Australian decided to have a go. As traditionally rural populations aged and declined, more city dwell-ers sought to reconnect with the bush in a spreading exurban fringe of soul-stirring views and plentiful land.[18] The Blue Mountains are a classic manifestation of this fringe, a place known in the international literature as a wildland/urban interface (abbreviated WUI, pronounced in Australia as "wooie"). Here Pyne sees "exurban expansion replicating the agricultural expansion, but freezing the transformation in the vulnerable phase."[19] Ex-urbanites don't totally clear the bush, as farmers do, because the bush is what they came for. In Australia, as in the western United States, Europe, and elsewhere, it's this new mosaic of flammable vegetation and exurbs that brings the most explosive disaster. Once communities are established, it be-comes imperative to defend them, and firefighting services expand in size and expense. Yet suppression remains impossible even for the best-equipped brigades when fires reach a certain size and intensity.[20]

Acknowledging this, authorities in the United States often dictate man-datory evacuation during times of high fire risk. But what seems sensible in California is highly contentious in Australia, where a bush culture of self-sufficiency persists. Most Australian states now offer residents a choice: leave early or stay and defend your own property. As explained by a member of the Tasmanian Fire Service in 2003, this is a departure from paternalis-tic emergency services and "a move back to the pre–emergency service era when communities were left to take care of their own safety."[21] And it spoke volumes about the Australian self-image.

The policy was put to the test in Victoria on February 7, 2009. The ex-traordinary, ruthless fires of "Black Saturday" killed 173 people and destroyed more than two thousand homes. This national tragedy led to appointment of a royal commission and a closer analysis of the leave-or-defend policy.[22] A group of researchers who were already studying coroners' records of Austra-lian fire deaths from 1900 through 2008 found that male deaths continued to happen most often while defending properties and livelihoods, but these numbers had fallen off in recent decades. Women, who mainly succumbed while sheltering or fleeing too late, were dying more—as would be expected with rising population in the bush. This revealed that the gender roles in-herent to traditions of bushfire protection were changing for men (fewer worked on the land or tried to defend their homes) but not as much for women (who stayed fatally excluded from fire knowledge relevant to both flight and defense).[23] Another group of researchers found these dynamics very much in play in Victoria as well.[24] The new exurban residents upheld the view of bushfire as "men's business," perpetuating a gender divide in risk awareness, knowledge, preparedness, and the belief in one's personal capacity to act. The women quoted in interviews spoke of tropes much older than their own residence in the bush, with one saying, "The mythical building up of the bushfire volunteer . . . it's very important that we always have that mythical icon. It has to be male. We cannot have, I mean, women bake the scones and sell them to raise money but we must have that icon. Bushfire always gives us that. . . . It gives us community. It gives us heroes. It gives us empathy."[25]

The leave-or-stay policy's apparently easy choice had ignored the gendered aspects of resilient bush culture, and, the researchers charged, in its sidelining of paternalistic emergency services it had ignored a deeper paternalism. Bushfire management "was the domain of men even in the pre–emergency service era, when the tradition of women staying behind to care and nurture for house and home was initiated," they wrote. In fact, we can trace it back as far into the past as we wish. Pyne describes a similar division of labor in Aboriginal society: there, too, he wrote, "the bushfire belonged to the male universe; the hearth fire to the female."[26]

In Victoria's 2009 catastrophe these roles had tragic echoes, both in differential mortality during the fires and in heightened domestic violence in the years after, inflicted by men who felt an acute failure to live up to their own legend. All of these outcomes were recognized in the unusually active field of research on the gender dynamics of fire following the 2009 catastrophe. The scholarship was also unusually effective, leading to the establishment of a broadly inclusive Gender and Disaster Taskforce. In 2015, Victoria's fire slogan was revised to a simple entreaty: "Leave and live."[27]

INSTITUTIONS OF FIRE

Paternalistic or not, there is no denying the power of the bushfire volunteer as a mythic figure. Rural brigades first sprouted up in outlying towns in the nineteenth century, a collectivization of the self-defense activities of bush dwellers. In the 1960s these were brought into a national system, and by the 1970s one rural Australian in ten belonged to a brigade.[28] The Blue Mountains are a quilt of local brigades who coordinate under the New South Wales Rural Fire Service to manage fires. The state provides the equipment, but most firefighters ("firies" in Aussie slang) are volunteers. Increasingly, both men and women are members, but their respective roles vary from one brigade to the next. Times are changing, according to David Howell, president of the remote Mount Wilson/Mount Irvine brigade. "There are some brigades which are still blokey brigades, where they all hang around and drink beer and sharpen chainsaws, stand with their legs apart," he told us. "But we've got lots of guys in catering, and we've got a lot of women on trucks." In fact, the brigade fought the mammoth State Mine fire in 2013 under the command of a female captain, Beth Raines. "It's a great advantage because her empathy is different from a bloke's," Howell said. "She's much more feeling of people's feelings."[29]

Along with its permissive evacuation policies and army of volunteers, Australia's reputation in the world of wildfire management has another, even more prominent element: enthusiasm for the use of fire. Picking up the firestick on a national level in the 1960s, the country engaged in prescribed burns to reduce fuel loads on a scale never attempted before or since. At a time when the United States and Soviet Union were dropping water, fire retardants, and smoke jumpers from airplanes, Australians were dropping aerial ignition capsules of their own invention, burning off huge expanses of bush from the air. The intention was to reset fire regimes to a rigid timetable and do away with huge, destructive fires for good. This proved not only ineffective at preventing large fires but also impossible to sustain in the face of changes in settlement and sentiment. The new exurbanites wouldn't stand for burning or the smoke it produced, nor would the environmentalists. The emergence of resilience ecology spoke to the danger of a widespread, homogenizing management system such as large-scale prescribed burning, which could endanger regime-sensitive species niches. Only natural fire regimes (with the occasional nudge from careful, knowledgeable ecologists) could sustain natural biodiversity, according to the latest research. The environmentalists' approach was more current, and, crucially, it promised to be cheaper.

Pyne summarizes how a unified strategy fragmented by the end of the 1970s into competing fiefdoms, with nature reserves taking over public land management and rural fire services taking over protection. Both practiced burning—"ecological burning" to sustain rare natural communities and "hazard reduction burning" to protect human communities—but on a much smaller, labor- and data-intensive scale, in both cases concentrating on the most valuable assets at risk.[30] Wrote Pyne, "The premier achievement of the Australian strategy had been its assertion that fire practices, and fire regimes, could be shaped consciously, that something could be done about

catastrophic bushfires. Now that legacy, too, fractured into a familiar fatalism. If large fires were not common, they were not a serious problem.
If bushfires became large, they were uncontrollable. Either way one simply
endured them along with taipans, floods, and other inscrutabilities of an-
tipodean Australia." [31]

THE CITY AND THE INFERNO

This fatalism reaches its peak in Hobart, capital of the island state of Tasmania. Australia's most endangered capital is situated along a harbor in the
shadow of towering Mount Wellington. The city creeps up into the bush
at every opening. It's a rich bush: Tasmania is green, lush, and wet, with
stands of a treasured ancient rain forest on its western half. Over to the east,
however, the climate can change dramatically in a matter of days. When hot
winds over on the continent blow out of the outback and across Victoria,
stoking fires all the way, they also cross the strait south to Tasmania. Any
moisture the winds pick up falls on the island's central plateau; by the time
they reach Hobart, the air is as desiccated as California's Santa Ana winds.
All of the vegetation that grows so bountifully in wet weather turns to tinder
in this continental blast furnace.

In 1967, this wind brought one of Australia's most traumatic fires to
Hobart. It swept the length of the city's edge in less than two hours, taking
entire suburbs with it. It burned 1,300 houses and 128 major buildings and
took sixty-two lives. The official inquiry that followed blamed the disaster on rural fire practices—that is, on the bush and its poorly controlled
regimes invading the city. [32] But like an Escher mosaic, this was only one
way of seeing the picture. Hobart and the bush reach deep into each other,
potentially carrying flames in any given direction. Decades later, with the
great blaze all but forgotten, Hobart's residents regard their tessellated disasterscape with a resolutely blind eye.

David Bowman is Australia's most outspoken fire scientist, in his position as a professor at the University of Tasmania's School of Plant Science.
"In terms of resilience, you're looking in the wrong place. Nobody in Hobart
has any understanding of what's going to befall them," he said in late 2014.
He foresees not just a repeat of 1967 but something worse, in a capital that's
become even more bush-bound. But he's had a hard time convincing most
residents of what's possible. The typical reaction to his warnings is disbelief:
"A city being incinerated? That doesn't happen. It happened in the nineteenth century but not now." [33]

As he tries to get through to complacent audiences, Bowman admits that his message to the city has become ever more strident. But the fire, if not the brimstone, is real. In February 2013, it very nearly happened. Amid intense drought and record temperatures, almost fifty thousand acres of Tasmania burned. The Fire Service threw everything it had at one large blaze near Collinsvale, over the ridge from Hobart, in the same path the 1967 inferno swept through. Across the harbor, another fire descended from the hills on the village of Dunalley. The fire front moved with such ferocity that it leapt in a fireball across mile-wide Blackman Bay to burn houses on the other side.

Dunalley's destruction was the greatest tragedy of that fire season, and, like the following season in the Blue Mountains, it was well documented. In one newspaper photo the sculptor Gay Hawkes stands in the ruins of her house and workshop, wearing overalls and a bright orange identity brace- let. She carries the ruggedness of bush culture, but she is clearly in shock.[34] Hawkes's own self-published book on the fires is very different from the Blue Mountains' *As the Smoke Clears*. Simply called *Time and Chance*, it is a document of loss. She flees the fire unprepared and lives among sixteen hundred people and hundreds of dogs—"burnt people, one tribe"—in an evacuation site. Edvard Munch–like paintings with red skies illustrate the text. The fireball passing over Blackman Bay is a supernatural blue orb, a medieval comet, observed by awestruck figures in the water below. Eventu- ally Hawkes returns and mourns her home: "They say 'no lives were lost' but many birds and animals died and what about the sawmill? So many liveli- hoods, whole lives of endeavor lost, the work of a dynasty. Life takes many forms, it's not just the little life to be snuffed out."[35]

The victims in Dunalley captured national headlines, but if not for the vagaries of the weather, the fire in Collinsvale could have been the one that got out of control, and then it would have swept out of the hills into Hobart itself. In Bowman's words, the cue ball was set up: "The real disaster is the absolute, catastrophic failure of the imagination. Hell, Dunalley got incin- erated. They saw balls of fire, buildings were exploding. Fire was jumping over water bodies. Bare ground was burning. It was a complete, catastrophic disaster. But you scale up from Dunalley to Hobart . . ."—and there, as he said, imagination fails.

Hobart's narrow escape triggered little action, only another inquiry. Among many generalized recommendations, the investigator heading the Bushfires Inquiry Report had one very specific concern. The day before Dunalley's destruction, the Tasmanian Fire Service had run a computer model that showed the fire spreading exactly as it would proceed to do, even hitting the village in the afternoon at the forecast hour. Surely this simulation should have been shared more widely and should have prompted evacuation calls, the investigator charged.

This public allegation has been a headache for the Fire Service's chief officer, Michael Brown. In essence, the model's accurate prediction for Dunalley was a lucky strike. "The modeling is all dependent on the weather just being spot on, and your fuel mapping and everything being absolutely current, and fire control measures being effective. Twenty-four hours out it's rarely accurate," says Brown. Indeed, that summer few of the other models were on target, which is why the Fire Service doesn't use them to order evacuations. The models on that one day looked solid enough to trigger evacuations only in retrospect. "Had we used our fire modeling to make our decisions about warnings, we would have evacuated half of Hobart twice," Brown explains, tracing the potential path of the Collinsvale fire on a city map. Of course, nobody minds at all that the models for that fire got it wrong.[36]

THE ANGRY SUMMER

The weather was unpredictable from hour to hour through the summer of 2012–13, but it made fire a certainty. Tasmania had fallen under a little corner of a continent-wide heat wave remembered as the "Angry Summer." Temperate Hobart, which sits as far from the equator as Buffalo, New York, reached its highest temperature on record, 107.2°F. The Australian Bureau of Meteorology added new colors to its weather forecasting chart, choosing purple and black to forecast temperatures beyond its previous hottest color, dark red.[37] Tasmania's fire season lasted for nearly six months, longer than ever before.

Sandra Whight, head of the State Fire Management Council, is no longer surprised by what she sees. Having worked in both Tasmania and New South Wales for many years, she has watched what was once a rhythm of bad fire seasons every six or seven years merge into an annual crisis. Since the Black Saturday fires in Victoria in 2009, there hasn't been a quiet season, she says. "The bigger seasons are more frequent, and because of climate change the seasons are longer. It might only be a week or two either side, but when you're talking about crew fatigue and people's clear thinking, it all starts to add up." And the longer the fire season, the less of the year is left to recover and take action to reduce next season's hazards. Whight has also tracked more subtle changes. Dry lightning storms that bring ignition without rain are sweeping over the island, and vegetation types that don't normally burn—such as parts of Tasmania's World Heritage rain forests—are drying out and burning like never before.[38]

With the Angry Summer's incredible new records raising interest in the climate community, two researchers at the University of Melbourne carried out one of the fastest analyses ever performed for a climate event, publishing their results the following July. They found that climate change models for 2006–20 raised the likelihood of such extreme temperatures fivefold, suggesting that "the human contribution to the increased odds of Australian summer extremes like 2013 was substantial." This was especially the case with the Angry Summer, which happened in a La Niña year, when ocean temperatures usually exert a cooling influence.[39] Modeling by another group of Hobart-based climatologists showed that under a high-emissions scenario Tasmanians would see a 120 percent increase in days of very high fire danger over a 250 percent larger area by 2100.[40] More and more, Australian summers would be provoked to the point of anger.

The following spring brought more moderate temperatures overall but another monster fire season, this time in the Blue Mountains. Now public officials, rather than climatologists, vied for the last word. UN climate chief Christina Figueres promptly stated that "the World Meteorological Organization has not established a direct link between this wildfire and climate change—yet," but that the science was absolutely clear on increasing heat waves in Asia, Europe, and Australia.[41] At the time, Australia's newly elected prime minister, Tony Abbott, was out volunteering with a back-burning brigade in the Blue Mountains. On his return he accused Figueres of "talking through her hat." The fires he'd seen, he opined, "are certainly not a function of climate change, they're a function of life in Australia."[42] By then, Abbott had backpedaled from his earlier reputation as a vehement climate change denier, but he had every reason to be on the defensive. The government he had defeated in the previous month's election had introduced a national carbon-pricing scheme (commonly called the carbon tax), and he had promised to repeal it. In fact he had put this promise at the center of his party's campaign, declaring the week before the vote that "more than anything, this election is a referendum on the carbon tax."[43]

The carbon tax was a pioneering effort by a country that ranked eleventh among the world's per capita emitters and that was already living the climatic future other countries feared. In the first year of the tax, an industry group recorded that emissions fell 7 percent on average among the 350 energy, petroleum, and transport companies large enough to be subject to the new pricing scheme.[44] Not surprisingly, there were other factors affecting the trend, and one year was too little time to settle the matter, but Abbott's government didn't wait for the verdict.[45] They chose a cold winter day, well after fire season, to follow through on their promise and repeal the carbon tax. Australia—the sunburnt country of droughts and flooding rains, the fire continent—became the first country to adopt and then scrap a carbon-pricing mechanism.

That year, with carbon unpriced and bipartisan support for renewable energy targets broken, investment in alternative energy sources plummeted 88 percent. Australia dropped back behind Panama, Sri Lanka, and Myanmar in renewable energy investment.[46] When countries were asked to file national commitments in advance of the Paris climate summit of 2015, Australia produced a conspicuously weak pledge that would soon leave it with the most carbon-intensive economy in the developed world.[47] If burning bush was, for Abbott, "a function of life in Australia," so was burning coal and gas. The vision of emissions reduction failed in the political arena, leaving only the dream of resilience.[48]

FLAMMABLE LIFE

In the paper that sketched out the concept of resilience ecology, C.S. Holling wrote of "the random perturbation caused by fires in Wisconsin forests," which bring "a sequence of transient changes that move forest communities from one domain of attraction to another."[49] This kind of perturbation is something alien and external to the forest community, and Holling likened it to climatic events. Today's fire ecology presents a different consensus. David Bowman, Stephen Pyne, and their co-authors argue in their recent book *Fire on Earth: An Introduction* that "the interactive nature of fire highlights the fundamental difference between fire and other large scale disturbances, such as tropical storms or tectonic activity, that are independent of vegetation. Rather, fire is more akin to a generalist herbivore"—an animal that eats all forms of vegetation—"as the effects are modulated by direct and indirect feedbacks; hence, fire genuinely has an 'ecology.'"[50]

The close relationship between vegetation and fire in environments such as Australia's eucalypt bush makes this perspective easy to see. There is even a suspicion, a morbid undercurrent in fire ecology since the 1970s, that some plant communities have evolved to catch fire. The authors of *Fire on Earth* call this phenomenon "flammability as an emergent property of plant communities."[51] Plants that are more flammable can torch competitors around them, grabbing space and nutrients for themselves or their offspring. In recent times invasive fireweeds, such as buffelgrass in central

Australia and America's Sonoran Desert, have shown how species can accelerate fire regimes to their own benefit.[52] In a survey of scientific literature with the provocative title "Have Plants Evolved to Self-Immolate?," Bowman and his colleagues show that proving this idea has been so far impossible, given the difficulty in establishing cause and effect between landscape fire and flammable plant traits. "It is more parsimonious," they conclude, "to view fire activity as a powerful filter that sorts plants with pre-existing flammabilities and hones regeneration strategies."[53]

Which came first, the fires or the trees? Like all chicken-and-egg questions, there may be no point to even answering it. If the bush didn't evolve to catch fire, then fire regimes shaped the bush to the point where it can't persist without burning. Adaptation became addiction. Unfortunately, as the re-greening forest mosaic became a popular metaphor for resilience in all our systems, something about this co-dependence between hazard and ecosystem was lost. The other message of the bush is that learning to live with fire means you can't live without it.

Dr. Guy Bannink, a palliative care specialist, lives far up on the deeply forested slopes of Mount Wellington. From his house the white stags of trees from the great 1967 fire can still be seen. He is one resident who is very aware of the danger that Hobart lives under, and he and his family have taken their safety into their own hands, constructing a concrete bunker in the hillside next to their house. What convinced them was the 2013 Collinsvale fire, which reached a point only five miles away—twenty minutes with the wind speed on that day. Bannink's philosophy is straightforward: "The house is going to burn down, we don't give a toss, there's nothing we can do about that." Instead they built the bunker. It has space for the whole family plus a neighbor or two and is equipped with a twenty-hour tank of medical oxygen. In nonemergency times, it doubles as a wine cellar. The family hopes never to use it as a fire refuge—it's a plan B, in case they don't have time to get down the mountain.[54]

Bannink is pleased with his bunker, in a backyard-engineer sort of way. He has also planted fire-resistant trees, dogwoods and blanket-leaves, around the edge of the property. Standing outside the fireproof door in the hillside, looking over the city and the bay, he admitted that he's starting to think about skipping plan A and going straight to plan B. "Maybe we need to rethink just rushing out, because what happens if we get down to Hobart and Hobart takes off? Where do you go then? Up here, you just go in the bunker, slam the door, turn on the Walkmans, and just relax for three or four hours. It'll all be ash when you come out, but you'll be alive."

With a bunker or without, planning—imagination—is everything. In the conclusion to her fire memoir, Gay Hawkes shares her own suggestions, drawn from bitter experience in the Dunalley fire:

> If we had any notice whatsoever of what could occur the following "fire plan" would be appropriate for a lone person without water, a pump and a generator:
>
> Just this—have at hand a list of the most important things to be taken at the point of evacuation. Under stress and extreme heat the brain does not work clearly and one needs such a list. With even one hour's prior notice that the fire was coming I could have acted more properly.
>
> Now I plant trees, knowing that I shall not live to enjoy their shade, but planting them on my black land for the sake of my children and their children, in honour of the beauty I have enjoyed there before and to cool the warming world.[55]

SEA OF CARBON

Determining the realities and implications of that warming world is a data-hungry business. To chart the short-circuiting carbon cycle, experts such as those on the Intergovernmental Panel on Climate Change (IPCC) need all the information the planet can supply. It flows from observatories that are carefully placed, generally remote, and preferably high up. Take, for example, the Zotino Tall Tower Observation Facility (ZOTTO). As tall as Sydney Tower or the top platform of the Eiffel Tower, ZOTTO soars above the middle of the Siberian forests in central Russia, much farther north than Tasmania is south. Instruments on the 991-foot mast sample greenhouse gas concentrations in the atmosphere at multiple elevations, far from human activity. They also monitor a nearly four-hundred-square-mile area of the taiga, the northern boreal forest belt that fills a vast zone of Eurasia and North America. About 10 percent of all carbon stored in the world's vegetation and soil is in Siberian forests, and it's impossible to predict what is happening to the world—even to Australia or California—without understanding this environment. And it's impossible to understand the taiga unless we understand the long, irresistible cycles in which it burns.

The climatologists at ZOTTO gained some understanding in 2012, when one of that summer's many large fires swept toward the tower. It burned more than a third of the monitoring area but fell just short of the mast and laboratory. The sensors carried on gathering data, with new data now measuring every wisp of carbon release from the burned area. Results streamed out to the Sukachev Institute of Forest in the distant regional capital of Krasnoyarsk, as well as to the Max Planck Institute for Biogeochemistry in Jena, Germany, where the mast's data feed into the projections and models of the IPCC.[56]

The great Siberian carbon sink developed in cold, dry environments where vegetation grows slowly but decomposes even more slowly. The forest piles up on itself, hoarding nutrients and freezing the ecosystem in place. Yet neither the more open "light" taiga of central and eastern Siberia nor the "dark" taiga of the south and west is as dry and welcoming of fire as Australia's eucalypt forests—unless drought descends. Within the Siberian taiga, the world's largest biome, a drought occurs somewhere every year. Eventually, on the order of decades in light taiga and centuries in dark taiga, the conditions are right for a fire to stir the pot.

The Soviet Union couldn't abide the capricious nature of the Siberian fire regime, so throughout the Cold War years the nation mounted a highly militarized aerial campaign against it. The strategy used the largest firefighting force on earth and imposed a total ban on burning for agricultural or protective reasons, a prohibition that would stamp out centuries of rural land practices. The fire regimes of remote Siberia were never, of course, under full totalitarian control, but for a time the government came perhaps as close to that as anyone could.[57] After the Soviet Union's collapse in 1991, the system of suppression weakened but persisted in its basic outlines. It ended conclusively only in 2007 when Russian president Vladimir Putin introduced a new Forest Code. This law stripped away the federal forest defense program and devolved responsibility to regional and local authorities, along with private timber leaseholders, while strengthening Moscow's monitoring of their activities. The plan fulfilled the Putin doctrine of central power through omniscience but brought a cascade of new problems. Foresters came to spend 75 percent of their time filling out paperwork to report on their activities. No one was allowed to help fight fires without a lengthy licensing process. And regions hit hard with fire in a given year couldn't easily call in help from idle resources elsewhere.[58] Meanwhile, the role vacated by foresters at the federal level was filled by the Ministry of Emergency Situations (EMERCOM), an agency with no wildland fire experience. EMERCOM shared the presidential emphasis on monitoring, spending two-thirds of its wildfire budget on watching fires happen from afar. In fire management, the Soviet culture of suppression had given way to a culture of surveillance.

The Sukachev Institute, on a bluff above the Yenisey River in Krasnoyarsk, hosts EMERCOM's monitoring unit for all of Siberia. Victor

Romasko is an army colonel and a programmer who writes the software that the unit, as well as the institute's own remote-sensing team, uses. They receive satellite data covering all of Siberia at a resolution of one kilometer per pixel. "We are monitoring for fires that are close to settlements or economically important infrastructure," Romasko explained. "We report those, but otherwise we don't. In these regions EMERCOM establishes boundaries at five kilometers [3.1 miles] around settlements. If it moves within the boundary we respond, but outside of that fires don't need to be managed." [59]

So far this strategy has worked for Siberia—and it may be the only strategy possible in a barely inhabited forest larger than the continent of Australia. West of the Urals in European Russia, however, Putin's 2007 reform seemed to call down disaster in very short order. The year 2010 brought incredible heat, setting new records in the history of Russian meteorology. This well-studied weather anomaly devastated Russia's wheat harvest that year, caused thousands of urban heat wave deaths in Moscow, and at its other end, perversely, brought catastrophic floods to Pakistan.[60] Fires incinerated dozens of villages and crept into the peatlands around Moscow, smothering the city in smoke. In the reinsurance company Munich Re's estimation, "at least 56,000 people died as a result of heat and air pollution, making it the most deadly natural disaster in Russia's history." [61] In a country where climate change is still rarely discussed, the weight of blame for the fire and smoke fell on the mismanagement that was a direct result of the 2007 reforms.[62]

The Russian government, like the Soviet state before it, passed some of the blame down to rural populations. Now, it charged, the problem wasn't that people were burning fields and forests in premodern ignorance; the problem was that they were leaving. In 2010 alone, nearly a hundred thousand mostly young Russians left the countryside for cities, part of an ongoing rural exodus that began with the collapse of collective farming two decades before.[63] An area of farmed land larger than the entire state of California was simply abandoned. This wasn't all bad news; the regrowth of wild vegetation on this land led to the sequestration of close to 50 million tons of carbon, helping Russia meet its Kyoto Protocol targets.[64] In climate change terms, the abandonment of almost a quarter of Russia's agricultural area could be the greatest act of carbon capture humanity has achieved in the past century. But in terms of politics and Russian identity, it was a shocking collapse. And to foresters and firefighters, all that captured carbon is just more fuel for wildfires. In the areas set ablaze in 2010, the rural exodus had brought the forest closer to villages, with abandoned fields and pastures acting as conduits for fire.[65] That's because the farmers who had remained, fewer and older, relied increasingly on the ancient (but illegal) labor-saving technology of fire. Poorly attended burns on fields and pastures had spread quickly into the wild.

The 2010 fires were a national disaster because of their proximity to inhabited regions of Russia, most significantly Moscow. But far larger areas of the country burned in 2011 and 2012, taking steps up the ladder of an unmistakable trend. These fires didn't raise the same alarms because they were in Siberia, where most fires are watched only from orbit. As the government and media ignored these fires, they also ignored their causes, and those might have had less to do with a declining farm sector than with a healthy logging sector.

Elena Kukavskaya, one of a new generation of scientists at the Sukachev Institute, has been researching the effects of logging on fire regimes in three districts of Siberia. In Angara, Krasnoyarsk region's big logging district, clear-cutting of stately pine trunks leaves huge fuel loads on the ground and substantially increases fire risk and carbon release. Most fires start in the logging camps but burn an area on average four times larger than the clear-cut itself. Kukavskaya has found that the relationship between logging and fire is a complicated one, just like the relationship between agriculture and fire. Yet it has received far less attention in the media and the research community. "It's politics, you know," Kukavskaya said. "The abandonment of agriculture is a big issue for the country. It causes an emotional response. Logging isn't a problem, it's an industry." [66]

THE COST OF A BOX OF MATCHES

We spoke with Kukavskaya after an international forest fire conference in Novosibirsk, the next stop west along the Trans-Siberian Railway from Krasnoyarsk (a mere twelve-hour journey).[67] Several of her Sukachev Institute colleagues had also attended, discussing fire experiments, fire management, and climate change with their national and international peers. At a parallel expo, companies from Russia and beyond displayed their firefighting tools to regional authorities who are suddenly, post-2007, in charge of their own budgets. Sales representatives demonstrated armored vehicles, fire-retardant liquids ("Within two months, objects sprayed could not be set on fire. . . . Can also be used as a nitrogen fertilizer"), automated camera stations, and access to satellite feeds. Behind every pitch was the best possible return on money and manpower for overwhelmed authorities.

The idea of value wasn't just on the expo floor. Increasingly, national and regional authorities are interested in prescribed burning and fire ecology to help manage the unmanageable. The organizer of the conference, Johann Georg Goldammer of the Global Fire Monitoring Center in Freiburg, Germany, is a name inseparable from these topics. A "missionary for fire ecology" since the 1970s and "the most widely traveled fire scientist in history," according to Stephen Pyne, Goldammer has made a life's mission out of trying to restore prescribed fire to Europe (still a project with severely limited success) and expand knowledge of fire regimes around the world.[68] He was one of the first Western fire scientists to enter Russia after the fall of the USSR, and in Siberia he got straight to work.

In 1993, Goldammer, the Sukachev Institute's Valentin V. Furyaev,[69] and a group of other scientists from Russia, North America, and Europe ended a conference on fire ecology by flying out to an uninhabited spot in the middle of Krasnoyarsk region.[70] Here a 140-acre expanse of dark taiga stood on an island surrounded by wet, treeless bogs. On July 6, under extreme fire weather, the scientists ignited the island. Helicopter-mounted cameras recorded a firestorm that consumed the forest in a convective vortex, creating a pillar of smoke three miles high. The aim was to foster an all-consuming stand replacement fire—to set the taiga's clock back to zero—and to watch it recover over the next two hundred years.[71]

The scale of the Bor Forest Island Experiment, in both size and centuries, was ambitious, and it signaled an intention to use Siberia as a vast natural laboratory. Goldammer's involvement in the country, its conferences, and its forests has continued on this scale. Along the way he has developed a close symbiotic relationship with the government, according to Alexey Yaroshenko, senior forest campaigner at Greenpeace Russia. "Definitely he is a person who is interested in large-scale fire experiments, and Russia is the one place where he is allowed to make such large fires," Yaroshenko told us.[72] In exchange, the administration gets the benefit of Goldammer's ecological message about prescribed burning and, in a more general sense, about the natural, beneficial role of fire in the ecosystem—a new message of resilience, and potentially a cheaper one.[73]

"Actually the big idea of our government is to justify a decrease in expenses for firefighting and to say that we should not fight all forest fires," Yaroshenko said. In practical terms this would mean enlarging the zone where fires are watched only from afar and flirting again with prescribed fire. In 2010 and 2011, the government ordered prescribed burning over an area triple the size of what was burned in the United States in the same years—although, under the decentralized system, these "orders" are really just suggestions to the regional authorities, who are left to burn wherever they can, however they can, he explained.

Properly done, of course, prescribed burning is anything but a cheap prospect. The U.S. Forest Service applies huge resources, expertise, and manpower to burn relatively modest patches of forest. But for put-upon regional and local authorities in Siberia, the cost of a burn may be no more than the cost of a box of matches. A common, stereotypical message of fire ecology—that fire regimes are for the greater good, and that the forest can take care of itself—becomes cover for a state eager to shed untenable responsibilities.

REGIME CHANGE

Can the forest take care of itself? In remote Siberia, where the sound of a tree burning is heard only by satellites, the answer may seem to have an impact on very few people. Yet through the carbon cycle, it ultimately impacts us all. And the taiga's future may have very little to do with the strategies of its would-be managers.

In most climate change models, the Russian and North American taiga becomes a mitigating force, a negative feedback: its upward creep into the warming Arctic will fill the far north with new carbon stocks. Granted, this will happen alongside the melting of permafrost (hastened by the heat of fires) that may release vast stores of carbon dioxide and even more potent methane, but the forest's advance is at least one reason for hope.[74]

On the other hand, a 2013 study by Charles Koven at the U.S. Department of Energy's Berkeley Lab drew attention to the taiga's neglected southern fringe. Simulating climate change as geographical shifts in climate properties from one region on the map to another, Koven found that "southern boreal carbon loss as a result of ecosystem shift is likely to offset carbon gains from northern boreal forest expansion"—in other words, the northward expansion of Siberia's and North America's taiga would look more like a northward migration, with the whole boreal belt shifting to higher latitudes. In its wake the grassy steppe of Central Asia and the Great Plains will creep north. This likelihood was missed by the majority of carbon-climate models, which are "typically without explicit simulation of the disturbance and mortality processes behind such shifts."[75]

Nadja Tchebakova, a climate researcher at the Sukachev Institute, is clearer about what these disturbance and mortality processes will be: "When the forest gives way to steppe, the trees don't actually get up and migrate. Something has to happen to them. What happens is they don't get enough water, so they dry up, and then the fires start." The southern taiga will burn out, not fade away. Under sufficient global warming this won't be a healthy rejuvenating fire but a true regime change. Where the forest can no longer live, the steppe moves in with its fast and shallow carbon cycles and frequent burns—offsetting, if not exceeding, the taiga's gains in the north.[76] Elena Kukavskaya has already seen this happen in areas of Transbaikal where logging and frequent fires have conspired to push forests over the edge.[77] Her landscape photographs of innocent waving grass might show the beginnings of a future landscape. Tchebakova has produced her own dramatic maps of reconfigured eco-regions based on the IPCC's predictions of Siberia's climate circa 2100.[78] She also found biomes shifting north by as much as 370 miles. The great expanse of central Siberia becomes mixed forest-steppe, and all of southern Siberia turns into pure steppe. "Here around Krasnoyarsk, the city, and the Institute of Forest will be nothing but grass," she says. "That is certain."[79]

At the top edge, the taiga will move through successive generations of new seedlings, tree by tree, taking perhaps a thousand years to cross the tundra. On the south edge, where the balance of zones is so delicate that in places you can see taiga on the north sides of hills and steppe on the south, it will happen suddenly, in flames. A fire regime will collapse in on itself, giving way to an accelerated world of grassland cycles and free-flowing carbon. This transformative fire will sweep beyond any notions of human control, because it will be just one front in a larger global carbon burn, one that touches every landscape.

2

LEAVE IT UP TO BATMAN

The Philippines

No single term has yet emerged that defines the areas where disasters are more commonplace: the media often sensationalizes a certain region as a "belt of pain" or a "rim of fire" or a "typhoon alley," while scientific literature makes reference to zones of "seismic or volcanic activity," "natural fault lines" or to meteorological conditions such as the El Niño–Southern Oscillation (ENSO). Whatever the denomination, however, there is an implicit understanding that the place in question is somewhere else, somewhere where "they" as opposed to "we" live, and denotes a land and climate that have been endowed with dangerous and life-threatening qualities.

—Greg Bankoff, *Cultures of Disaster: Society and Natural Hazard in the Philippines,* 2003[1]

You know Filipinos. They will tell you, "We're still okay," even though, of course, they're not.

—Juber Lugas, volunteer rescuer in Davao del Norte, 2012[2]

Two months after the strongest storm landfall ever recorded, Judith Buhay stood on a balcony at the point of impact, overlooking her community. There was no wind today, but rain was cascading from the deep gray sky, running off Buhay's polka-dotted umbrella and the forest of tarp roofs surrounding her. Everything, including this balcony atop the community hall of Barangay 37, Tacloban City, Philippines, had been blasted by a nineteen-foot storm surge when Typhoon Yolanda (known internationally as Haiyan) landed on November 8, 2013. The super typhoon's record-setting winds of 195 miles per hour, earning it the unofficial status "category 6," had had little direct impact here.[3] The waterside neighborhood had simply been washed away.[4]

Buhay's barangay—the smallest administrative division in the Philippines, but also a word that replaced, and absorbed some connotations of, the Spanish word *barrio*—was home to hundreds of families, many of whom ran stalls in the city's covered vegetable market. The market complex, now collapsing, trash-filled, and unusable, lay a little way south on the shoreline.

Sales had moved into the open air, with vegetable dealers by the dozens setting up shop on the dockside out front, using their own tarps and old umbrellas to keep off the rain. Fresh produce had returned, and business, despite everything, was good. To the north, a canal separated Barangay 37 from the other barangays along the waterfront, an area that goes by the name Seawall. The ruined dwellings stretched out of sight along the curve of the shore. Near the point of disappearance, three huge container ships were stranded where Yolanda's surge had deposited them, between fifty and a hundred yards inland. The crews were still living aboard, their laundry drying on the deck railings. The government had asked the ships' owners to remove their vessels, but the necessary process of scrapping them piece by piece would not begin for many months to come.[5] Beside one of the ships, someone had constructed a tree out of plastic water bottles and decorated it with Coke cans; it was less than three weeks after Christmas.

HARD RAIN, HARD WORK

The Philippines' deadliest typhoon on record killed 6,300 people in the central islands of the country, with more than a thousand others recorded as missing, and on this January afternoon cash-for-work crews in United Nations Development Programme (UNDP) helmets were working through the rain, digging out more victims along the shore. Their efforts would eventually bring the number of recovered dead in Tacloban alone to more than two thousand. In the early days, Buhay had helped with the removal of bodies from her barangay and sprayed about one hundred of them with disinfectant as they lay, lined up, on the main road for days. "Most of my neighbors stayed put during Yolanda," she said. "A dentist and his family lived opposite us. He said he'd been through typhoons before, and it hadn't been so bad. I told him, 'This one will be different! You must evacuate!' He stayed, and he, his wife, and his daughter all died." Buhay and her family did evacuate, but they returned home as soon as they could. "I cannot describe what it was like to come back to this barangay and not even to be able to find our street, much less our house. There was no food in the city. We took the bus to Ormoc"—three hours to the other side of the island in ordinary times—"but they too had been hit hard and it was all panic buying there, so we had to take the ferry all the way to Cebu"—another island and province—"to bring back supplies."[6]

Buhay has always been community-minded. She is the president of a local organization representing workers in the informal sector, a category including many market dealers. She also helps organize occupational training

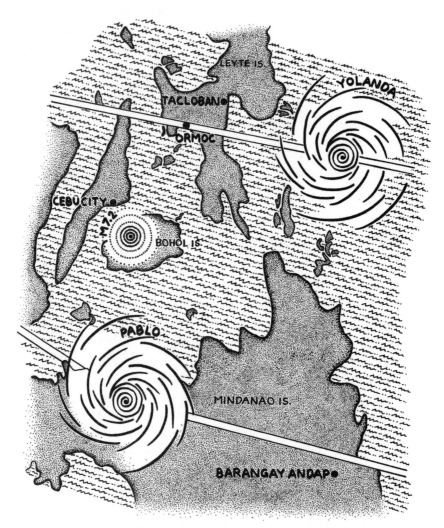

Figure 3. The central Philippines, showing the position of Super Typhoon Pablo at about 5:00 p.m. local time, December 4, 2012; the position of Super Typhoon Yolanda at about 3:00 a.m. local time, November 8, 2013; and the location of the magnitude-7.2 Bohol earthquake that occurred at 8:12 a.m., October 15, 2013.

for young out-of-school residents to learn hair, nail, and foot-massage skills. But at the moment all of that was on hold, and right now everyone was a volunteer. In that, Buhay had found a silver lining amid the chaotic ruins: "It turns out that everyone has some skill to contribute. I've never seen the community work together like this. All the old arguments disappeared." Rain muffled the sound of hammering and sawing down below as construction continued despite the weather. Behind the barangay headquarters, some of Buhay's neighbors with carpentry skills were building her a new house on what remained of the foundations of the old one. While the original had been a family home solidly built of cinder block, the new one was a simple frame of coconut palm lumber—a lightweight material, but one that was plentiful now that the typhoon had felled millions of palms. Once the structure was done, she planned to put a photovoltaic array on top to supply her home as well as help out neighbors; the barangay as a whole was still without power. "But that is all in the future. There is so much to do first," she said.

Palm lumber formed the skeleton of the new waterfront architecture, with mostly thin plywood as cladding, and on top a few sheets of highly sought-after corrugated iron and tarps of all colors and origins. Prominent in the seafront area, as viewed from higher ground, were numerous identical advertising banners. Very large and made of thick vinyl, they had been repurposed as waterproof roofing. Somewhat cruelly, the bright pink banners pictured a woman with freshly conditioned hair flowing in the wind and the slogan "Ultimate Blow Out."

The people of Barangay 37 had largely returned from the evacuation centers, driven out by the cramped conditions and the need to restart their lives and businesses at the food market. For the same reasons, they had almost unanimously rejected the idea of moving into the flimsy, barracks-like "bunkhouses" being built by the government far on the outskirts of the city; there was nothing for them to do there. As Buhay's neighbors worked on her new house, they had already finished constructing shelter for themselves on the sites of their old homes. As a result, the barangay had become something more than a tent city but much less than what it once had been.

Most of the debris had been cleared through countless hours of communal work. The last of it was heaped in the former neighborhood gymnasium and basketball court below the balcony where Buhay stood. In front of the junk pile Sarah, another volunteer, was handing out and collecting forms on green cardstock. These were for households to record emergency assistance given by the Department of Social Welfare and Development; however, two months on, very little had been written on the forms. Almost no government or private aid had reached the neighborhood. Instead there were neighbors helping neighbors. They included Jalyn, an elderly doctor who had been tending to barangay residents' health for the past twenty years. "In any case, we don't need relief supplies," said Buhay. "We need our livelihoods and help in rebuilding. We can't go live in some other place where there is no good work for us and just live on relief."

The downpour intensified, but whole blocks of ramshackle palm-wood residences, most around twelve by twelve feet, were serving as dry havens for families cooking and laughing together. Rain ran off tarps into plastic tubs, replenishing the only on-hand source of drinking water. The tubs were overflowing, and rainwater ran inches deep through the pathways of the barangay, toward the shore. The rebuilding effort had reached the waterline of the harbor and even beyond, with the last row of houses perching on platforms inches above the high tide. But the future of this shoreline construction was uncertain. That morning, word had come of a tensely anticipated decision handed down by the national government on what sort of permanent rebuilding would be allowed. After weighing interests and risks, officials had decided to enforce the little-used Philippine Water Code to ban structures within forty meters (130 feet) of the shore. Everyone agreed that it was a pointless compromise—the storm surge had reached much farther inland than that—and it would cut the living space of Barangay 37 by half. Judith Buhay's house and youth training center were on the right side of the line; Jalyn's rebuilt clinic was not. For now, however, it made little practical difference. Families on the wrong side of the line carried on their lives in those rough sketches of their old homes, trusting in the state of general chaos and neglect to keep them safe from eviction for a little longer.

The hard, steady rain that fell on Tacloban that January day was a consequence of a huge low-pressure area churning just off the coast of the Philippines. Three hundred miles due south, on the country's big southern island of Mindanao, the rain fell even more heavily. Floods and landslides were driving more than ten thousand residents to evacuate, destroying thousands of homes, and ultimately taking seventy lives.[7] The torrential rains were a sad footnote to the story of Typhoon Yolanda, a new disruption for relief efforts, a source of compounded misery for typhoon survivors—and a perfectly ordinary disaster for the Philippines.

Two names stand out among the seventy fatality reports from the 2014 rainstorms down in Mindanao: Estanislao and Antonio Manla, who died in a landslide in the mountain village of Barangay Andap in Compostela Valley. What was remarkable was that anyone was still residing in Andap at all. Eleven months before Yolanda, on December 4, 2012, the village had been destroyed in a massive debris flow triggered at the height of Typhoon Pablo (international name Bopha). It was the worst single incident caused by Pablo, a super typhoon that left more than a thousand dead and caused more than a billion dollars in damage. Buried under a field of boulders half a mile wide and five miles long, Andap was spoken of in the past tense for quite a while. When President Benigno Aquino III pledged to reporters after Pablo that he would mandate relocation of hazard-prone communities, he mentioned the village only as a warning to others. "Are there other places like Barangay Andap?" he asked, fearing for the living, not the dead.[8]

But Andap wasn't gone. As in Tacloban's Barangay 37, the will to stay put, rebuild, and persist is strong, regardless of official policy and future risks. The city of Ormoc was building back in place as well. The reason Judith Buhay could find no food in Ormoc just after Yolanda was because the storm's eye had passed directly over the city, causing severe destruction. Sadly, this did not even come close to being Ormoc's worst disaster. That had come back in 1991 when a weak tropical storm called Uring dumped six inches of rainfall in three hours on the mountain slopes above Ormoc— slopes that had been stripped of their tree cover by heedless commercial logging. Pouring out of the mountains, the Anilao River instantly filled the city in floodwaters up to a twenty-foot depth. Eighty-five percent of buildings were destroyed, 5,365 people were killed, and another 2,046 were recorded as missing.[9] For more than two decades, the name Ormoc has evoked the spirit of struggle in the face of tragedy that the names Tacloban and Andap will be evoking for decades to come.

COPING AS CULTURE

Historian Greg Bankoff has studied the Filipino approach to living on the world's most dangerous archipelago for years, most thoroughly in his book *Cultures of Disaster: Society and Natural Hazard in the Philippines*. For Bankoff, such cultures "question the basis on which the whole concept of disaster has been constructed":

Hazard is treated as a constant feature of the physical environment, one that for most people is a frequent life experience, and consequently one where the chronic threat of such events has been normalised as an integral part of culture. Within this interpretive framework or schema, societies are characterised by the development of specific coping mechanisms that permit particular communities to reach permanent accommodation of the effects of these occurrences and that is manifest in the behavioural and normative values of those cultures. . . . Filipinos have developed their own specific coping mechanisms . . . visible in the historical records of architectural adaptation, agricultural practices and migration patterns, and in the popular manifestations of calculated risk assessment, resignation, mysticism, self-reliance and reciprocity common to many cultures in the archipelago.[10]

The most proudly Filipino of these values is something called *bayanihan*. The country's great cultural anthropologist F. Landa Jocano, whose slum studies underlie Bankoff's analysis, defined *bayanihan* as "coming together voluntarily and helping one another accomplish certain tasks requiring group action."[11] For Bankoff, it connotes "toiling on another's behalf, and assuming another's burdens."[12] Unlike patriotism or town pride, *bayanihan* can be invoked on any scale, from the national level down to one-on-one exchanges between neighbors. On full display in places such as Barangay 37 after Yolanda, it is a feeling of unity and togetherness, but one that must be expressed through action and labor. It appears similar to disaster communitas around the world (see pages 116–21); the key distinction is that it is not limited to times of disruption but is basic to all sorts of social organization. Historically, it was how many communities would coordinate labor-intensive agricultural or building work.[13]

Another trait that Filipinos recognize in themselves but don't celebrate as often is expressed as *bahala na*. This is often heard in a phrase such as *bahala na si Lord*, "leave it up to God"—or, when the situation seems to be beyond even divine intervention, *bahala na si Batman*. This trait is sometimes defined as fatalism and sometimes, paradoxically, as determination. Jocano described it as "a willingness to accept reality, especially in situations where one has no control over the consequences of his/her actions."[14] Bankoff maintained that it is as much about courage, daring, and a finely calculated assessment of the odds, "providing cover for risky action, such as taking to sea during a storm, as well as inaction." To Bankoff, this come-what-may sense of destiny provides a "formidable armour against the suffering brought by disasters."[15] Indeed, it's hard to imagine living in as unpredictably tempestuous a land as the Philippines without a highly developed sense of *bahala na* or at least of *que será, será*.

In Barangay Andap after the debris flow and in Barangay 37 after the storm surge, it was with an air of *bayanihan* that residents opened their homes to one another and came together to rebuild. And it was with a shrug of *bahala na* that they rebuilt in some of the same places where the next storm system could claim more lives. Of course, it is important to distinguish between attitude alone and any number of other economic and social forces that prevent people from moving to safer ground. Notions such as *bahala na* are born of necessity, as a response to situations of chronic and inescapable risk. They are an acceptance—often a very brave acceptance—of risk, not a cause of it.

This is the Filipino way, a heroic collective spirit of perseverance and faith that has served the country's two thousand inhabited islands well through a history punctuated at close intervals by disasters. To live anywhere in the Philippines is truly to live life between storms—and between volcanoes, earthquakes, and landslides.[16] Nevertheless, Pablo and Yolanda pushed the resilient Filipino way of life to its limits. Pablo was the world's deadliest disaster of 2012. Yolanda was the world's deadliest disaster of 2013. Pablo was the costliest disaster in the history of the Philippines, until Yolanda doubled its toll. If the rest of us want to catch a glimpse of what it might be like to live between storms in the decades to come, it is worthwhile at this point to examine how events unfolded in the Philippines during those eleven months between the departure of Pablo and the arrival of Yolanda.

2013: THE EYE OF TWO STORMS

Pablo was particularly jarring because it struck Mindanao, which is traditionally well outside of the country's typhoon belt. While an average of eight or nine tropical cyclones make landfall in the Philippines every year, the highest rate in the world, Mindanao is seldom at risk; cyclonic storms have trouble forming so near the equator.[17] Pablo was the second most southerly typhoon on record to achieve Category 5–equivalent strength and the most powerful ever to hit Mindanao. The next morning's headline in the *Philippine Daily Inquirer* was a direct quote from a resident: "So, That's What a Typhoon Is Like."[18]

The eighteenth annual Conference of the Parties to the 1992 UN Framework Convention on Climate Change (known as COP18) happened to kick off in Qatar just as Pablo was tearing through the Philippines. Some climate activists had already been hoping that Hurricane/Superstorm Sandy, which had hit the Caribbean and United States scarcely a month earlier, would serve as a wake-up call at the conference. But now Naderev "Yeb" Saño, commissioner of the Philippines Climate Change Commission, had an opportunity to sound an even more desperate alarm to the foot-dragging delegates in Qatar:

We have never had a typhoon like [Pablo], which has wreaked havoc in a part of the country that has never seen a storm like this in half a century. And heartbreaking tragedies like this is not unique to the Philippines, because the whole world, especially developing countries struggling to address poverty and achieve social and human development, confront these same realities. . . . I appeal to the whole world, I appeal to leaders from all over the world, to open our eyes to the stark reality that we face. I appeal to ministers. . . . I ask of all of us here, if not us, then who? If not now, then when? If not here, then where? [19]

In the first sentence Saño represented Typhoon Pablo as something unprecedented; in the second sentence, it also seems somehow commonplace—not just to the Philippines but to all developing countries. This contradiction was heard everywhere after Pablo, and then again after Yolanda. In it is voiced the attempt of an inveterate "disaster culture" to absorb, explain, cope with, and patch over a particularly destructive event, while contending with the possibility that this one really was the start of something different. In Qatar, the world ignored both facets of Saño's wake-up call.

A typical news report from stricken Mindanao found *bayanihan* welling up from the lowest levels of village life:

Rescue squads that descended on [New Bataan] were by turns astounded and inspired by the open-arm welcome from the residents, whose actions bespoke their resilience. . . . At a school grandstand serving as an evacuation center in Barangay Cabinuangan, hundreds of villagers poured in on hearing word that fresh supplies of relief goods were to be distributed. They crammed the bleachers, as if they were about to watch a sports spectacle. Sporadic laughter broke out as residents from villages far and wide chatted among themselves, munching on bananas handed to them by social workers. . . . Police officer Cabuñas cited several instances when the villagers opened their homes to neighbors in need of shelter as the typhoon raged on Tuesday. That, he said, was why there were still people who survived in Barangay Andap. [20]

From the urban centers of Mindanao, groups such as the Digos Joggers Club and the Dog Owners of Tagum stepped up after Pablo to help the search and retrieval efforts. [21] Meanwhile, the rural electric cooperatives in charge of repairing power lines in remote areas engaged Task Force Kapatid, called "a modern day *bayanihan*" in the *Philippine Daily Inquirer*, through which technicians pooled their skills and worked for free with any cooperative in need.[22]

Given Pablo's early December timing, the shared burdens of *bayanihan* were felt most emotionally at Christmastime. Following a directive from President Aquino for all government agencies to hold only subdued celebrations, the national police and armed forces canceled their Christmas parties nationwide, donating the budgeted money to relief.[23] Many other agencies and companies followed suit, even though parties also had been canceled the previous year after Tropical Storm Sendong (international name Washi)—and would be again the next year, after Yolanda. A Catholic bishop joined hands with an environmental group called the EcoWaste Coalition to urge ordinary citizens to join in with "an ecological and simple celebration," allowing money to be set aside for charity while reducing the environmental impacts of holiday trash, or "holitrash."[24]

These appeals were laced with imagery from the field of devastation, reported on December 24 under headlines such as "Christmas Among the Ruins":

> Dominga Daipan's only wish this Christmas is to serve pansit (noodles) as noche buena to her seven children. The problem is she doesn't have this special Filipino dish for Tuesday's special occasion. . . . With their local chapel destroyed, they have been holding [mass] on a tennis court. . . . "We only have canned sardines and NFA (National Food Authority) rice but we do not care. What is important is that we were able to survive Pablo. We should be really thankful that we are breathing right now," the 49-year old Mandaue said.[25]

The spirit of giving began with relief pledges from companies and foreign governments, but everyone soon joined in, with one telethon raising more than $22,000 in text message donations as small as 11¢.[26] In February, as rebuilding commenced, environment secretary Ramon Paje announced a partnership with the Department of Tourism to bring money back to stricken parts of Mindanao through "voluntourism," a new concept whereby Filipinos could plan their summer vacations around *bayanihan* in disaster-hit areas. He suggested Pablo-struck regions as places where tourists could go to help build houses for villages destroyed by flash floods.[27]

WAKING UP

Bayanihan and *bahala na* are two parts of a cultural immune system that heals over wounds such as Typhoon Pablo, allowing life to carry on. Not everyone thought the harm should be brushed off, however. The idea of the disaster as a wake-up call also resonated widely, urging something other than just relief and recovery. Yeb Saño expressed this to the climate change conference, and he would renew that call a year later after Yolanda. Meanwhile, within the national discussion, there rang other alarms calling the Philippines to action.

Secretary Paje, the man behind "voluntourism," was a strong voice of warning over the contribution of illegal logging and mining to the devastation. Despite a total logging ban in all natural forests, imposed by President Aquino after deadly floods and landslides in 2011, the industry continued to operate with impunity. It had done so through repeated ineffective and temporary bans over the years,[28] even after the country's forest cover shrank from 70 to 20 percent over the course of the twentieth century.[29] Out of thirty-one illegal logging hotspots tracked by Secretary Paje's department, 80 percent were in the disaster-hit regions of Mindanao, including in Compostela Valley, where Barangay Andap had been buried. "This [devastation] is now proving that a total log ban is right," he urged as soon as Pablo had passed. "Several quarters are criticizing the declaration of a total log ban but look at what happened."[30] He further suggested that landslides in mountainous areas of the province were made worse by illegal small-scale gold miners.

The governor of Compostela Valley, Arturo Uy, came to the region's defense, maintaining that neither mining nor logging was currently going on in the affected barangays. The last big mining company to operate in the area, the Canadian-owned Sabina Gold and Silver Corporation, had pulled out years before. As for the state of the forest, "there's no illegal logging here anymore, because obviously, there are few trees left to cut," Governor Uy said.[31] Little notice was taken of such claims made by politicians—it was widely assumed that they all benefited in some way from both gold and timber. For the columnist Neal H. Cruz, the evidence was right there in the disaster photos coming out of Compostela Valley: "Many logs can be seen scattered among the ruins or floating in the waters around them. They are not trunks of trees uprooted by Typhoon 'Pablo.' The trunks have been neatly cut by chainsaws. They are logs. Washed down the mountainsides by the floods, they served as the battering rams that wrecked the houses left standing by the strong winds of Pablo. There is a total log ban nationwide.

So why are there still so many logs?"[32] Another columnist, Ramon Tulfo, pressed harder: "Most of the masterminds are politicians, a congressman, a governor and a mayor, among them. [Paje] knows who they are. If he doesn't, then he's *tanga*"—that is, an idiot.[33] Two months later, Paje found an apolitical and constructive response. His Department of Environmental and Natural Resources earmarked $5.25 million for a cash-for-work program that would employ more than four thousand residents of Compostela Valley and Davao Oriental to replant trees.[34]

Journalist and sociologist Randy David weighed in with a broader story of the region's vulnerability. "The poor have a natural affinity with disasters," he wrote. "More than anything else, therefore, what the devastation wrought by Pablo has revealed is the reality of persistent poverty in Mindanao." He continued:

> In the banana town of New Bataan in Compostela Valley, most of the affected families are workers or, as an Agence France-Presse report puts it, "sharecroppers in an industry that has grown to become the world's third largest exporter of bananas—after Ecuador and Costa Rica." . . . Its economy has been shown up to be as frail as the wooden homes of its residents, its prosperity as shallow as the roots of the banana plant. . . . The veneer of progress it has installed is visible in the road networks that connect the plantations to the ports from where boxes of bananas are shipped. What Typhoon Pablo has done, in effect, is strip away that veneer, exposing the reality of the poverty underneath.[35]

The auxiliary bishop of Manila heard yet another alarm bell, one that was imperceptible to almost anyone else. Bishop Broderick Pabillo had long been embroiled in the fight against a reproductive health bill being pushed by President Aquino that would provide free contraceptives across the country. In a radio interview, the bishop remarked, "I don't know if it's just coincidence, or God is sending a message that every time [the bill] is being discussed seriously a lot of sufferings happen to us." Pabillo's quote caused a minor brouhaha and drew a rebuke from the presidential palace, whose spokesperson asked Bishop Pabillo instead "to join hands with us to pray for a successful search for those who are still missing and to pray for the people who perished in the tragedy."[36]

More joining of hands took place in Davao City, the metropolitan center of the Pablo-hit region, at the end of January when activists, thinkers, artists, and musicians gathered for the Liwanag World Festival on Creativity and Sustainability. The luminaries who converged to highlight the lessons of the typhoon included organizers, social entrepreneurs, and inspirational speakers from around the country as well as a few international headliners, such as an Occupy Wall Street participant from New York and California's Barbara Marx Hubbard, "the world's leading exponent of conscious evolution and planetary transformation." Festival director Nicanor Perlas, a longtime activist and 2010 independent presidential candidate, announced the gathering not with a wake-up call but with a promise: "There is another world and another Philippines emerging in the very midst of decay and pain."[37]

RECOVERY BY NUMBERS

Meanwhile, the United Nations took on the immediate consequences of that decay and pain with the more traditional response: a series of international appeals. These followed a script reassuringly familiar to both the Philippines and UN agencies. In the days after Pablo, the need was pegged at $65 million to cover food, temporary shelter, and clothing for nearly half a million people and a projected six months of sustained assistance—about $130 per person.[38] A month later, the United Nations reported that the number requiring food assistance had doubled, and funding gaps were already slowing down the effort.[39] The appeal was revised to $76 million at the end of January for "immediate life-saving assistance" to the nearly 1 million people still living in temporary shelters (i.e., $76 per person), though only $27 million had been raised.[40]

While the global news media remained focused on the recovery from Sandy in New York and New Jersey, there were bright spots of international enthusiasm. A few weeks after the new UN appeal, former heavyweight boxing champion Evander Holyfield landed in Manila to advocate for affected children—a showstopper in a country where boxing is second only to basketball in popularity.[41] Then in May, the Philippines won the New Frontiers Award at the tourism industry's Arabian Travel Market. The annual prize recognizes destinations that have overcome great catastrophes to revive tourism, and the country came out ahead of New York and flood-ravaged Pakistan to take the year's prize. Tourism secretary Ramon Jimenez responded by reiterating the spirit of voluntourism: "What makes the Philippines more fun is its people. The *bayanihan* spirit, amplified during crises, is a unique social phenomenon that will forever drive the Filipinos to remain steadfast and dependent on each other for strength, through good times and bad."[42]

As 2013 wore on, though, bad times persisted in most of the region. Six months after Pablo, in Barangay Andap a new school year began for four hundred students under a tattered United Nations Children's Fund (UNICEF) tent and other scrap building materials. The teachers took turns cooking meals for the construction crew as they finished up the makeshift structure.[43] They hoped it would be a temporary fix, but many temporary buildings had already become permanent. The United Nations released another revision to its appeal, raising the target to $91 million to see the region through the year—more than twice what had been offered so far.[44]

This goal would not be met before other disasters took the stage late in 2013. By December, the UN World Food Programme (WFP) warned that malnutrition had become entrenched among children in the region. The agency, which had appealed for $21.6 million after the typhoon, had distributed $2.5 million in food and cash-for-work programs in the Pablo-hit region, but that was coming to an end at the one-year mark.[45] This was a timely deadline, for the WFP's resources were by then urgently needed by Yolanda survivors. The trucks had already moved north. On the other hand, the government itself barely met the one-year deadline to release $24 million of Pablo rehabilitation money from the budget of the Department of Public Works and Highways, which would be used to rebuild ten bridges on major transport routes. This topped off a national rehabilitation budget of nearly a quarter of a billion dollars.[46]

A modest share of the budget went to improving technological preparedness for future storms. The Department of Science and Technology (DOST) immediately unveiled plans for an advanced wind tunnel to test the resiliency of structures against Pablo-strength gusts.[47] This department had already undertaken a suite of technological projects after other recent storms, including geohazard mapping of vulnerable areas, an online public information and alert platform, and hundreds of automated water level sensors along rivers. In his July 22 State of the Nation Address, President Aquino praised these advances at length and pledged that the geohazard maps would be expanded to cover every corner of the country. On the same day, DOST announced the launch of a specially developed tablet computer for accessing the maps and related information, promising to eventually distribute one unit to each of the country's 42,028 barangays.[48]

The most painful recovery was faced by Mindanao's agricultural industry, reliant as it was on banana and coconut production that would take years to restore. Banana farmers and exporters asked for millions in assistance to reestablish plantations, 40 percent of which were destroyed. Along with the banana plants, some 3.5 million coconut palms were shredded.

Total damage to agriculture and fisheries was calculated at well over half a billion dollars.[49] The impacts on livelihoods were sudden and absolute: many farmers were landless tenants working on plantations, and their jobs disappeared overnight. Those who had their own land and money to re-plant faced the prospect of years without a harvest (new banana plants take on average eighteen months to bear fruit, while coconut palms often take five years).[50] Some joined government and UN cash-for-work programs for a time, while others simply migrated to the cities in destitution.

In the following months, exporters had to abandon an estimated $23 million in contracts for supplying bananas to the United States and other trading partners, losing much ground to competitors in Latin America.[51] The larger economic legacy, however, was not immediately clear. With the typhoon having struck less than a month before the end of 2012, the country charted strong annual growth for the year. The agriculture secretary suggested that the blow to agricultural output would show up only in the coming year and might be balanced overall by public spending on the res-toration of Mindanao.[52] As it turned out, the economy outstripped even the most optimistic forecasts: for the first quarter of 2013, the Philippines was the fastest-growing economy in Asia, beating even China's GDP growth rate. The boom was attributed to the manufacturing and construction sectors, as well as increases in government spending. It had the look of a silver lining.[53]

THE PRICE OF PORK

Government spending also became the source of the year's juiciest scan-dals. The bulk of Mindanao's rehabilitation package came out of the na-tion's $170 million calamity fund, emptying the fund. Beyond the reach of this package, local rebuilding and social services were paid for through other channels. A controversial, and soon to be downright notorious, source was the Priority Development Assistance Fund (PDAF). This discretionary money was allocated to each member of Congress every year to be spent in her or his district. Most people—and all newspapers—simply called it "the pork barrel." In July, the *Daily Inquirer* broke the story of business-woman Janet Lim-Napoles, who had been successfully diverting money from the fund through a syndicate of crooked contractors and politicians.[54] Over the past ten years they had scammed the government of an amount roughly equal to the quarter-billion-dollar cost of the Typhoon Pablo re-habilitation. Allegedly, Napoles even had one of her own cousins, Benhur Luy, kidnapped to prevent him either from revealing her misdeeds or from setting up a competing scam of his own. Once rescued by National Bureau of Investigation agents, Luy told them how he had handled bank deposits and withdrawals for his cousin, at times bringing so much cash back to her condo that it had to be stored in the bathtub. NBI called it "the mother of all scams." (In April 2015, after a two-year trial, Napoles was convicted of kidnapping and given a life sentence. She still faced charges of "plunder" in connection with the pork barrel scam.)[55]

As the scandal spread, the Supreme Court put a halt to all PDAF payouts. Local governments in Compostela Valley and Davao abandoned rebuilding projects, and social services were left half funded by their representatives.[56] Outrage over pork spread to another program recently created by President Aquino, the Disbursement Acceleration Program (DAP), which was billed as a tool to fast-track public spending and push economic growth. Despite protests that it had contributed in large part to the country's stellar growth in the first half of the year, along with bringing quick aid to victims of Pablo and other disasters, the DAP began to look like just another pork barrel— this one for the personal use of the president himself. A series of cases filed against the program in October brought its constitutionality into question, and it too came before the Supreme Court.[57] The court scheduled oral arguments on the DAP to begin on November 11, just four days after Napoles, now crowned by the media the "Pork Barrel Queen," would appear before a Senate committee. As it happened, both showdowns coincided with the arrival of Typhoon Yolanda, but neither process was sidetracked by the fresh catastrophe. On November 19, the Supreme Court ruled the PDAF unconstitutional. Soon after, the government backed down on DAP, stating that the program had done its job in growing the economy and was no longer needed.[58]

The unraveling of the pork programs put an end to a flow of resources that could have—and occasionally did—come to the aid of communities recovering from and preparing for disaster. It also more than vindicated the ingrained cynicism that serves as the default political lens for ordinary Filipinos. Politics as usual is, after all, the corrupt norm against which *bayanihan* gets its particular shine. "Confidence in the local community's ability to cope with trauma does not extend much beyond the confines of family, friends and neighbors," wrote Bankoff. "There is a deep-seated mistrust of politicians and bureaucrats who are regarded as largely impotent and venal, and disregard for a political system that is seen as woefully ineffective when it comes to disaster preparedness and largely reactive when it comes to disaster management."[59] The same dismissal of power can be heard in sighs of *bahala na*: if the community can't handle things on its own, only God (or Batman) can provide.

The alternative reaction is opposition. Sometimes it's a matter of words: behind every assertion of a "wake-up call" is a blame-casting project, accusing leaders of being asleep at the wheel. Other times it's a matter of all-out insurgency. The southern islands of the Philippines have a long history of armed rebellion, and the three largest modern militant groups can all be found in Mindanao: the communist New People's Army (NPA), the secessionist Moro Islamic Liberation Front, and the fundamentalist Abu Sayyaf Group. The communist insurgency is one of the longest-running in Asia, flaring up on different islands since the 1960s, its peaks closely tracking increases in poverty, inequality, and popular dissatisfaction.

At the time of Pablo, such conditions prevailed in Compostela Valley. The Philippines Army considered Barangay Andap part of a "red area"—a zone of strong communist influence—and had twenty-one soldiers stationed at a temporary command post in the village. Eleven died in the debris flow, and all were honored for acts of heroism.[60] Hundreds more soldiers soon rushed to the area to help with rescue work. The original twenty-one had been in the village under the banner of peace: the military was engaged in a new peace and security plan with a softer touch, pointedly called Operation Bayanihan. This strategy was launched in 2011 to replace the earlier Operation Freedom Watch, a more aggressive plan of eradication inspired by the United States' Operation Enduring Freedom, a global effort one branch of which had assisted Manila in fighting communist and Islamic insurgents. Whereas the old plan had become synonymous with political killings and torture, Operation Bayanihan promised a more holistic approach tying development and relief projects to the suppression of outlaw groups. The inspiration this time was the U.S. government's 2009 Counterinsurgency Guide, with its greater focus on winning hearts and minds away from militants.

The military's role in post-Pablo relief, rebuilding, and rehabilitation could not have been a better opportunity to showcase Operation Bayanihan. The new face of the armed forces was summed up in a Christmas Eve headline, "Santa Comes to ComVal" (shorthand for Compostela Valley).

They arrived immediately after Typhoon "Pablo" hit Mindanao. Most of them are soldiers from the 10th Infantry Division, who, along with rescue workers, were among the first to come, braving floods, mud and fallen trees to save lives in the provinces of Compostela Valley and Davao Oriental. For the soldiers, there was no need to know who had been naughty or nice, even if the village of

Andap, which was hardest-hit New Bataan town in Compostela Valley, is known to be influenced by the communist New People's Army.[61]

The holiday spirit apparently was not contagious. The next month, the New People's Army abducted a soldier and policeman in the valley and released a statement vowing to "frustrate" the charm offensive, calling Operation Bayanihan a ploy to continue "human rights abuses accompanied by more publicity gimmicks."[62] As the year of rehabilitation wore on, continuing clashes in the valley drove typhoon survivors to re-evacuate, and villagers alleged that soldiers were using house-to-house "needs assessments" to squeeze them for information on rebel movements.[63]

Just as troublesome were mass protests against the slow pace of relief. The leaders of these were not the communists but a wide variety of local alliances, from the survivors' collective Barog Katawhan (People Rise Up) to the relief group Balsa Mindanao. On January 15, nearly six thousand victims barricaded the national highway through Compostela Valley demanding more help for their remote villages. Facing down the police and military, they forced social welfare secretary Dinky Soliman into a personal negotiation in which she promised the survivors ten thousand sacks of rice.[64] When both rice and relief failed to appear after more than a month, thousands of survivors came to her department's regional office in Davao City, knocking down a gate and taking distribution into their own hands. There followed a forcible recovery of the rice by the police, a riot, and a three-day occupation of the office. Soliman again negotiated an agreement to accelerate relief but afterward filed lawsuits against the protest leaders.[65] One of those leaders was Christina Jose, village councillor from the mountain community of Baganga. On returning to the village, she was told by the military contingent in the area that any households who had participated in the protest would be removed from aid lists. The threat sent Jose back to Davao City to lodge a complaint. On the way to board her bus, however, she was shot and killed by motorcycle-riding assassins.[66] Her murder remains unsolved.

In this violent climate, poorly concealed behind declarations of post-disaster unity, perhaps the most divisive actor was the U.S. military. More than two decades before, Philippine legislators had voted to close all U.S. bases in the country, and at the time the Pentagon had seemed almost grateful to get the boot, its bases having become nearly uninhabitable under ashfall from the eruption of the Mount Pinatubo volcano.[67] America has not

been allowed to establish a permanent presence since, with military involvement in its onetime colony limited to that of an adviser. This was displayed
in annual joint training exercises with the Philippines military code-named
Balikatan (meaning "shoulder to shoulder"). As American interest in reestablishing a foothold grew, these war games became massive spectacles of
cooperation, staged to soften the popular opposition that had closed the
bases. When Pablo struck in late 2012, the two militaries happened to be
meeting in Davao City to draw up plans for the following April's Balikatan
exercise. Setting the conference aside, the Americans joined in the rescue.
Within hours U.S. aircraft were arriving to transport aid supplies; soon after, water purification units brought in by the Marines were pumping out
clean water for evacuees. But few skeptics were convinced by these actions.
Unease about the U.S. role revived in advance of Balikatan in 2013, when
a rumor spread that the exercise would be staged in the midst of Pablo-
devastated villages. Bishop Antonio Ablon, another organizer behind the
mass protests, warned that the war games would "only intensify the imple-
mentation of [Operation Bayanihan] in the hinterlands of Mindanao, especially in Davao regions where American-controlled mining corporations
are presently devouring mineral deposits and resources."[68]

The American ambassador, Harry Thomas Jr., tried to put this rumor to
rest at the Balikatan opening ceremony. The eight-thousand-troop exercise
took place in the north, far from the site of U.S. relief efforts. Thomas was
not, however, shy about drawing a connection. "Through that tragedy, we
saw the best of our partnership," he said in his speech. "It is incumbent upon
us to discuss, refine and promote humanitarian assistance and disaster relief
interoperability among our multinational regional partners. Balikatan provides the opportunity and commitment for this dialogue."[69]

In August, the mayor of Davao City sparked new controversy when
he loudly rejected a request from the Americans to use the city's old airport as a drone base. While the overture was guaranteed to raise alarm, the
United States had couched this proposal in terms of disaster assistance.
The embassy revealed that the military had already deployed drones over
the Philippines, but only at Manila's request and only for humanitarian
operations—including an aerial survey of Typhoon Pablo's devastation.
Amid local outrage, the Presidential Palace had to defend the use of U.S.
drones in such "special cases."[70] Meanwhile, the Philippines's Department
of Environment and Natural Resources was drawing up plans to use drones
in its own campaign against illegal logging in the region. The communists
thought it more likely that the drones would be looking for them.[71]

THE GREAT WALL OF BOHOL

Although we are tracing the eleven months between Typhoon Pablo and Typhoon Yolanda as an emblematic time between storms, it was not a disaster-free interlude. In fact, it wasn't even a typhoon-free interlude. Like the United States, the Philippines's naming system for tropical storms is alphabetical, so Yolanda's name is an indication of what a busy year it was. Among others, Typhoon Labuyo left eleven dead and caused $23 million of property damage in Northern Luzon in August. Tropical Storm Maring flooded much of metro Manila and surrounding provinces the same month, killing eight people and forcing 1,744,000 to evacuate—the second largest displacement of the year worldwide, to be surpassed only by Yolanda. In September, Odette, a super typhoon like Pablo and Yolanda, brushed the country on its way to a $4.3 billion strike on the southern coast of China; in the Philippines, it brought floods, landslides, and a tornado that tore through a village. Typhoon Santi cost $70 million and fifteen lives in Luzon in October, followed closely by the $6 million Tropical Storm Vinta, which killed four. The week after Yolanda, Tropical Depression Zoraida finished out the alphabet and the year, complicating relief efforts and causing two more deaths in Mindanao.[72]

Life outside of typhoon season wasn't much smoother. January and February saw various episodes of flooding and landslides in six different provinces of Mindanao and the far western island of Palawan. Another flash flood hit Mindanao in June, and landslides came in June and July. March and June brought flooding to Manila. A waterspout in March sank a riverboat with ten passengers in Maguindanao, while other destructive tornadoes struck Bulacan in April, South Cotabato in May, and Cebu City in June, the last destroying fifty houses.

A peculiarly Filipino—and decidedly unnatural—disaster claimed four lives on April 19. In one of the vast landfills surrounding metro Manila, heavy rain set off a trash slide on the side of a mountain of garbage, burying two and a half acres.[73] Yet this was nowhere near the scale of the tragic Payatas trash slide of 2000, which had crushed an entire barangay. Three hundred and thirty people, mostly scrap gatherers who had settled on the edge of the Payatas landfill to pursue their livelihood, are confirmed to have been buried alive in that tragedy, though in such an informal settlement the true number was impossible to establish.[74] Metro Manila authorities had made quick promises to reform the way the capital dealt with its trash, but subsequent incidents like the one in 2013 cast doubt on how much had changed and could change. The rivers of trash flowing daily out of Manila's 12-million-person urban conglomeration are their own sort of unstoppable physical force.

Calamitous 2013 was also punctuated by seismic bursts. Volcanic activity was not remarkable in 2013, but on May 7 the country's most temperamental volcano, Mayon, did generate a surprise ash eruption that cost five climbers their lives. The incidence of earthquakes was more typical—for most of the year. From January through the first half of October, the Philippine Institute of Volcanology and Seismology measured several earthquakes per month, varying between magnitudes 5 and 6.2, with none of them causing much damage.[75]

Then on the morning of October 15, every needle jumped. On their quiet family farm on the small central Philippines island of Bohol, a brother and sister looked up from their work to see the world break in half. All that Momerto Bautista could see at first was his prized water buffalo being launched into the air as the ground thrust upward. As he threw himself to the shaking ground to offer prayers to the Virgin (he says that at the time, "I actually believed the world was ending"), his sister, who had been making bread in the kitchen, looked out the window to see a ten-foot wall of earth rise up before her, pushing the grass of her family's pasture up and out of her line of sight.[76] A hitherto unknown geologic fault running through the farm had revealed itself, in the process triggering a magnitude-7.2 earthquake. As it slammed Bohol violently up and down, the quake moved the entire island nearly two feet to the west while tilting it slightly eastward. Bohol—50 percent larger and 20 percent more populous than Rhode Island—was devastated. The extreme vertical motion pancaked multistory structures, including houses, municipal buildings, and schools. Several large adobe churches, national treasures dating to Spanish colonial times, crumbled completely. Neighboring islands also shook, including the Philippines's second-largest metropolitan region on Cebu. With no way of knowing that Typhoon Yolanda lay three weeks in the future, the country faced what looked like the mega-disaster of 2013. The Bohol governor's office estimated the damage at $57 million, and 222 deaths were recorded, making it the Philippines's deadliest and most expensive quake in twenty-three years.

Bohol boasts an enchanted image featuring pirates and conquistadors, the dreamlike landscape of the conical Chocolate Hills, and the uncanny Philippine tarsier, perhaps the world's strangest-looking primate. Here can be found everything that makes Philippines tourism tick. The government had worked hard in recent years to make Bohol one of the country's most popular destinations, an effort that included winning the island back from a tough communist insurgency. The Armed Forces declared Bohol insurgent-free in 2010; in fact, it was that development-linked military success that would become a template for the national Operation Bayanihan three years later.[77]

The earthquake was a blow to all of Bohol's development plans. Iconic churches crumbled; landslides marred the slopes of the Chocolate Hills and swept away viewing platforms; roads, bridges, and other infrastructure collapsed. The disaster threatened not only the homes and safety of Boholanos but also the carefully nurtured tourism industry and the peace it had been counted upon to secure. The Department of Tourism rushed to declare Bohol "back in business" before the end of October, as roads to the Chocolate Hills and Tarsier Sanctuary reopened.[78] Quick financing came from the U.S. Agency for International Development and UN World Tourism Organization. Governor Edgar Chatto went even further, announcing that "Bohol's tourism has not been diminished, it has actually been added with so many attractions." The governor was an inexhaustible wellspring of optimism. He had visited Bali after the 2002 nightclub bombing and studied Thailand's Phuket tourist zone after the 2004 tsunami; he knew what it took to ride out a disaster. In Chatto's vision, Bohol would seize the moment and add geoscience tours to the itinerary. The three-mile fault line running through the Bautista farm would be rebranded as the Great Wall of Bohol. Visitors to the Chocolate Hills could get a glimpse into the famous cones where the landslides had exposed the raw "white chocolate" limestone within. A patch of uplifted seafloor that added forty-five acres to the island's west coast would become a must-see curiosity. Of the obliterated Spanish churches, Chatto declared, "The ruins are tourist attractions by themselves."[79]

When we took our own informal geoscience tour of Bohol a couple of months after Chatto's announcement, the sense of wonder seemed to have receded a bit. Along freshly repaired roads, residents still camped outside crumpled concrete houses and attended class in the playgrounds of fallen schools. Yolanda had passed safely north of the island but had diverted attention and funds away from the earthquake, so little rebuilding activity was obvious.[80] The UNESCO World Heritage churches of Maribojoc and Loon were not even recognizable as ruins; most of the centuries-old masonry had been reduced to yellowish powder by the quake. (Signs nonetheless warned souvenir hunters that the heaps of dust and rock and everything in them were still protected as cultural treasures.) Green grass was overgrowing the white filling of the Chocolate Hills. On the island's west coast, the barren shelf of brand-new land created by uplift of the seafloor drew a few interested tourists, but the local women selling government-issued tickets to the site said they were dissatisfied. Formerly they had made their living collecting shells in the shallows and selling them to tourists; now all sea life was gone. (Soon after our visit, some locals constructed huts on the new landscape. Authorities demolished them, but there would eventually come a three-way tug-of-war among two local governments, which were looking to turn the new land into a tourist area with attractions such as an amusement park and sports facilities, and the national government, which decreed that the area be preserved as a "geological monument." Bohol University scientists note that it will be greenhouse warming that eventually steps in to resolve the dispute when, possibly as soon as 2025, the new land will be submerged once again by rising seas.)[81]

The Great Wall of Bohol itself could still be found at the end of a long series of rough country roads, with hand-painted signs marking each turn. The fault face was not quite vertical, having slumped in the rains, and weeds were colonizing the soft black soil. It was clear that the aboveground evidence of the fault, so striking in early news photos, would soon become an unremarkable feature of the already hilly landscape. There was, however, one permanent effect here at the epicenter: the fault had created subterranean cracks that swallowed up all the water from the aquifer below the Bautistas' field. The family who had watched the birth of the Great Wall now had a dry well, with no other water source available nearby.

YOLANDA

On the night of November 6, 2013, eleven months after Typhoon Pablo and only three weeks after the Bohol quake, the Philippine weather agency gave the name Yolanda to a fully formed super typhoon bearing toward the country's east coast. It was still far out on the open Pacific, but experts could already see that this would be a bad one. Using satellite measurements, the Japan Meteorological Agency reported maximum ten-minute sustained winds of 145 miles per hour, and then one-minute sustained winds of 195, securing Yolanda a place among the strongest tropical cyclones ever witnessed; it may have been *the* most powerful ever, at least until 2015's Hurricane Patricia roared through the eastern Pacific. And unlike other record storms, which had spent at least some of their energy out at sea, Yolanda looked set to hit the Philippines at near-peak strength.

The day of the storm's christening, President Aquino activated all local and municipal disaster councils and tasked them with readying hundreds of thousands of people for evacuation. The following night he addressed the nation, as he had during Pablo. "I decided to speak before you so that you'll be aware of how serious the danger that our countrymen will face in the coming days," he said, "and to call for *bayanihan* and cooperation."[82] The government had pre-positioned $4.5 million worth of emergency relief along the exposed eastern edge of the country, and the navy and air force were ready to deliver it. The U.S. military was again standing by for support. The secretary of presidential communications, Herminio Coloma, put on a brave face. The government was better prepared this time, he told the press, having learned the lessons of Pablo and other storms. "Over the past three years, there has been a very substantial upgrade and modernization of our meteorological forecasting equipment and capability," he assured the press. Aquino had set a "zero casualty" goal, and through its improved data-gathering and -sharing systems, the country had the tools and "mind-set" to avoid unnecessary loss of life.[83]

With the path of the monster storm still uncertain and a third of the country under threat, the government was especially concerned for the areas hit by earlier disasters. The Mines and Geosciences Bureau warned that the freshly shaken earth of Bohol and Cebu was primed for landslides, in areas where tens of thousands of people were still living in tents.[84] Mindanao was much less likely to take a direct hit, but officials there weren't taking any chances. All of the previously Pablo-hit provinces scrambled to organize preemptive evacuations.[85] As the typhoon's path began zeroing in on Leyte Island in the historic typhoon belt, well north of Mindanao, Davao Oriental governor and Pablo veteran Corazon Malanyaon had just one piece of advice for his neighbors on Leyte: "Pray hard."[86]

On Leyte, the residents of Tacloban, not yet knowing they were fated to be at ground zero, began preparing. All fifty-four public schools, the state university, and the convention center (better known to locals as the Tacloban Astrodome) were designated as evacuation centers. Actual evacuation was, as usual, only partial. On the south side of town in the coastal village of Candahug, a reporter found soldiers evacuating women and children. Most of the men were staying to brave the night and guard their homes. Elegio Altis, a sixty-five-year-old gardener tending to a beachside memorial, shrugged off the reporter's questions. "I have encountered many storms in my life and I've always survived," he said. "If this storm will take me, then I leave it to God."[87]

Then Yolanda hit.

"WE CAN STOP THIS MADNESS"

It was five violent days later, at the COP19 climate negotiation in Warsaw, Poland, that climate delegate Yeb Saño had his opportunity to make a second annual appeal to the world. In it, he repeated verbatim his words from the previous year's conference: "If not us, then who? If not now, then when? If not here in Warsaw, where?" he pleaded. "We can stop this madness."[88] This was only the first of many repetitions.

Invocations of Filipino values began with no less than U.S. president Barack Obama, who wrote in his condolences, "I am confident that the spirit of Bayanihan will see you through this tragedy."[89] For a third year running, Christmas parties were canceled or scaled back around the country. Op-ed writers argued that this was it, the big one for sure, the true cyclonic wake-up call—to poverty, corruption, climate change, the precarious environment. Most prominently for environmentalists, where Pablo had exposed the scandalous treatment of mountain forests, Yolanda revealed the sad state of coastal mangroves.[90] The Philippine government began 2014 with a $926 million budget allocation for recovery and ended the year with a $3.8 billion rehabilitation plan for the years ahead.[91] The United Nations' aid appeal climbed to $788 million, ten times the request after Pablo. International donors ultimately committed 60 percent of that amount.[92] Agriculture was crushed, with 34 million coconut trees damaged, all but collapsing the industry, and hundreds of square miles of rice flooded.[93] Yet overall economic growth had touched 7.2 percent in 2013 and was headed for 6 percent in 2014, still carried forward by the industry of construction and reconstruction—and heavy doses of fossil energy.[94] Much like Australia, the Philippines shrugged off climate vulnerability to double down on coal power, with fifty-five new coal-burning power stations announced or under construction in late 2015.[95] Dissatisfaction with the response and recovery was more rampant than ever, with much of the criticism revolving around a suite of flimsy, tiny, overpriced, and possibly kickback-laden government bunkhouses. (A year later, 20,570 survivors would still be living in these, sharing ninety-three square feet to a family. Two years later, only 342 permanent housing units had been built in Tacloban, out of the 12,345 targeted.)[96] Despite lurid rumors of New People's Army attacks on typhoon-hit towns and aid convoys, the communists called a cease-fire in the region and joined in relief drives; all was quiet until the following July, when the NPA ambushed and killed a mayor who was delivering medical aid in collaboration with the national army.[97] Retaliation followed, and not just against the guerrillas. By late 2015, a local human rights group blamed the

Philippine military for episodes of torture and extrajudicial killings of thirteen Yolanda survivors and community leaders, including members of the vocal new survivors' alliance called People Surge.[98] Throughout 2014, Judith Buhay visited Yolanda-shattered barangays presenting seminars on disaster risk reduction and preparedness. As Typhoon Ruby bore down on Tacloban in December, she helped in the mandatory evacuation of Barangays 36 and 37; by January 2015, Buhay was buying seeds of okra, beans, eggplant, tomatoes, peppers, and luffa in preparation for an organic food production network of twenty-five households.[99] But in Buhay's Barangay 37, the ban on rebuilding within forty meters of the water evolved fast. Two weeks after we'd visited, the government had decided to amend the "no-build zone" to a "no-dwelling zone," citing the possibility that future investors might want to build tourist resorts and other structures that would be "resilient enough" to handle a future storm surge.[100] Within months, seventy-two rebuilt houses in the barangay were demolished under the zone rule to make room for the new $1 million Tacloban Fish Port Complex.[101] As a next step, the government reconsidered its plan to return the forty-meter band along the waterfront to mangrove forest, instead seeking Japanese assistance to construct a seventeen-mile, $160 million concrete seawall to protect the waterfront. Inevitably, this controversial plan came to be known as the Great Wall of Tacloban.[102] U.S. military involvement in relief efforts redoubled after Yolanda, peaking at 13,400 troops, 66 aircraft, and 12 naval vessels.[103] The following year's Balikatan cooperative military exercises focused on disaster situations. This was a prelude to the signing of the long-sought Enhanced Defense Cooperation Agreement, which finally secured the first firm rotational presence of American soldiers in the country since the Pinatubo eruption year of 1991, "desiring to enhance cooperative capacities and efforts in humanitarian assistance and disaster relief."[104] Meanwhile, the next Liwanag World Festival came to quake- and storm-hit Cebu City with the theme "Disasters: From Chaos to Rebirth."[105] And voluntourism returned with a bang, this time on an international scale. David Beckham and Justin Bieber were the most famous visitors—until January 2015, when Pope Francis visited Tacloban.

Leading an open-air Mass at the airport, the pope spoke to 150,000 worshipers: "Let us respect a moment of silence together and look to Christ on the cross. He understands us because He endured everything." The pope wore a yellow poncho over his vestments as rain lashed the gathered faithful. After he'd been on the island just four hours, however, the papal visit was forced to conclude abruptly with an early departure. Tropical Storm Amang, the first of the new year, was on its way.[106]

3

NEIGHBORS TO THE SKY

New York City

From the edge of the horizon furiously comes to them
The most terrible of the progeny
Which the North has till then contained within it.
The tree holds up well; the reed bends.
The wind doubles its trying;
And does so well that it uproots
That, the head of which was neighbor to the sky,
And the feet of which touched the empire of the dead.
   —Jean de La Fontaine, *The Oak and the Reed*, 1668, after Aesop [1]

Disaster survivors can try to restore their world as it was on the day before, or they can hit fast-forward, attempting to speed over the rough patch to a better tomorrow. Cities can build back, or they can build back better. Communities seeking light can look to the past or to the future. Either way, the present must come first: they have to restore the basic machinery of life, ameliorate the hurt, feed the hungry, clear the debris, and mourn the lost. But where does the story go from there? That depends entirely on well-laid plans and the forces that break them.

DISASTER AS FIASCO

As the waters began to recede, a thirteen-ton crane boom seemed to hang by a thread over the city. Every TV network had at least one camera fixed on the swaying wreckage, a thousand feet above Midtown, where the winds had peeled it away from the frame of a seventy-five-story skyscraper under construction. Fifty-Seventh Street and neighboring buildings had been cleared; nobody could predict where steel falling a thousand feet into the canyon of a Manhattan street might ricochet. Now everyone just watched. Amid the dark, swirling chaos brought to New York City by Superstorm Sandy, here at least was one simple problem: the crane would hold or it would fall.

The twisted sword of Damocles hung over the city for seven days before

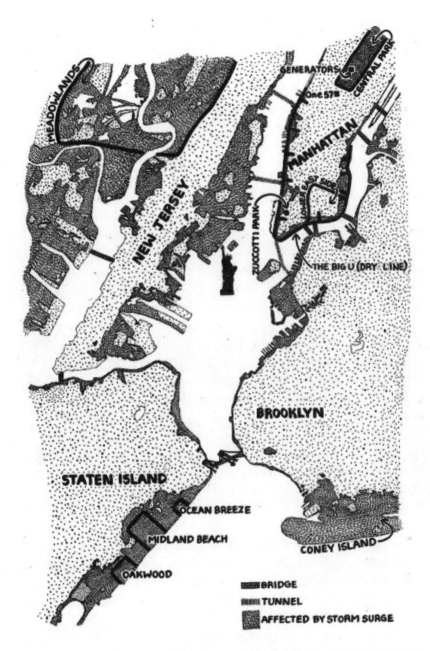

Figure 4. Areas of New York and New Jersey affected by Superstorm Sandy in October 2012. Darker-stippled areas were flooded by storm surge. The "Dry Line," formerly the "Big U," would, if built, surround the southerly part of Manhattan; hash marks along the Dry Line indicate "resilience districts" that would be compartmentalized by smaller barriers. Residents of Ocean Breeze and Oakwood on Staten Island requested and accepted property buyouts and moved away after Sandy.

crews carefully secured it to the side of the tower and declared the area safe. Fifty-Seventh Street was reopened. Another six months later, almost to the day, a second crane was hoisted beside the tower in order to remove the failed appendage and finish the construction job. The building's developer, Gary Barnett, shrugged off a failed insurance claim for the wayward crane and a lawsuit from two Fifty-Seventh Street dentists over lost business. It all proved to be small change. Within two years of the storm, the new tower, called One57, stood as the tallest completed residential building in the city, rising to a height just forty feet shy of the Chrysler Building. A five-star hotel operated in the first twenty-five floors, and above that, Barnett told *Bloomberg Businessweek*, he had sold ultra-luxury condominiums worth a total of $1.5 billion and counting. The magazine made no mention of the crane collapse, its memory and cost erased by a single, record-setting $90 million penthouse sale. "He's the developer of what looks like the most successful condominium in New York," a real estate tracker told the magazine.[2] Another year later this was still true. By Sandy's third anniversary, One57 claimed the most expensive square footage in the city and had inspired a crop of even taller condo towers planned along the new "Billionaire's Row" of Fifty-Seventh Street: 1,396-foot 432 Park Avenue, 1,428-foot 111 West 57th, and Barnett's own follow-up act, the 1,550-foot Central Park Tower.[3] Once completed, these would become the three tallest residential buildings in the world.

Had every story after Superstorm Sandy gone this way, life could have carried on much more smoothly: a terrifying night, a nail-biting week, and then—everything under control. Get a crew in, take care of it. Just the cost of doing business.

Michael Bloomberg, mayor of New York City and the tenth-richest man in America at the time of the storm, had no doubt that this was the way his city worked. He spent the days after Sandy making appearances in affected neighborhoods, meeting the standards of a leader during crisis, speaking of hope and resilience. Then the issue of the New York City Marathon came up, and things got messy. The world's largest footrace was scheduled to take place only six days after the storm, snaking through all five boroughs of the city. Many of the 47,000 registered runners had already arrived, waiting for the marathon in their hotels. In the streets they would be joined by 2 million spectators, eight thousand volunteers, and the throngs of city police, emergency services, and sanitation workers required for an event of such scale. City services were already stretched to breaking point but, Bloomberg decided, if it could happen, it must happen. Symbolically and economically, the city needed the marathon more than ever. "It's a great event for New

York," the mayor said the day after Sandy, "and I think for those who were lost, you know, you've got to believe they would want us to have an economy and have a city go on for those that they left behind."[4]

The backlash began immediately and grew with each day that passed. The event seemed an extravagance at a time when unknown thousands of city residents were without power, shelter, and basic necessities (as the mayor spoke, more than 6 million East Coast residents remained without electricity, including most of Manhattan below Thirty-Fourth Street).[5] Rumors spread of marathon volunteers being pelted with eggs while setting up equipment. Worse still, the run's traditional starting line was on Staten Island. In ordinary years, it would be one of the rare occasions when the least populous borough makes it on television; this time, television crews were already on the island, documenting some of Sandy's worst devastation, and marathon organizers feared that they would broadcast images of a hostile crowd at the starting line, eggs in hand. Yet it was a trio of giant generators sitting in Central Park, at the other end of the race, that did the marathon in before it began. Two were powering the event's media tent near the famous Tavern on the Green restaurant, with the third standing by as a backup. The press, primed for outrage, were more interested in the generators than anything going on in the tent.[6] By the New York Post's calculations, each one of these three machines in the park could instead have been powering four hundred homes in the large swaths of the city still cut off from the grid. A police source, angered by diversions of New York Police Department (NYPD) resources, told the paper, "You know what this is about? This is all so Bloomberg can stand at the finish line Sunday and tell the world we bounced back."[7]

Two days before race time, Bloomberg was still in no mood to compromise on the marathon as a necessary beacon of hope. He told reporters he had been speaking with his predecessor Rudy Giuliani, who encouraged him. On a trip to Sydney, Giuliani told a receptive Australian press that Bloomberg should, in local terms, have a go: "I hope beyond hope they have the marathon on Sunday, because we end up with about thirty thousand runners and a couple million people on the streets of Manhattan, which reaffirms the fact that we're tremendously resilient and will overcome anything," the ex-mayor said. "One of the proudest things I participated in after September 11th was making sure the marathon took place. There were more people on the streets than maybe ever before, and I think that we'll probably see the same thing again."[8]

In the end, even with the spirit of 9/11 invoked, the generators chugging away in Central Park were too much to bear. "The conversations around it had become caustic," a mayoral adviser told the *Times*.[9] It was too much celebration too soon—it didn't look like resilience so much as blind disregard. The plug was finally pulled, less than forty-eight hours before the starting gun was to go off. Unused PowerBars, bottled water, and T-shirts were quickly handed over to the Sandy relief effort, and thousands of runners headed to Staten Island to join the cleanup effort and run supplies up the stairs of blacked-out high-rise blocks. The disputed generators stayed put in the park for a while, until the rental companies came to collect them. Whether any of them got a chance to power four hundred homes is unknown. The vendors who owned the machines took them back to fill new orders from other customers. According to the mayor's office, some left the city for New Jersey. The clumsy impropriety of the marathon could now be left behind, but the misdirection of resources ran deeper. The power that would have supplied the event streamed back into the circuits of the free market, where generators were in sky-high demand.

OCCUPYING SANDY

The storm had made its hardest landfall "out there" in the Rockaways, Staten Island, Coney Island, the Jersey Shore: devastated zones that were hard for many New Yorkers to imagine. They were far away even for people now living "south of power" in the dark lower end of Manhattan, where people's needs were also great, but most of their homes—at least those above the ground floor—were intact. Here neighbors and community groups did what they could for one another and waited for the power to come back. Where life had to go on, New Yorkers had their own *bahala na* outlook. In the words of one FedEx courier making a delivery to a Grand Street warehouse while the storm bore down, "I had a shift. What are you gonna do?"[10]

Some of the neighbors had particularly recent training in urban survival and mutual aid. It was less than a year since the Occupy Wall Street movement had been crushed and evicted by the NYPD from Zuccotti Park. Unable to claim a new public space, the Occupiers had scattered to all corners of the city and a multitude of new undertakings. Goldi Guerra, an activist and musician who had been with Occupy from day one, was back at home on the Lower East Side. In the first days after Sandy, with much of that low-income neighborhood flooded and all of it without power, Guerra helped out however he could at a Sixth Street aid center coordinated by the community group Good Old Lower East Side (GOLES).[11] Elsewhere in the city, two other Occupiers joining in early aid efforts started up an online fund-raising campaign under the title Occupy Sandy. The name stuck fast, and it struck a chord: mutual aid was back. Eager volunteers—old Occupy faces and new—streamed to the hubs listed on the movement's website. GOLES was the only hub designated in Manhattan. That weekend a thousand volunteers showed up. The power had just come back on for the Lower East Side and there wasn't enough work for so many people. Guerra handed out an address on Staten Island that he had gotten from Occupy Sandy organizers and promised to meet them there.[12]

The first day was pandemonium. Sixteen percent of the island had been inundated by the sea. "Thousands of people just walking around taking pictures," Goldi recalls. It was the day of the canceled marathon, and runners were dashing around with supplies—some with destinations, some without. The address Goldi had been given, it turned out, was a church that had offered a lobby as a staging space but otherwise had neither connections nor suggestions. The Occupiers started with daily rounds, going where people were gathered outside and getting contacts. "Just kept moving and seeing who's doing the best stuff," Guerra said. This eventually came down to six hubs that were "right in the middle of the action," where Occupy sent volunteers and supplies. A week on, they got their own hub in another church. "As great as it was, it took us a month to realize we weren't getting a cross section of the population coming in. We were getting their parish and neighbors and that was it." They moved into a warehouse and reached out to a larger community.

On Staten Island the volunteers stepped into dense networks of local groups. Yet as long as crisis was in the air, anyone willing to get their hands dirty was welcomed in. For six months straight, the task was simple: "mucking out" flooded houses to remove wet furniture, drywall, and general filth. Mucking out was the first step that came before anything that might look like recovery. It took more than just volunteers; by an estimate from the National Day Laborer Organizing Network, at least four thousand day laborers—mostly undocumented immigrants from Latin America—worked on cleanup in the New York area, enduring dangerous, sickening conditions for a few dollars an hour.[13] But at least they, too, benefited from the spirit of the times, if only briefly. One such Staten Islander, Miguel Ángel Piñeda, told a reporter two years later, "There wasn't racism then, not like now or like before. There was a lapse of time when Americans didn't see us as immigrants, but as other people here dealing with the same things."[14]

Occupy Sandy drew more media attention than anyone else responding to the storm. For the Occupiers, the positive coverage was something new. On the decimated Rockaways in the immediate aftermath, *New York Times* reporter Alan Feuer discovered that the only relief groups operating were the National Guard, the Federal Emergency Management Agency (FEMA), the police and sanitation departments, and Occupy Sandy. "Maligned for months for its purported ineffectiveness," he wrote, "Occupy Wall Street has managed through its storm-related efforts not only to renew the impromptu passions of Zuccotti, but also to tap into an unfulfilled desire among the residents of the city to assist in the recovery." A hot-and-cold critic of the movement, the *Times* was won over anew: impressed by Occupiers' ability to organize chaos at their Brooklyn distribution points, charmed by their DIY attitude, and inspired by a force "capable of summoning an army with the posting of a tweet." Judged the paper, "It is only with Hurricane Sandy that the times have conspired to deliver an event that fully calls upon the movement's talents and caters to its strengths."[15]

Even the Department of Homeland Security and FEMA complimented Occupy through their information-sharing network, albeit in a report with a title that read like a condescending pat on the head: "Youthful Energy and Idealism Tackles Real-World Disaster Response." In the first week after Sandy, the report stated, the movement was distributing ten thousand meals a day through the work of fifteen thousand volunteers. Within four months Occupy could claim sixty thousand volunteers and was on the way to raising its first million dollars in online donations.[16] The NYPD was also paying attention: it was discovered a year later that one regular visitor at the Brooklyn distribution points was an undercover detective tasked with keeping tabs on Occupy's new plans.[17]

The Occupiers on Staten Island planned to stay out of trouble, but that sort of thing didn't come naturally. They soon linked up with other energetic responders, such as the immigrant organization Make the Road New York and the homeless charity Project Hospitality—groups that, despite the common cause, were treated with suspicion by residents who still subscribed to the image of a borough of conservative Italian American homeowners. Here reality diverged from representation, as the low-lying coastal neighborhoods had actually become some of the most ethnically diverse in the city, and many of those homeowners were devastated by the mortgage crisis long before the storm hit. Nevertheless, the traditional self-image had endured.

Veterans of Occupy Wall Street were well aware of how they were being viewed. Their ideas of community-powered recovery extended well beyond mucking out shoulder to shoulder and came a little too close to politics for Staten Island's tastes. Once, walking into a meeting with three civic groups, Guerra was greeted with the question "So, you're a communist? Tell us about it." After they left their original church location and pushed their partners to work in more diverse areas, Guerra says, Staten Island's borough president referred to the Occupiers in another meeting as "Iraqi insurgents."

HIDDEN ROT

As the work dragged on into spring, it turned out that the real enemy was living under the floorboards. With the arrival of warmer weather, mucked-out houses erupted with mold. Bill Sothern, a mold researcher who heads the remediation company Microecologies, estimates that virtually all of the seventy to eighty thousand single-family homes that were flooded around the city experienced mold problems, in most cases significant ones. The invasion began on the paper backing of drywall, but where the wallboard was removed—an important part of the mucking-out process—mold was also rampant on exposed structural wood. And it didn't go away. Two years on, Sothern was still concerned about five to ten thousand homes of mostly poor New Yorkers where mold remains a health danger.[18]

"The moment the weather hits 70 degrees and rainy, the mold pops out like in a horror movie," said Guerra. Like many others, he got sick from constant exposure to the organisms. There were days when spores would hang in the air over the beach like a mist. Recalling the "Katrina cough" heard everywhere in post-hurricane Louisiana, a local Fox affiliate identified the sound of "Far Rockaway cough."[19] It did seem like a horror movie for residents who felt beset by an alien organism they couldn't fight. Many organizations recommended bleach, but this was both caustic and ineffective. Proprietary chemical blends looked promising but were hugely expensive, not priced for such a large cleanup. (Sothern believes that soap, water, and elbow grease work as well as anything.) It was even harder to find good health information. After decades of legal wrangles over sick-building syndrome and toxic black mold, researchers still haven't come to any airtight conclusions on the dangers of spores. But the creeping invader wasn't as alien as it appeared. Sothern's company detected none of the dreaded black mold in properties but overwhelmingly found ordinary green *Penicillium* molds, along with some equally familiar but more dangerous *Aspergillus*—common household molds that flourish on wet wood and drywall. The invader had been in the house all along; it was just the environment that had changed. Not all residents were shocked by this. A survey of nearly five hundred public housing residents found that 45 percent had visible mold in their apartments after Sandy, but 34 percent had already had it before.[20]

The mold was one of many problems that had been lurking in the walls long before Sandy. Everyone cleaning up got a close look at what Sandy had washed in and what it had simply uncovered. The storm brought volunteers and workers together in a certain understanding of urban disaster—something that may prove far more enduring than the heady solidarity of the mucking-out days and quite unlike the view from the mayor's office or from atop the One57 tower. The Superstorm Research Lab, an academic collective organized after Sandy, documented both of these perspectives in an interview-based study called *A Tale of Two Sandys*:[21]

> Responses to Hurricane Sandy consistently cluster into two types according to how the issues have been defined and understood . . . on one hand, the crisis was seen as a weather event that created physical and economic damage, and temporarily moved New York City away from its status quo; on the other hand, Hurricane Sandy exacerbated crises which existed before the storm and continued afterwards in heightened form, including poverty, lack of affordable housing, precarious or low employment, and unequal access to resources generally.

Interviews revealed that affected residents, community-based groups, and volunteer responders tended toward this second interpretation. They knew the existing problems in already struggling and racially divided neighborhoods, and the fault lines became clear as some people recovered while others didn't. Meanwhile, city officials and large organizations managing the recovery favored the first and simpler understanding. Many of them just seemed confounded by people who couldn't bounce back. One experienced disaster manager balked at the interviewer's mention of race, class, and gender, calling the question "horse shit":

> I have never been involved where you think of poor versus rich, black versus white. I've been telling people for years storms do not differentiate between Republicans and Democrats. You respond the same way no matter who they are. The problem is in the poorer neighborhoods it's a little more difficult to get back on your feet because they don't have the resilience, the money in the bank, the borrowing ability to go out and get the money you need to rebuild your house, buy new stuff for your apartment. It's not there. And that, I think, is the fundamental issue.

A city official, asked about shelter in the depths of the post-Sandy winter, sounded even more heartless. "This is three freaking months, if you haven't found a place to be warm in three freaking months you don't need the city's help; you've already figured it out," he griped. "At some point the responsibility of government has to stop somewhere and you as Joe Blow citizen need to figure out how to do your own thing."[22]

FANTASIES OF ACTION

Understanding disaster as a temporary interruption of the urban machine, the authorities had made plans to fix it. Lots of plans. At city, state, and federal levels, planning for disasters like Sandy is an intense and never-ending process. In another study, Superstorm Research Lab members learned something interesting about these plans: when Sandy struck, they were mostly left on the shelf. Officials and first responders described to the researchers how Bloomberg sidelined the city's Office of Emergency Management, with its detailed hurricane plans and exhaustively mapped-out hierarchies of command, and brought emergency functions directly into the mayor's office. Larger, intricately coordinated plans drawn up by the Regional Catastrophic Planning Team of emergency managers from New York, New Jersey, Connecticut, and Pennsylvania also went out the window.[23]

The administration wasn't alone in this failure. Abandonment of emergency plans seems to be the norm during events like Sandy. Authorities who spend millions of dollars on planning have their own reasons for ignoring the results; often, they're seen as too complex to be effective when the system itself is under pressure. This tendency is so pervasive that the sociologist Lee Clarke has called such emergency management plans "fantasy documents."[24] Fantasy documents emerge, Clarke says, from efforts to deal with unpredictable situations in which the ability to know what constitutes "effectiveness" is terribly low or nonexistent. For those who claim authority over disaster, "rationalistic plans and rational-looking planning processes become rationality badges, labels proclaiming that organizations and experts can control things that are, most likely, without the range of their expertise. Planning then becomes a sign that organizations hang on themselves advertising their competence and forethought, announcing to all who would listen, 'We know what this problem is and we know how to solve it. Trust us.'" By jockeying for expert status, planners put themselves in charge even when their plans aren't. "Since claims to expertise are always claims that somebody should be left out of the decision loop," he writes, "planning is deeply, unavoidably political."

Some organizations themselves began to look like fantasies in the weeks after Sandy. The day Guerra arrived on Staten Island, the first person he met, the pastor at the church where Occupy was supposed to set up, was cursing the name of the Red Cross: "He was saying, 'Fucking Red Cross, they don't know what they're doing. You guys know what you're doing. You guys know what it's like to not have food, to be in a park without light. The Red Cross doesn't have a clue.' And I was thinking to myself, really, it's that bad already? We're only five days into it and this guy hates them?" But the pastor's expletive would become a familiar refrain. Two days earlier American Red Cross CEO Gail McGovern had been on the island for a press conference, her emergency response vehicles lined up behind her. As the Red Cross personnel struggled across the region to launch a coordinated response, fifteen of their vehicles had been diverted for this exercise in pure public relations. More generally, the charity was notable for its absence from the worst-hit areas, often for weeks after the storm. Volunteers had come from all over the country, and two-thirds had never participated in a major disaster operation. They were seen wandering the city lost and falling into a series of blunders, like delivering flashlights without batteries and bringing a truckful of pork lunches to a Jewish retirement home. A ProPublica investigation later obtained minutes from a post-operation assessment in which top Red Cross officials admitted the organization was "not good at scaling up" to a Sandy-sized disaster, and "multiple systems failed" under the challenge. "We didn't have the kind of sophistication needed for this size job," noted a vice president.[25]

On Staten Island the Occupiers finally met up with the Red Cross, along with FEMA, three weeks after the storm, in a meeting of so-called Voluntary Organizations Active in Disaster. Of about three hundred responders present, seventy-five were from Occupy Sandy. "The guy from the Red Cross said, 'You know, we didn't know where to go. You all didn't call us,'" Guerra recalls. "At the time they had a hundred million dollars. We had a hundred thousand." Red Cross did scale up its fund-raising and ultimately raised $312 million for Sandy. An irate press was unable to obtain information on how the organization used these funds, however. ProPublica's request for public records on this spending was intercepted by the Red Cross's lawyers, who asked that details be redacted as trade secrets. They argued with unintentional humor that if the information were disclosed, the charity "would suffer competitive harm because its competitors would be able to mimic the American Red Cross's business model for an increased competitive advantage."[26]

In perhaps the most memorable scene of the Red Cross's Sandy drama, the organization booked supermodel Heidi Klum for a photo op on Long Island in December, where cameras looked on as she handed a few boxes of supplies out of the back of the truck. The former Victoria's Secret model and TV talent judge was perfectly suited to the fantasy. Her words to the cameras, however, were a little too real: "People still are suffering here and people still need things. It's not like, oh, you send a truck full of things and then everything is fine again. This will take a while to rebuild and for people to get back on their feet."

BUILDING IT BACK

If Bloomberg had incurred bad karma with his marathon push, it all came back around with the city's federally funded rebuilding effort, called Build It Back. In the initial aftermath of the storm the city had first opted for a strategy of hiring contractors to quick-fix basic damage on many homes, helping residents to "shelter in place." Pleased with this so-called Rapid Repairs program, the city made direct deals with contractors to lay the foundation for permanent recovery under Build It Back. Two years later, that recovery hadn't reached many.

In 2014, Amy Peterson, already the third director of the ill-fated effort, told us, "Our program was set up in a way to avoid the problems of Katrina and other programs where a lot of money was given out, but the end result was that the housing stock wasn't rebuilt, because the homeowner didn't use the money, or there was contractor fraud, or for whatever reason." [27] Peterson had the advantage of two years' perspective as she worked hard under a new mayor to finally kick the program into gear. The number of construction contracts signed for new houses had just passed a measly eight hundred, already two years late. Just months before, in September, Build It Back hadn't built back a single wall. The spectacular failure resulted from a good intention to follow the "build back better" philosophy—specifically, to prevent the waste, corruption, and poor work that had occurred when federal funds were handed out too freely in New Orleans. Bloomberg chose a hands-on approach in which the city itself would pay contractors to rebuild, and where necessary elevate, houses after a thorough application process to determine the real needs of homeowners.

Before the construction crews, the city brought in a herd of consultants. Top management consulting firm Boston Consulting Group billed $8.3 million over the first six months for dispensing ineffective advice to the project's planners. Another company won $50 million to run intake centers that were equipped with untrained temp workers and faulty custom software. As the months dragged on, no such money was being spent on actual construction. The weight of responsibility was on homeowners to bring document after document into the intake centers—if they could recover these from their flooded-out homes. After more than a year, not a single one of the twenty thousand applicants had made it to the end of the obstacle course. The optimistic Build It Back name became a punch line. (*Gothamist*: "'Build It Back' Sandy Recovery Program Has Built Nothing Back"; WNYC, more imaginatively: "If Tolstoy were alive today, he might say: Every unhappy Build It Back family is unhappy in its own way.")[28] Thousands of applicants withdrew or simply stopped returning calls.[29]

Bloomberg's successor, Bill de Blasio, brought in Amy Peterson a few months after his swearing-in to finally drag the program across the starting line. By the second anniversary of the storm, 762 constructions and elevations were under contract. But to achieve this milestone, the new management had to make a compromise by scrapping income priority levels. Because higher-income groups traditionally have an easier time navigating these programs and locating their documentation, the original designers of Build It Back had dictated that the first thousand constructions should go to low-income households. This turned out to be a substantial roadblock, because low-income homeowners struggled to provide the extensive documentation required (with little help from city staff) while everyone else waited in line behind them. With the restart, equitable intentions were ditched for the sake of achieving any progress at all. "Now everyone who's applied is eligible, and that's a big deal to be able to move people forward," Peterson told us. Once the program was finally under way, she stood by the decision made by her predecessors to put rebuilding under a microscope. "Sure, there are some homeowners where sending them a check works best," she said. "But I think the lesson learned, and the reason the program was set up as it was, is because if you really want to ensure you're rebuilding the housing stock and you're making it more resilient for the future, then you want to take a hands-on approach."

By the third anniversary of Sandy in 2015, the mayor's office could claim that Build It Back was finally hitting its stride, with more than five thousand reimbursement checks sent out, more than five thousand design starts, and more than eighteen hundred construction starts. All three counts had stood at zero at the start of 2014.[30]

STATEN ISLAND STRONG

Occupy Sandy finished most of its hands-on action six months after the storm. Volunteers started moving on to other activities, right around the time that mucking out was fading into rebuilding—or rather, into waiting to rebuild. The need for helping hands gave way to more expensive material requirements. Goldi Guerra did join the board of the Staten Island Long Term Recovery Organization (LTRO) as an Occupy representative, however. LTROs are bodies set up by FEMA, as dictated in the National Disaster Recovery Framework, to empower communities to take care of themselves after FEMA moves on. Occupy Sandy and its partners thought the LTRO was going to be "the holy grail of organizations," Guerra recalls. They also thought it would get them federal funding. FEMA quickly explained that no, the group wouldn't get federal money; instead, it would need to raise private funds. With federal Sandy aid tied up in the Build It Back program and the lion's share of private money directed to the Red Cross, there were few places for the group to turn. Occupy itself pitched in the first $100,000 to kick-start the group's bank account.

While Guerra still hopes the LTRO will help Staten Islanders recover, he resigned from the board in 2014. Occupy was back to being attacked by the press, and the recovery organization was suffering by association. "We were sweethearts for six months. Then the conservative nature of Staten Island took over and we became the enemy once again," Guerra says. He decided the other organizations in the group would fare better without the Occupy name attached.

On the night of Sandy's second anniversary, Guerra stood at the sidelines of a commemoration event on the borough's Midland Beach, greeting old friends. "Looking around now, it's amazing we were able to come over here and do anything at all. Everything was so different after Sandy," he said. Under a line of tents, servers from local restaurants were heaping pizza and eggplant parmigiana onto paper plates. A Boy Scout troop performed color guard, and a police officer sang the national anthem. On stage, the homeowners who told their stories were all women, and the heroes they thanked from the bottoms of their hearts were all men—not just men but, in the words of one local organizer, the "beefy guys" who are something of a Staten Island specialty.[31] Behind them hung banners representing the volunteer rescue services and local aid groups to which the heroes belonged. One group's banner had been quietly moved from the stage before the crowd arrived, however: that of the New York State Nurses Association, whose volunteers had treated survivors for months, delivered prescription medicines, and administered tetanus shots on street corners. The nurses had retrieved their banner from a periphery fence, and it now hung near a table at the back of the event where they were handing out health information.[32]

The event concluded with an appearance by congressman Michael Grimm, days away from reelection. With twenty federal indictments on fraud, perjury, and obstruction-of-justice charges hanging over his head, he nevertheless was ahead in the polls. A Republican Party outsider with a tight campaign budget and a short fuse, Grimm knew all about Staten Island and all about Sandy. He had stood by the borough through the storm and later co-sponsored a law to head off large increases in flood insurance rates. But Grimm's speech this night was heartfelt and humble, almost to the point of tears. He asked Patricia Dresch, a survivor whose husband and daughter were lost to the storm, to join him on stage. With her congressman's arm on her shoulder, she spoke of everything he and others had done for her: rescuing her from the wreckage, helping her move into a new house, even—in Grimm's case—showing up on her doorstep with a dachshund to keep her company. (Two months later the federal charges would catch up with Grimm and he would leave his hard-won seat. In July 2015 he was sentenced to eight months of prison and one year of supervised release for tax evasion.)

The event's theme was a narrow one: celebration of the ordained heroes. This was not the time to dwell on the crushing difficulties still facing residents or on the vulnerabilities that remained. As the crowd formed a procession to place candles atop a newly constructed sand dune facing out to sea, there was a sense of equilibrium restored and a chapter struggling to close.

BUILDING BACK RICHER

Staten Island's hard trek back to the status quo shares features with countless other post-disaster recoveries, as we will see in the chapters ahead. Some residents bounced back, some didn't, and many fought back just far enough to find their old problems again. The story might end here, with, to borrow Sigmund Freud's words, hysterical misery fading into common unhappiness. But Superstorm Sandy wasn't like any storm that had hit the city, or the country, before. It belonged to a future that America was finally ready to imagine. Where Hurricane Katrina had merely fanned the debate over climate change, Sandy, just seven years later, rolled into a different era, a time when it was safe for everyone from your next-door neighbor to the president of the United States to call the superstorm a climate change phenomenon.

In 2005, *Time* had asked, "Is Global Warming Fueling Katrina?" In 2012 the headlines still bore question marks, but the questions were more pointed. "How Can Cities Be 'Climate-Proofed'?" asked the *New Yorker*. "Are We Ready for a Superstorm Sandy Every Other Year?" asked *Mother Jones*.[33] Sandy was so destructive because its cyclonic winds had joined forces with a strong cold front slumping down from the Arctic. And a *Scientific American* article by Charles Greene of Cornell University, published just days before the superstorm struck, had discussed research showing that such monster cold fronts are generated by Arctic melting associated with human-induced global warming, and that therefore we should expect destructive winter weather more and more frequently in coming decades.[34]

This was an issue that Mayor Bloomberg was ready to address. However tarnished his post-Sandy record became, he knew he would be remembered as a climate visionary. Breaking with much of the GOP, he had been talking about climate throughout his three terms, and in 2007 had tasked his then-deputy Daniel Doctoroff with developing a plan for the city. The so-called PlaNYC brought together twenty-five city agencies and 127 points of action to develop New York City in directions that would be more sustainable, lower-carbon, and ready for a future climate. Mitigation of carbon emissions was central to the plan, positioning Bloomberg as a pioneer. Shortly after, he became chair of the C40 Cities Climate Leadership Group, an international group of mayors taking action against climate change.

Following Sandy, Doctoroff published an op-ed stating that without PlaNYC the superstorm's damage would have been much worse.[35] The reasons had nothing to do with New York's so far unimpressive efforts to cut its own greenhouse emissions but everything to do with its development. "Mayor Bloomberg's most lasting legacy won't be a single park or cultural site—it will be his vision for a city prepared to survive and thrive in a world of new competition, threats and dangers," Doctoroff predicted. Among the changes of the previous five years, a series of high-profile waterfront redevelopments stood tall after Sandy, and Doctoroff cited several that had weathered the storm well. "A new approach to waterfront development allowed the city to do more with this valuable resource—safely, and despite the potential for flooding," he wrote. "Indeed the performance of waterfront areas influenced by PlaNYC during Sandy shows that we should be advancing, not retreating, in the face of danger."

After the storm, Bloomberg asked the president of the NYC Economic Development Corporation to lead the creation of a new agenda for PlaNYC, titled "A Stronger, More Resilient New York."[36] Like many such documents, it began with a definition of the word "resilience"—but this time, it had a New York accent.

re•sil•ient [ri-zil-yuhnt] adj.
1. Able to bounce back after change or adversity.
2. Capable of preparing for, responding to, and recovering from difficult conditions.
Syn.: TOUGH
See also: New York City

The word "tough" hadn't appeared anywhere in the pre-Sandy PlaNYC. The 2013 plan, in contrast, was infused with toughness, trading away talk of green growth and environmental stewardship for battle-ready language that surely would appeal to the beefy guys of Staten Island.

> In our vision of a stronger, more resilient city, many vulnerable neighborhoods will sit behind an array of coastal defenses. Waves rushing toward the coastline will, in some places, be weakened by offshore breakwaters or wetlands, while waves that do reach the shore will find more nourished beaches and dunes that will shield inland communities. In other areas, permanent and temporary floodwalls will hold back rising waters, and storm surge will meet raised and reinforced bulkheads, tide gates, and other coastal protections. Water that makes its way inland will find hardened and, in some cases, elevated homes, making it more difficult to knock buildings off their foundations or knock out mechanical and electrical systems. And it will be absorbed by expanded green infrastructure, or diverted into new high-level sewers. Meanwhile, power, liquid fuels, telecommunications, transportation, water and wastewater, healthcare, and other networks will operate largely without interruption, or will return to service quickly when preventative shutdowns or localized interruptions occur.

In its *Tale of Two Sandys* report, the Superstorm Research Lab confirmed that the tough new attitude had taken hold everywhere. Interviewees from policy-oriented sectors repeated that the storm had "changed the conversation, putting climate change at the top of the public agenda." Yet in doing so, the conversation became all about armoring the waterfront for future superstorms, not cleaning up emissions.[37] This persisted into the de Blasio administration until, just in time for the massive People's Climate March of September 2014, the new mayor picked up the fallen mitigation banner and pledged to scale back 80 percent of the city's carbon footprint by 2050.[38]

While the Build It Back program stalled out, hundreds of millions of dollars from the federal aid package were fast-tracked into Bloomberg's "tough" vision. At first, it seemed like the city (and states) would pay any price. The Dutch engineering firm Arcadis designed a $6.5 billion barrier that would have crossed the Verrazano Narrows, closing off all of Upper New York Bay. But this was quickly turned down. Bloomberg knew that the waterfront could protect itself, with sufficient investment. The important thing was to keep building. Launching his new plan at the Brooklyn Navy Yard, the mayor made his resolve clear, stating, "As New Yorkers, we cannot and will not abandon our waterfront. It's one of our greatest assets. We must protect it, not retreat from it." But in reality, New York had abandoned its waterfront long ago—now it was simply in the process of rediscovering it. As shipping traffic moved out of the city in the early twentieth century, the low-lying shores became a home for the poor. They moved in first on their own, and later were brought in larger numbers by master planners including Robert Moses, who sited huge public housing developments along the shore in the 1960s and '70s.

Only over the past two decades had the waterfront been gaining a new image. In a process familiar to many American and European port cities, developers began negotiating choice waterside lots for luxury developments, and harbor or river views became real estate gold. For the most part, these new, well-fortified condominiums held up to the storm better than buildings in low-income neighborhoods. Bloomberg, who had waved through zoning changes to make many of these ventures possible, could prove that they didn't just bring money to the waterside—they brought resilience.

"How is it possible that the same winding, 538-mile coastline that has recently been colonized by condominium developers chasing wealthy New Yorkers, themselves chasing waterfront views, had been, for decades, a catch basin for many of the city's poorest residents?" asked *New York Times* writer Jonathan Mahler just after Sandy. Telling the history of waterfront public housing, Mahler called it "a perverse stroke of urban planning," a bad move by the dictatorial Moses and his predecessors.[39] Now there was a chance to reverse the mistake. If the poor were vulnerable and the rich resilient, why not let the rich hold the front lines? Old charges of neighborhood gentrification seemed silly when survival was at stake. Bloomberg (and then de Blasio, in equal measure) green-lit larger and larger luxury developments and boosted them with tax breaks, with the only requirement that they rent a modest percentage of units at affordable rates. For homeowners tired of waiting under the Build It Back program, meanwhile, the recovery office began offering to buy up damaged houses at post-storm value under a program called Acquisition for Redevelopment. Now developers could fortify the coastline as they recolonized it, making the waterfront safe for those who could afford it.

DESIGNS ON THE WATERFRONT

When a relief appropriations act made its way to President Barack Obama's desk three months after Sandy at a final price of $50.5 billion, the federal government was facing the same pressures to make the recovery climate-proof that New York had faced. The answer was a bold experiment: the Department of Housing and Urban Development (HUD) and the president's Hurricane Sandy Rebuilding Task Force launched a competition called Rebuild by Design to elicit fresh thinking from the design world. The competition brought top firms, many from Europe, together into teams to generate resilience ideas while consulting closely with the communities in their neighborhoods of choice. After a yearlong competition, HUD awarded $930 million from the federal Sandy funds to six winning ideas.

For Staten Island's south shore, one group won $60 million for a line of offshore living breakwaters that would slow storm surge while creating new underwater habitats. Another largely Dutch team successfully proposed a $150 million band of flood-blocking new development to stand between wetlands and surrounding towns in New Jersey's Meadowlands. But the biggest ticket was a $335 million plan called the Big U—later redubbed the Dry Line—designed by the New York office of Denmark's Bjarke Ingels Group (BIG).[40] It proposed wrapping the entire foot of Manhattan with a ten-mile continuous line of storm surge barriers, which would change in form from one neighborhood to the next and, it was hoped, fit into the needs of those neighborhoods in ordinary weather. Planning the Dry Line piece by piece also allowed the plan to chop up risks. "Like the hull of a ship, it's built in compartments, so if one is ruptured only that compartment will flood," explained BIG designer Daniel Kidd, sitting in the firm's Chelsea offices. Each of the districts would be separated from its neighbors by barriers reaching back into the floodplain to keep any breaches contained.[41]

The eastern side of the U would make use of FDR Drive, an existing freeway along the East River shore. Where its downtown end runs elevated above the waterfront, BIG proposed installing flip-down steel shutters that would deploy from under the highway during storm surges. A continuous "big bench" under the structure would combine lower-level flood protection with recreational space. Where the freeway drops to road level, an East River Park berm would widen the storm wall into a sloping parkland between road and river that, in the final stage of the plan, could bury the FDR itself underground. Together these protections would more effectively seal off the flood-prone Lower East Side.

As Kidd tells it, protecting the Lower East Side was the team's first priority, but that alone would have been a tough sell in the design competition. "Ultimately it's politics. It was clear very early on that you can't propose to handpick your favorite neighborhoods, but at the same time you can't protect everyone equally because the neighborhoods aren't equal." Thus a Big U of ten miles, extending all the way around the financial nexus of Lower Manhattan and up the West Side, past quarters such as fashionable Chelsea, home of BIG's own offices. In these big-money districts the designers envision private-public partnerships coming together to fund the new construction.

As required by the Rebuild by Design brief, the BIG team worked through a long series of community consultations in the Lower East Side. "We had this big grand Robert Moses–esque gesture of protecting the entire island, but it required a Jane Jacobs bottom-up grassroots approach to get support from the people who live there," Kidd said.[42] Good Old Lower East Side was one of the community organizations engaged by the project. Lilah Mejia, a lifelong resident who joined GOLES as disaster relief coordinator after Sandy, was pleased that their participation was invited as the design came together, and then overjoyed when their ideas won $335 million of funding. Yet in a neighborhood that is already gentrifying faster than any other in the city, the Dry Line's success is also a dangerous prospect. "Ever since this happened there is a focus on the East River, where there never was in this community," Mejia said. "So now our community is saying OK, we want something to protect us, but we're scared that if you bring all these amenities to our community, you're kicking us out. And that's something we've been fighting for a really, really long time, so it's a scary feeling." Mejia recalled that Damaris Reyes, the director of GOLES, was never shy about sharing these feelings with the BIG team. "Damaris is very open in saying to all these people, 'Yeah, I'm on board, but the minute you try to take us out, we're gonna be on that front line fighting you guys. As long as you're cool with that, then yeah.'"[43]

Kidd understands the concerns, but in a neighborhood with the most and the oldest public housing projects in Manhattan—mostly packed along the waterfront—there are limits to what developers can do. At the same time, he admitted that BIG does get e-mails from developers all the time, looking for an in.

Gary Barnett, who had developed One57 of dangling-crane fame, was one of those who already had his eyes on the Lower East Side. He had bought up a large low-lying site next to the FDR, home to the neighborhood's only full-size supermarket, where residents had flocked before Sandy to stock up on emergency supplies. Two months later, the supermarket closed for demolition to make way for an enormous eighty-story condo tower and, to fulfill the developer's bargain with the city, a separate thirteen-story affordable-housing building. The fancifully named One Manhattan Square, which will soar hundreds of feet above anything ever built in the neighborhood, is initially being marketed to pied-à-terre buyers in China, Singapore, and Malaysia. And now the sales offices can promise buyers the protection of the fold-down storm gates under the FDR. Yet current neighborhood residents expressed outrage at the "economic segregation" of the separate buildings and skepticism at Barnett's promise that a new supermarket on the ground floor of the towers will cater to the same class of customers who had been served by the old one.[44]

Whose priorities will guide the protection of downtown Manhattan and the other Rebuild by Design districts? The answer won't be known for a while—Kidd expected that the first construction along the Dry Line wouldn't start until year five after Sandy, if another storm doesn't interfere— and it won't be in the hands of the winning designers. "We're essentially here to channel money from the federal government into their neighborhood, and that's all we can do," Kidd explained. HUD's $930 million was slated to be parceled out to grantees within local government, which in the Dry Line's case is the City of New York. The city then received a new round of proposals from engineering firms, who are the prime contract holders to carry out BIG's vision. These firms then hire their own teams of consultants and architects. "We were an ideas competition. I think a lot of the teams are really nervous that they're essentially going to be cut out of the whole process in the real world," Kidd said. But he tries not to worry. "It pays to be optimistic a lot of the time that some of the stuff is going to happen and some of it is going to look like we hoped it would and that the citizens who helped us design it hoped it would."

RETREAT, NO SURRENDER

The results of rebuilding by design are years away in all six project regions, but one fact is clear: something will be built. As Bloomberg insisted so strongly, the waterfront will not be abandoned—not by the city and not by the federal government. But what of the people who live there? History shows that residents of an area are often the most adamant of all in their desire to stay in place. A survey six months after Sandy found that a majority

of Americans supported government-assisted rebuilding over managed retreat, and the majority was even larger among individuals affected by Sandy.
This was despite the fact that affected respondents were more inclined than
other Americans to agree that another similar disaster is likely and that
these disasters are becoming more severe.[45]

Retreat is a difficult thing to imagine, most of all in a city such as New
York. The Lower East Side certainly isn't going anywhere soon. Out on the
far shore of Staten Island, however, some residents did begin talking about
doing the unthinkable—and then fought to make it a reality. Not only
were these neighborhoods low-lying, but in some spots they actually sat in
depressions behind a raised coastal boulevard that had held back Sandy's
surge just long enough to let it all rush in at once, knocking houses right off
their foundations. Residents of these former marshlands were no strangers
to flooding, but Sandy was unprecedented. This is where the majority of the
borough's twenty-three deaths occurred.

The call for retreat began in Oakwood Beach. Here homeowners were already well organized into a Flood Victims' Committee, the legacy of a storm
in 1992, when they had banded together to lobby for coastal protection. A
few weeks after Sandy, two hundred residents met to discuss another plan
of action. The near-unanimous decision was that protection wasn't enough.
The community instead began lobbying to unmake itself.[46] The new Oakwood Beach Buyout Committee began by researching the precedents for
public buyouts of disaster-exposed homes, most notably the case of a FEMA
buyout of Tennessee houses after a historic flood in 2010 and a buyout under way upstate in Jay, New York, after 2011's Hurricane Irene. They made
proposals to various officials up to the state level. Their perseverance, more
than any outside initiative, brought the question into the offices of power.
The city was unresponsive, but the following January, New York governor
Andrew Cuomo visited the island and took notice. He subsequently announced a state buyout program with Oakwood Beach as a pilot site. Under
the plan, the state would buy properties in the area at their pre-storm values,
and the land would never again be built upon. "There are some parcels that
Mother Nature owns. She may only visit once every few years, but she owns
the parcel and when she comes to visit, she visits," the governor said.[47]

Bloomberg was unenthused about the idea and took more time to give
his assent. It went against his vision and, in essence, against the whole history of New York City, where nothing is ever demolished without blueprints
for something bigger to replace it.[48] The city was soon courting homeowners with its own Build It Back program, dividing opinions in other Staten
Island neighborhoods. A short drive up the shore, residents of the tiny

neighborhood of Ocean Breeze took to the streets. Their buyout committee set up a tent in front of their skewed houses and debris heaps every
weekend, gathering signatures for letters of interest in a buyout, and making
their opposition to building back clear to journalists and visiting politicians.
Eventually Ocean Breeze also won a buyout, followed by Graham Beach. After these the governor capped off the program and ended requests in progress from at least four other neighborhoods.[49] By October 2014, nearly five
hundred Staten Island households had filed applications, and demolitions
were proceeding.

In Ocean Breeze, the second anniversary of the storm was quiet, with
many of the remaining residents having handed over their houses just a few
days before. Wild turkeys flocked the empty streets. Where the raised coastal
boulevard passes the neighborhood's edge, a cute wooden house bearing the
name "The Little House in the Gully" stood boarded up. Joe Hernkind, one
of the departees, had left signs hanging outside for passing cars: "Mother
Nature thanks the Gov. for giving her back her land," read the first. "For giving my neighborhood and I the opportunity to move on and find closure,"
continued another. The rest of the neighborhood was a mix of empty lots,
empty houses, and a very few homes where residents had chosen to stay. Tall
Phragmites grass was moving in from the surrounding wetlands.[50] At the
reeds' edge, one patch had been kept clear, decorated with plantings and a
granite slab. It bore the names of Ella Norris and James Rossi, the two Ocean
Breeze residents who drowned in their homes when fourteen feet of water
rushed into the hollow. Ronnie Loesch, their onetime neighbor, brought
flowers on the anniversary. "They're just fake flowers, but I wanted to have
something here. One for Jimmy, one for Ella," she said.[51]

Loesch handed over her house in the buyout. To stay would have meant
elevating it nine and a half feet. "I'm old," she told us by way of explanation.
"And besides, nobody else is here." For now Loesch was still living in a tiny
one-bedroom apartment with her daughter and grandson, but she had just
signed on a new house on a higher part of the island. Still, she came down
often to sit on her old porch and feel the sea breeze. She also left behind the
memory of the house where her mother, Veronica Weiler, had raised her and
her five siblings. They watched a crew tear it down. Weiler was eighty-four at
the time of the storm and has since passed away. Her children almost hadn't
convinced her to leave on the night of Sandy, and when they returned they
found a giant hole in the middle of her roof. The water had poured in from
above, right onto Weiler's bed. "We brought her to see it, and she said she'd
never be back there again. It would never be rebuilt in time for her," Loesch

recalled. And it didn't matter. "When I visited her in the nursing home, she said she just wanted to remember the neighborhood as it was."

On the anniversary, Loesch was joined by a handful of other former residents for a yearly memorial gathering. A visiting priest said a prayer for the two victims and the neighborhood. Bill Johnsen—an outspoken Staten Islander, a Vietnam veteran, and an early Occupy Wall Street organizer—stopped by to pay his respects to old friends. The neighbors discussed upkeep of the memorial site, which was already well tended by Frank Moszczynski, head of the Ocean Breeze Civic Association and one of the organizers of the buyout.

Those who left now regarded the advancing reeds warily. Given the chance, *Phragmites* could swamp the area as deep as the storm surge did, and in dry periods it has a tendency to catch fire, melting the siding off houses. Yet they have also seen muskrats, raccoons, and deer moving in. A hundred years ago the area was a prime fur-trapping marsh, Moszczynski said. They talked of a ten-point buck that several of them had sighted on recent visits.

But with most of the old residents now living elsewhere on Staten Island or in New Jersey, the most urgent order of business was to see to it that the granite memorial stayed in place. "That thing weighs over nine hundred pounds; it's not washing away," Moszczynski assured the others. "This way the name of Ocean Breeze is always going to be here, and Jimmy and Ella's names will always be here."

4

EVERY SILVER LINING . . .

Would it console the sad inhabitants
Of these aflame and desolated shores
To say to them: "Lay down your lives in peace;
For the world's good your homes are sacrificed;
Your ruined palaces shall others build,
For other peoples shall your walls arise;
The North grows rich on your unhappy loss . . . ?"

—Voltaire, "Poem on the Lisbon Disaster;
Or an Examination of the Axiom, 'All Is Well,'" 1755[1]

"How cruel it seems for the sky to be so blue and the ocean so calm!" Kino said.
But his father shook his head. "No, it is wonderful that after the storm the ocean grows calm and the sky blue once more. It was not the ocean or the sky that made the evil storm."
"Who made it?" Kino asked. . . .
"Ah, no one knows who makes evil storms," his father replied. "We only know that we must live through them as bravely as we can, and after they are gone, we must feel again how wonderful is life. Every day of life is more valuable now than it was before the storm."

—Pearl S. Buck, *The Big Wave*, 1948[2]

In a speech he gave the year before he was elected president, John F. Kennedy put the perfect name on the perfect quotation about crisis. "When written in Chinese the word *crisis* is composed of two characters," he said. "One represents danger, and the other represents opportunity." We haven't stopped quoting this optimistic bit of ancient-Chinese-by-way-of-Massachusetts wisdom since. It's not important that it's totally untrue; based on a misunderstanding of how the language works, it started out as the 1930s equivalent of chain e-mail fodder among American missionaries in China. Even if it isn't ancient or Chinese, we all understand the sentiment. The contemporary world is a glass half full of nitroglycerin.

After missionaries and politicians, the saying found its most enduring home among corporate executives (who might also be the first to tell you the market price of nitroglycerin). We expect inhabitants of the business world to make the most of misfortune, the only question being the level of tact with which they do so. Yet to single them out would be a little myopic. We are all deeply attuned to the silver linings of life, most of all when the clouds are darkest gray. The silver beams burst into view right at the crossroads in Holling's figure-eight cycle of life (see Figure 2), as we struggle along the path that, we're told, leads from destructive crisis to resilient reorganization. But silver linings don't shine for everyone, and even metaphorical precious metals can be magnets for conflict.

Not only in Australia, Siberia, the Philippines, and New York, but almost everywhere disaster strikes we see the rise of new opportunities propelling three sorts of actors into webs of cooperation and competition. Each of these groups—business, government, and the rest of us—sees opportunities either to disrupt or to reinvigorate the status quo in ways that further their own goals. It is the disaster that opens up the space for potentially sweeping change, but it is largely the relative strengths of different factions, as they existed at the time disaster struck, that influence how the struggle plays out. In a seminal 1993 book, *The Political Economy of Large Natural Disasters*, J.M. Albala-Bertrand wrote:

> Setting aside well-reported short-term co-operative behavior, a society's pre-existing dynamics and structure appear to be revealed more strongly after a disaster. Disasters also provide new opportunities for social sectors. Regional or ethnic identity can appear to be reinforced. Businesses enjoy new opportunities associated with reconstruction contracts (a seed of corruption in more developed countries). Planners use the occasion to propose more ambitious development plans which go beyond the narrow disaster region. And general property and power patterns as well as socio-political conflicts may be brought to the fore. Whether this can be sustained in the longer term or bring about a more significant social change has little to do with the disaster itself, but with pre-existing economic and socio-political conditions.[3]

How do we know a disaster has thrown society onto a new path? Joern Birkmann of United Nations University and his colleagues equate change in this context with "a critical juncture—a change that sets into motion a new trajectory for action, policy, or institutional regime."[4] But how to identify and measure that alteration or new trajectory is not always obvious. To do it perfectly would require access to data from a parallel universe in which the hurricane missed, the fault held steady, the rains came on time. Instead of this exercise in educated fiction, experts usually just try to measure the impact of a disaster—the difference between what was there before and what is there after. In theory it's a simpler problem, but it remains open to many questions.

Scholars, analysts, governments, corporations, insurance companies, media outlets, and aid groups measure impact in three numbers: the number of deaths, the number of people affected, and the value of property and business losses. Obviously, none of these, alone or even combined, tells the whole story. In her diary from the typhoon-ravaged Philippines in 2013, British aid worker Sandra Bulling grappled with the inadequacy of numbers: "A lot of the questions circle around the death toll. I want to scream: 'But we can't measure the disaster simply by the number of dead people!' Entire communities have been wiped out. As many as 14 million people could be affected. The casualties are a tragedy but those who survived have found their lives in tatters."[5] In this book, too, we cite many loss and mortality figures, but we also try to give ample attention to the harder-to-measure impacts of a "life in tatters"—disruption of communities and the ecosystems on which they depend, the production of misery, and vicious cycles. Change, in turn, must be evaluated not only in terms of modifying the difficult conditions of life (already far too widespread, with or without a disaster) but also in terms of the political, social, and economic structures that help create these conditions—and that seek opportunity everywhere. Markets, states, and affected populations measure the impacts of a disaster differently, and beyond the pain and shock each dreams of a different kind of change, or at least a glimpse through the clouds.

SILVER LININGS FOR THE MARKET

With the waters of Superstorm Sandy still draining from the East Coast, analysts at Goldman Sachs attempted to predict the storm's economic consequences by looking back at how America's gross domestic product (GDP) reacted to the eighteen strongest post–World War II hurricanes that had struck U.S. coasts up to that time. They found that production and sales in general "show a clear dip in the month of the disaster, followed by a significant recovery within one to three months that typically takes their growth rate above that seen prior to the disaster." *Forbes* saw a rosy outlook as well: "Sandy has left many casualties in its path and has shaken the United States to its roots, but it could have a positive economic impact in the near-term. Companies in the housing and construction sector, like Home Depot, Loews, KB Home, and Lennar, could see a meaningful boost in the coming months from the rebound in activity . . . beyond the human tragedy, there may be a silver lining to Sandy."[6] The *San Francisco Chronicle* also looked forward to a happy economic impact from the calamity, quoting Bernard Baumohl, chief global economist at the Economic Outlook Group: "Construction costs to rebuild all that was lost will be more than simply replacement, because a lot of the work will also involve fortifying structures. . . . We'll see construction ramped up, and that's going to bring in jobs and an increase in demand for material of all sorts, and that's going to further stimulate the economy."[7] With so many cars flooded out and ruined, the struggling automotive industry was also hopeful of stimulus. Industry site Edmunds.com reported that "Hurricane Sandy blew away some October car sales, but, as expected, the payback was big in November." Dealers rolled out special promotions, focusing on luxury cars and pickup trucks, and sales in the Northeast achieved a volume not seen since the eve of the recession nearly five years before.[8]

Stated in its crudest form, the argument that large disasters have economic silver linings is fairly easy to visualize. Predictions of accelerated growth can be found all over the media after any major storm, flood, or earthquake, and sometimes growth does seem to perk up, as it did in the Philippines in the months between Pablo and Yolanda. As early as 1850, the economist Frédéric Bastiat had tired of hearing such cheerful logic, so he decided to spoil the fun in an essay titled "That Which Is Seen and That Which Is Not Seen":

Have you ever witnessed the anger of the good shopkeeper, James B., when his careless son happened to break a square of glass? If you have been present at such a scene, you will most assuredly bear witness to the fact, that every one of the spectators, were there even thirty of them, by common consent apparently, offered the unfortunate owner this invariable consolation—"It is an ill wind that blows nobody good. Everybody must live, and what would become of the glaziers if panes of glass were never broken?"

Now, this form of condolence contains an entire theory, which it will be well to show up in this simple case, seeing that it is precisely the same as that which, unhappily, regulates the greater part of our economical institutions.

Suppose it cost six francs to repair the damage, and you say that the accident brings six francs to the glazier's trade—that it encourages that trade to the amount of six francs—I grant it; I have not a word to say against it; you reason justly. The glazier comes, performs his task, receives his six francs, rubs his hands, and, in his heart, blesses the careless child. All this is *that which is seen.*

But if, on the other hand, you come to the conclusion, as is too often the case, that it is a good thing to break windows, that it causes money to circulate, and that the encouragement of industry in general will be the result of it, you will oblige me to call out, "Stop there! your theory is confined to that *which is seen*; it takes no account of that *which is not seen.*"

It is not seen that as our shopkeeper has spent six francs upon one thing, he cannot spend them upon another. *It is not seen* that if he had not had a window to replace, he would, perhaps, have replaced his old shoes, or added another book to his library. In short, he would have employed his six francs in some way, which this accident has prevented.[9]

The economic optimist's silver-lining argument focuses on economic activity and growth, ignoring both the destruction of wealth that has to be replaced and the preexisting maldistribution of power and wealth. Disaster pessimists contend, as did Bastiat, that in calculating the net impact, it's necessary to subtract not only the value of wealth and property that were destroyed but also the goods and services that never get produced because resources were redirected toward rescue, repair, and rebuilding. In any case, once a disaster has happened, whatever the losses, businesses do as much as they can to make the best of the situation. In that, wealthy and well-insured institutions and corporations have the advantage, and, as Naomi Klein demonstrated in her book *The Shock Doctrine*, they can press their advantage to its cruelest limits.

The idea that disasters are "good" for economies has persisted partly because of the focus on growth rather than improving people's capability or quality of life, and partly because predictions that a specific disaster will boost overall growth typically are not checked later to see if they were accurate (chiefly because that is not an easy thing to do—but, as we'll see, it can be done).[10] Statistical analyses of a large number of economies that have suffered differentially from disasters can reveal what individual anecdotes do not, however. When economists started doing that, it seemed at first that silver linings might be the rule after all. In 2002, the economists Mark Skidmore and Hideki Toya published a paper that surprised many economists. Their analysis, which followed the economic fortunes of eighty-nine countries, showed that countries encountering larger numbers of climatic disasters between 1960 and 1990 had higher per capita GDP growth during those years. (Seismic disasters had no effect on growth.) Skidmore and Toya offered one explanation: "We infer that disasters provide opportunities to update the capital stock and adopt new technologies. We also suggest that disaster risks necessitate adaptability, so that cultures experiencing disasters may be able to adopt new technologies more readily."[11]

The idea that geoclimatic disasters can boost economic growth has been associated with the concept of "creative destruction" as articulated by economist Joseph Schumpeter in the 1940s, based on his reading of Karl Marx.[12] Because of overproduction and sagging demand, mature capitalist economies tend to stagnate; however, argued Schumpeter, the periodic introduction of new technologies sweeps away old, established equipment and infrastructure in capitalism's "perennial gale" and makes way for more productive ones. "Economic progress, in a capitalist society, means turmoil," he wrote.[13] In the Silicon Valley milieu of the following century, this bracing account merged with another of Schumpeter's innovations—the positive rebranding of the word "innovation"—to inspire the doctrine of disruption.[14] Self-described disruptive tech start-ups celebrate their annihilative role as they are poised to scour away established business "ecosystems." With a little imagination, the disruption regime of California's tech valleys shows affinities with the disturbance regime of its fire-swept hills.

But is all destruction creative? Schumpeter clarified that this process in capitalism "is not merely due to the fact that economic life goes on in a social and natural environment which changes and by its change alters the data of economic action; this fact is important and these changes (wars, revolutions and so on) often condition industrial change, but they are not its prime movers."[15] In other words, the perennial gale comes from within, not without. What Schumpeter called creative destruction is an internally triggered process of innovation inherent in economies, so it's not immediately relevant to disasters that strike out of the blue. For one thing, the timing of upgrades of production systems is usually determined by business owners' and managers' judgment, not by random chance. Unplanned devastation of property and productive capacity by a hurricane shouldn't be expected to accomplish the same thing. There is only a small chance that a geoclimatic hazard will get lucky and disrupt just the right economic targets—that is, those whose demise would make way for upgrades right at the time they are due for replacement. Silver-lining optimists like to rebut this rebuttal by claiming that business owners and managers often become too comfortable and set in their ways, and that, to borrow the economist Joseph

Berliner's unforgettable term, disaster can give them a swift "invisible foot" to the backside, motivating them to update and upgrade their means of production.[16]

Theoretical objections notwithstanding, a few empirical studies seem to confirm Skidmore and Toya's conclusions, but overall the results are mixed. For example, European companies in regions hit by severe flooding in 2000 showed faster growth in total assets and employment than did those that did not suffer flooding.[17] On the other hand, most other published studies found no stimulus. When 1970–2005 growth rates of all U.S. coastal counties along the Atlantic Ocean and the Gulf of Mexico were examined, hurricane strikes slowed growth by about half a percentage point in the year they made landfall and had no effect at all on growth rates in subsequent years.[18] All told, studies of post-disaster growth during the 2000s presented a mix of positive, neutral, and negative results that taken together can provide only one answer to the question of whether disasters provide stimulus: it depends.[19]

But the most serious shortcoming of such analyses is that they track only overall changes in national income following disasters, ignoring the distribution of wealth and income within and between countries. The underlying assumption is that the rising post-disaster economic tide will lift all boats (even if the disaster sank some) and that acceleration of aggregate growth is good, period. But when researchers did look at how geoclimate hazards interact with existing conditions to shape economic futures, that clearly was not the case. For one thing, the stimulus effect, when it does show up, is usually evident only in wealthy countries or in more prosperous regions of less wealthy countries.[20] Large, prosperous economies can be expected to better tolerate and even take advantage of the unexpected destruction of productive capacity by making technological improvements (while welcoming larger infusions of government reconstruction funds), while poorer economies cannot. Analyzing forty-five years of data from countries around the world, a 2009 report found disaster had positive effects only in wealthier countries.[21] The poorest countries—which are home to one-third of the world's population and, according to economist David Strömberg, suffer almost two-thirds of disaster fatalities—not only miss out on any economic stimulus but also take longer to recover.[22] They can even fail to recover at all, instead falling into "poverty traps": vicious cycles in which poverty deepens in the harsh post-disaster environment, rendering people more vulnerable to environmental degradation and future disasters.[23]

Examples abound. The post-Pablo growth spurt in the Philippines certainly didn't bring prosperity to Barangay Andap. In Mexico, the poverty rate has been found to rise in areas affected by droughts, frosts, catastrophic rains, and/or floods, and the affected areas lost two years' worth of gains in their human-development score over a five-year period.[24] In Nicaragua, households experienced downward mobility following Hurricane Mitch in 1998, falling from the middle class into the lower class at a higher rate.[25] Hardships can be especially severe in small island nations such as those of the Caribbean, which face more than their share of disasters.[26] Haiti is one of the starkest examples of an island crippled by frequent, diverse geoclimatic hazards, not just in the Caribbean but in the world.

Theoretical and empirical research on economic impacts of disasters almost always applies to large, one-time shocks. But economies like those of Haiti and the Philippines that are repeatedly battered over time by large, medium, and small disasters (both geoclimatic and economic) may not have sufficient time between shocks to recover fully. After successive assaults, the ecosystems and built environments on which those economies depend can be crippled. As a result, a society can become trapped in a downward spiral. The World Bank's Stéphane Hallegatte and colleagues have shown theoretically why disasters fail to bring creative destruction and predict that as poorer countries face more frequent climatic disasters in coming decades, they will get no shots at silver linings. This, they wrote, "may partly explain why some poor countries that experience repeated natural disasters cannot develop."[27]

Large, wealthy economies, with ample resources to direct toward rebuilding, have proven more capable of bouncing back. An oft-cited example of a strong comeback is the case of Kobe, Japan, which was struck by an earthquake in 1995 that killed 6,400 people, caused damages of $100 billion (2.5 percent of that year's gross domestic product), and left Japan's largest port and most of its vaunted industrial capacity in ruins. Within seven months of the quake, utility services, railways, and roads had been restored. Within fifteen months, manufacturing was almost back to its pre-quake output. At eighteen months, retail sales were already three-fourths of what they had been before. Import-export trade was almost back to normal within a year.[28] Many economists marveled at Kobe's rapid recovery of the status quo. A decade and a half after the disaster, however, William DuPont and Ilan Noy examined the city's post-recovery growth pattern, comparing Kobe's actual economic trajectory from 1995 to 2010 with a statistically constructed model of its economy as it would have performed had the quake never happened. In per capita GDP, the real Kobe was a consistent 15 percent poorer than the counterfactual, quake-free Kobe and showed no signs of catching up even fifteen years after the disaster; if anything, the real Kobe seemed to be falling farther behind. DuPont and Noy noted that even though "the housing stock was quickly rebuilt and people chose to return to the city soon thereafter," they returned to a different city, where "many of the employment opportunities that had existed before were no longer available, so that residents came back to a depressed economy." All of this ran counter to "received wisdom," which tells us "that modern market economies recover completely, and rapidly, from adverse shocks, especially if property rights and market operations are maintained."[29]

The received wisdom that DuPont and Noy were debunking also holds that geoclimatic forces can be absorbed by the human economy along with the rest of nature, and that dealing with hazards is simply a matter of incorporating them into profit-and-loss balance sheets. The view holds that nations of the global North not only have managed to reduce dramatically the mortality resulting from major geoclimatic hazards; by fostering highly developed infrastructure and social stability buffered by strong insurance and complex financial mechanisms, they have learned to ride out even unimaginable disasters and not be thrown into long-term decline.[30]

The trajectory of recovery from the Great East Japan earthquake and tsunami of 2011 has convinced many that there is no disaster so big that it can bring down a major economy, even a sluggish one such as Japan's. The magnitude-9 quake released *six times* the energy of the previous largest quake ever recorded in the region. Japan had seemed to be ready; its nuclear plants had been designed to withstand a magnitude-9+ quake. But the tsunami triggered by the quake, combined with the unfortunate placement of the country's major nuclear facilities, produced a disaster that many argued was much more than unprecedented or unexpected. The Japanese word *sōtegai*, roughly meaning "unimaginable," was repeated often by officials of the government and the nuclear utility company TEPCO in the aftermath of the tsunami as a way of rationalizing their failure to defend against a nuclear catastrophe of the type and scale that had actually occurred. They argued that no one should blame them, because no one could possibly have anticipated the *sōtegai* series of events that amplified the disaster and led to the destruction of the Fukushima reactors and release of radioactivity. After all, they argued, they had taken great care to ensure that strong seismic shaking, which they had believed to represent the worst-case scenario, would not damage the reactors.[31] While Japan recovered from even this unimaginable disaster, it may not have learned the lessons it brought. Three and a half years after the Fukushima crisis had prompted a nationwide nuclear shutdown, the government's Nuclear Regulation Authority tentatively approved the restart of four reactors in other regions. The agency's chairperson told reporters, "I am not saying that the [Takahama plant in Fukui prefecture] is safe, or that safety was confirmed or that we determined that it is not safe. Rather, we completed an evaluation into whether it meets the necessary requirements to operate."[32] Court challenges delayed the Takahama restart for more than a year, but it finally happened at the end of January 2016. In the interim, two other plants had been fired up.[33] The first was the Sendai nuclear power station, on August, 11, 2015. There, the chief concern had been that the plant is situated only thirty miles from one of the country's most active volcanoes, Sakurajima. So there could be no appeals to *sōtegai* when, just four days after the restart, officials issued a level 4 emergency warning for Sakurajima. On February 5, 2016, the volcano erupted but did no damage to the nuclear facility.[34] The post-Fukushima shutdown of the nuclear industry had exposed the country's deep dependence on those power plants. Japan's economy could handle a nuclear catastrophe but could not deal with the prospect of voluntarily and permanently idling 30 percent of its electricity supply.

Because we can't control when and where destructive geological and atmospheric forces strike, the resulting disasters are usually not seen as discriminating for or against any socioeconomic sectors. Instead they have at times been viewed as "great levelers." This erroneous argument ignores the long history of actual hazards, which conclusively demonstrates the greater exposure of disadvantaged groups to hazards, their greater vulnerability when hazards strike, the more formidable obstacles they face when trying to rebuild their lives after a disaster, and the recently recognized role of lopsided economic growth in creating the seemingly natural hazards themselves. A high degree of inequality can greatly worsen the situation.[35] Even in the United States and other wealthy countries, where catastrophe can have an economic silver lining—though only at times and for some—benefits to the more well-off regions and sectors can come at the expense of others. A large body of research shows that America's low-income families and communities are the most vulnerable in the face of storms, floods, heat emergencies, and wildfires and that they are the least likely to see any economic silver linings should they appear.[36]

The rich can, of course, make the case that they stand to lose the most from disasters, since their higher-value property and business losses weigh most heavily in the calculation of total losses. Meanwhile, the destruction of a slum or mobile-home park represents a much smaller wealth loss and may even prompt some hard-hearted observers to conclude that their community is better off without old or substandard housing anyway. The deadly tornado that hit Joplin, Missouri, in 2011 (see pages 292–97) destroyed a blighted business district; according to local wags, the storm had thereby wrought "millions of dollars in improvements."[37] But disaster losses of a size that would seem trivial to an affluent "one percenter" can trap families already living in precarious circumstances in inescapable poverty. An extensive review of research showed that, after a disaster, those who have "suffered from the intensification of previous economic stress and problems" take a bigger proportional hit to their wealth, homes, and livelihoods than do the affluent; have a harder time getting emergency relief; and encounter higher barriers to recovery, especially in the critical area of housing. They have a harder time getting access to emergency repair help; insurance benefits; health care; home or small-business loans; and decent, affordable housing and transportation. In low-income areas, government agencies or repair teams from charitable organizations are sometimes overly careful to avoid fixing damages that might be "preexisting conditions" not caused by the disaster. Suddenly jobless people are hit with foreclosures, housing shortages drive rents and home purchase prices up, landlords cheat on damage deposits or use the disaster as an excuse to evict residents from rent-controlled housing, contractor scams proliferate, and affordable housing is often the last to be rebuilt.[38]

The effect of poverty does not only increase a country's vulnerability to geoclimatic hazards. Even more important, poverty increases vulnerability to predatory governments, corporations, and lending agencies. In *The Shock Doctrine*, Naomi Klein relates how, in the aftermath of the 2004 tsunami, Sri Lanka's coastal residents set aside old religious and historical grievances and spontaneously united in an inspiring outbreak of mutual aid and cooperation. That style of recovery was choked off in short order, soon after the hastily appointed Task Force to Rebuild the Nation, drawn chiefly from powerful business interests, issued a report recommending development of high-end tourism and industrial fishing infrastructures that would make permanent the tsunami's erasure of coastal villages and livelihoods. Noting the outpouring of anguish and assistance for victims of the tsunami among people throughout the world, Klein wrote, "But the World Bank and USAID [U.S. Agency for International Development] understood something that most of us did not: that soon enough, the distinctiveness of the tsunami survivors would fade and they would melt into the billions of faceless poor worldwide, so many of whom already live in tin shacks without water. The proliferation of these shacks has become as much an accepted feature of the global economy as the explosion of $800-a-night hotels."[39]

Sri Lanka was hardly unusual. By far the most common prescription being offered to poor countries and regions is that they grow their way out of disaster vulnerability. And highly inequitable growth is always the fastest and easiest to achieve. *The Economist* concluded from its review of the year 2011, "As societies develop they can afford the human and physical infrastructure needed to protect against, and respond to, natural disaster. In time, last year's earthquake and tsunami and floods will be mere blips in the GDP of Japan and Thailand, thanks to the rapid reconstruction made possible by the same wealth that meant the disasters were so costly to start with." Their conclusion? "The lesson for poorer countries is that growth is the best disaster-mitigation policy of all." One of the most predictable consequences of disasters is big business's call to accelerate unfocused growth. The "lessons" that the world should take from a catastrophic event always seem to align fully with the pre-disaster goals of business interests.[40] But research shows that decreases in deaths from floods, windstorms, and landslides followed rising incomes only in already rich or middle-income countries. From the mid-1970s through mid-1990s, an individual's likelihood of being affected by a geoclimate disaster doubled even though real per capita income increased by one-third worldwide.[41]

Trying to reduce disaster vulnerability by boosting aggregate growth and capital accumulation leads to intensified resource exploitation and ecological degradation, as has happened in logging- and mining-plagued Mindanao in the Philippines. That degradation, in turn, amplifies not only the destructiveness of disasters but also the ecological impacts of efforts to prevent disasters. In less wealthy parts of the world, that ecological degradation can reinforce the spiral into poverty traps. The disaster geographers Mark Pelling and Katherine Dill cite studies of environmental degradation and climate change that "show a vicious cycle where human insecurity (limited access to rights and basic needs) generates vulnerability to environmental change and hazard, the impacts of which undermine livelihoods and capacity to adapt and survive future threats."[42]

Statistics have given us a much better understanding of the consequences of hurricanes, floods, and earthquakes. But numbers can never tell the whole story. When you ask people who actually have to live and work in places that have been struck by disaster, whether they're in Staten Island, Ormoc, or the Blue Mountains, few if any will say that a flood or fire or landslide has improved their daily life. Adding to the physical damage done to homes, livelihoods, and communities is of course the personal aftermath. The economist Aaron Popp writes, "Natural disasters affect people first and foremost. Even if a disaster does not kill anyone, the emotional and physical effects of the disaster on the population still harm people well after the actual disaster. Some survivors will likely suffer permanent physical disabilities and psychological conditions, such as post-traumatic stress syndrome. Workers who suffer from psychological conditions and disabilities will not be as productive as they were before the disaster." [43] The market is ill-equipped to deal with such trauma, either personal or societal, and that leaves an opening for governments and citizens to come to the rescue.

SILVER LININGS FOR THE STATE

Many scholars of the Enlightenment regard the great Lisbon earthquake of 1755 as history's biggest shakeup in the understanding of disasters, and of much else besides. The event almost completely destroyed one of Europe's largest cities and killed tens of thousands in a paroxysm of earth, fire, and water. The clergy went about the work of interpreting God's message in the catastrophe, while Enlightenment thinkers faced a bigger task: they were forced to ask whether it really was within their power to understand nature's ways and whether theirs really was the best of possible worlds. In Voltaire's 1759 novel *Candide*, it is in the rubble of Lisbon that the indefatigable Doctor Pangloss extols for the last time his belief that "all that is, is for the best"—a member of the Inquisition, hearing heresy, has him hanged to appease God and prevent another quake. [44] For men of reason like Voltaire, both Pangloss and the Inquisitor were looking in vain for signs of benevolence or punishment in the disaster. When news and broadsheet engravings of the carnage first spread through Europe, Voltaire had written,

> What crime, what sin, had those young hearts conceived
> That lie, bleeding and torn, on mother's breast?
> Did fallen Lisbon deeper drink of vice
> Than London, Paris, or sunlit Madrid?
> In these men dance; at Lisbon yawns the abyss. [45]

A fissure had opened in the modern mind between natural evil and moral evil, and the category of the indiscriminate, unfeeling, *natural* disaster was born. [46]

For those who followed naturalistic explanations, there were no great villains in Lisbon—but there was a hero. Sebastião José de Carvalho e Melo, Marquês de Pombal, had been appointed prime minister by King Joseph I just before the quake. As soon as the king assigned him to oversee the relief, recovery, and rebuilding efforts, Pombal is said to have issued a simple order to "bury the dead and feed the living." [47] With the efficient accomplishment of those tasks, disease and famine were prevented. At the same time, Pombal cracked down hard on crime within the city, fended off invading pirates, and aggressively quashed economic speculation and the Inquisition itself. Pombal has been seen as a human link between the philosophical revolution triggered by the quake and modern conceptions of effective disaster response. Philosopher Susan Neiman, for example, writes, "Pombal was explicit in supporting naturalist explanations of the earthquake. The more earthquakes were viewed as normal events, the easier it would be to incorporate them into a normal world—or to view the return to normalcy as a merely practical problem."[48] Pombal handled the rebuilding of Lisbon with an efficiency unburdened by moral questions, setting a technocratic standard for disaster management and recovery authorities ever since.

"Nature is dumb, in vain appeal to it," wrote Voltaire after the Lisbon quake. His foil Jean-Jacques Rousseau had a quick rejoinder: "It was hardly nature who assembled there twenty thousand houses of six or seven stories."[49] In contemporary terms we would commend Rousseau for recognizing the contribution of urban planning to seismic risk. Pombal recognized it, too, and introduced totally new, more seismically secure building methods to his blueprint for the new Lisbon. Now, as then, disasters routinely present local and national governments with a chance to "build back better" and thereby display their foresight and competence. "Better" may mean that the rebuilt area is more resistant to future disasters, more structurally sound and technologically advanced, less likely to contribute to climate disruption and disasters, or simply a better place for people to live and work.

A long tradition in disaster lore involves local leaders huddling among the ruins in the immediate aftermath, resolving to build their community back better than ever. The late *Atlanta Journal* reporter William Brice recounted such a meeting that took place the evening after an EF-5 tornado struck Gainesville, Georgia, on April 6, 1936, destroying the city's downtown square, touching off fires that obliterated what was left of many buildings, and leaving behind what is still the fifth-highest death toll of any U.S. tornado.[50] Two days after the storm, reported Brice, the president of Gainesville's Brenau College, a local bank president, a "prominent oil dealer," and the Chamber of Commerce president met in the Citizens Bank building and conceived a commission, composed of government officials and citizens, that would lead an ambitious build-back-better campaign.[51]

For them, the terrible storm at least had fortunate political timing. The federal government was struggling to pull the country out of the Great Depression, and the Roosevelt administration saw Gainesville as an opportunity to show how the decidedly visible hand of the government could turn even the worst of tragedies into something positive. Reconstruction funding flowed in, and Franklin Roosevelt himself visited Gainesville twice.

Emergency relief and restoration were accomplished with help from members of the federal government's most prominent New Deal agencies: five companies of the Civilian Conservation Corps and seven hundred workers from the Works Progress Administration. With funding from federal, state, and local governments, Gainesville got a new city center with a federal building, courthouse, and city hall, along with a hospital, a public library, a school auditorium, a bus terminal, and new, wider streets.

Other, less obvious factors were important to the recovery. Johnny Vardeman was the district editor and then editor at the *Gainesville Daily Times* (now *The Times*) from 1957 through 1998. As one of the city's key journalists for four decades, he had extensive contact with members of the community who had lived through the tornado and the rebuilding. He believes that local leaders' determination to turn the disaster to the city's advantage had its roots in a previous storm. A tornado that hit Gainesville in 1903, with a death toll of ninety-eight, had killed many child workers when a textile factory called Gainesville Mill collapsed. The factory had been promptly rebuilt, but the community continued to feel considerable shame and guilt over a tragedy tainted by the employment and endangerment of children. When the 1936 tornado hit, Vardeman says, Gainesvillians were determined to redeem their image as a city.[52]

Recovery was fast. The bright new city center, with its stately arrangement of county courthouse, city hall, and federal building, he says, "was the kind of thing that few Georgia communities had at that time." Much of the central business district was built back new as well. In a 1937 *American Journal of Nursing* article, registered nurse B.M. Sigmon described how on the day of the storm the city's hospital staff went to work "heroically and industriously without much sleep or rest," in their now roofless building, in steady rain, to care for throngs of patients who were coming in "bloody and wet, . . . faces and clothes blackened beyond recognition, crying, screaming, dying." She praised the selflessness that the entire community showed in the weeks that followed, but she also saw a more enduring legacy of the disaster: "'Every cloud has a silver lining,' and so we feel that out of the tragic aftermath of this disaster has come recompense—a new hospital."[53] Gainesville's comeback earned it a favorable reputation across the region. "People saw Gainesville as a spirited town," says Vardeman, one that could handle both adversity and prosperity.[54]

Gainesville may have triumphed over adversity and put its child labor stigma behind it, but the city remained solidly embedded in the Jim Crow

South. In *A City Laid Waste*, names and photos of white residents killed by the tornado are displayed yearbook style near the end of the book, while names of black victims are given within one sentence buried in the main text. On the same day as the Gainesville storm, a tornado hit Tupelo, Mississippi, with a body count of 216; as a result, April 6, 1936, remains the deadliest single day in the history of U.S. tornadoes. However, Tuplelo's official death toll is a distinct underestimate; many of the black victims were not included in the official count.[55]

Much like military conflicts, headline-making geoclimatic disasters open a door through which political leaders can push much-needed legislation that might otherwise meet stiff resistance. Often this can involve changes that government officials had been wanting to make anyway, but only in times of disaster do they see a chance to obtain votes to support the large infusion of funds needed to carry them out. Gainesville was not the only beneficiary of such state-sponsored silver linings in the 1930s. When Roosevelt and his allies set out to create the New Deal programs that would provide relief for millions of Americans left unemployed and destitute by the Great Depression, they met strong political opposition. They managed to clear those hurdles only by linking the New Deal programs (which applied nationwide) with two disasters: the towering drought-fueled dust storms that hit the country's midsection in the 1930s and the Mississippi River floods of 1938. Calling the programs "disaster relief" made them more palatable, because that placed them squarely within a history of federal government relief for disaster victims that stretched back as far as proponents cared to trace it—arguably to the First Congress in 1790, a time when the Lisbon quake was a living memory.[56]

In 1935, the two-year-old federal Soil Erosion Service, created as a stopgap measure to deal with the disaster in the plains, was on the verge of running out of funding, and its director, Hugh Hammond Bennett, was pushing to make it a permanent agency. On March 21, Bennett was to testify before a congressional subcommittee that was contemplating his plan. Having gotten word from his colleagues in the Midwest that a gargantuan dust storm was then making its way rapidly east, he dragged out his testimony as long as he could, giving the storm enough time to descend upon the District of Columbia. In a biography of Bennett, Willington Brink tells what happened when the skies finally darkened:

> The group gathered at a window. The dust storm for which Hugh Bennett had been waiting rolled in like a vast steel-town pall, thick and repulsive. The skies took on a copper color. The sun went into

hiding. The air became heavy with grit. Government's most spectacular showman had laid the stage well. All day, step by step, he had
built his drama, paced it slowly, risked possible failure with his in-
terminable reports, while he prayed for Nature to hurry up a proper
denouement. For once, Nature cooperated generously.[57]

Bennett had successfully leveraged a disaster. The new Soil Conservation
Service was created and eventually transformed American agriculture; it exists today as the Natural Resources Conservation Service.

When political leaders have performed well in the wake of disaster, like
Pombal in Lisbon, New Jersey governor Chris Christie and congressman
Michael Grimm after Sandy, and Benigno Aquino III in the Philippines after Pablo and Yolanda, they have garnered citizens' support. The legacies of
leaders who have faltered, as did George W. Bush in his response to Katrina,
have suffered stains that can never be washed out.

Unlike solid response and recovery, disasters averted seldom earn ac-
colades, because they never come to pass. The exception comes with a rare
combination of spectacular hazard and spectacularly effective preparation.
In 1999, the strongest tropical cyclone ever to hit India killed more than ten
thousand people in the state of Orissa. As the state recovered, its government put in place extensive disaster response and management plans. When
the nearly as powerful Cyclone Phailin threatened in 2013, Orissa was ready.
Almost a million people were successfully evacuated. According to the *Financial Express* newspaper,

> Adopting a strategy of "zero loss of life," 247 cyclone shelters were
> pressed into service in Orissa. However, what made the difference
> was the 10,000 specially constructed school buildings that housed
> a major part of the 912,848 people who were evacuated from the
> affected areas. These buildings had been constructed as part of a spe-
> cial plan after the 1999 super cyclone disaster. "All resources were put
> to use. This was the largest-ever evacuation at such short notice. We
> not only provided food and water to all but a week's supply of both
> was also kept ready at the shelters," a senior state government func-
> tionary said. Special rapid action force teams were placed at strategic
> locations in districts and fuel was stocked.[58]

When it was all over, there were eighteen deaths. The state government was
showered with praise by the media, national and international, and the ruling party swept the next spring's elections.[59] As we will see in Chapter 11, the

governments of Bangladesh and Cuba have achieved even greater progress in reducing mortality from tropical cyclones, and have also built legitimacy on governing disaster.

In elections that come soon after disasters, the state can find both dark and silver linings. Voters punish incumbent politicians for disasters from geoclimate hazards that strike during their term in office, whether or not the politicians could have done anything about them. Analysis of U.S. county-by-county electoral results show that, depending on how much damage it causes, a disaster can knock 0.5 to 2.5 percentage points off an incumbent governor's vote share in the next election.[60] But a well-received response to disaster also can pay off politically. By bringing in federal relief funds, governors can easily cancel out that "automatic" disaster penalty and even get a boost at the polls. While any government appropriations going to a particular district or region can help politicians garner votes and campaign contributions, disaster-response spending produces an especially large bang for the buck.[61] Republican Florida governor Jeb Bush's approach to hurricane relief aid in 2004 met broad approval, energizing his supporters and knocking the wind out of his political opposition; his post-disaster performance was even credited with boosting Republican voter turnout by 5.1 percent and decreasing Democratic turnout by 3.1 percent on Election Day later that year. Support via the disaster-response route doesn't come cheap, however; in that election, it amounted to an estimated $21,000 per Republican vote gained.[62] A nationwide study over five election cycles gave a similar result: one vote gained per $27,000 spent on disaster relief aid.[63] Other research shows that when a governor sends the U.S. president a successful disaster-declaration request, thereby triggering a flow of relief funds, materials, and disaster-response teams, voters in the counties covered by the declaration give the governor a boost worth an average four percentage points in the next election. Even if the president turns down the request for a disaster declaration, the governor gets an A for effort, with a 2.7-percentage-point bonus. (If they turn down a request, presidents who run for reelection stand to lose a point in the affected counties.)[64]

Importantly, those electoral benefits come only from relief spending; politicians get no payoff at all from spending on disaster preparedness. The relief aid that flows in after a disaster is highly visible, whereas prevention measures are easily forgotten—after all, when such efforts are successful, nothing newsworthy happens. Closer examination reveals another reason for elections being influenced by relief but not prevention spending: the former comes primarily (but not exclusively) in the form of direct payments to individuals, whereas preparedness spending is mostly in the form of public projects. Even with relief, voting patterns are influenced only by that portion of spending that comes in the form of individual compensation rather than collective relief programs such as the U.S. Public Health and Social Services Emergency Fund.[65] The benefits of spending on environmental initiatives such as reducing greenhouse emissions that could help lessen disaster threats worldwide are regarded as even more distant from the lives of individual constituents, so such actions are even less likely to be rewarded.

It's not surprising that when politicians receive such signals from voters they respond. If they can afford to do so, governments buy themselves some popularity with disaster declarations and other domestic relief mechanisms. The stimulus thus provided is amplified when insurance companies pour billions of dollars of their accumulated cash back into the economy. (Poor countries can afford far less "countercyclical" spending; in fact, they often are forced to cut their spending after a disaster, just at the time the economy needs it most.)[66] In the United States, legislators tend to steer more relief spending toward counties or regions that gave more support to their own party in the previous election. The federal government has been highly responsive to the electoral preference for relief aid, ignoring the fact that preparedness is far more cost-effective: $1 spent on preparedness reduces disaster damages by a whopping $7. Per capita federal disaster relief spending in current dollars shot up from about $5 annually in the 1988 election cycle to $68 in the 2004 cycle. Over the same period, per capita preparedness spending dropped from an already measly $1.41 down to $0.53.[67]

The economists Thomas Garrett and Russell Sobel reported in 2003 that "states politically important to the president have a higher rate of disaster declaration by the president, and disaster expenditures are higher in states having congressional representation on FEMA oversight committees." They estimated that nearly half of all disaster relief "is motivated politically rather than by need."[68] Presidents Ronald Reagan, George H.W. Bush, Bill Clinton, and George W. Bush all issued their largest numbers of disaster declarations in the year they ran for reelection, and declarations go disproportionately to the Electoral College's "swing states."[69] The number of disasters declared per year has grown rapidly since the mid-1990s. Barack Obama set a record for federal major-disaster declarations in 2010, with eighty-one, and then broke his own record in 2011, issuing a total of ninety-nine. Then, contrary to tradition, the declaration count plummeted to forty-seven in his reelection year of 2012, rose to sixty-two in 2013, and fell again to only forty-five in the midterm election year of 2014.[70]

In America, the regional nature of disaster damage combined with the increasing role of federal money in relief efforts complicates politicians' incentives. When Congress voted to direct $9.7 billion to the national flood insurance program to bring relief to homeowners after Sandy, some of the sixty-six dissenting votes came from representatives of Louisiana, Mississippi, Alabama, and Florida—states that had received some of the huge federal payments following Katrina. Back in 2005 when he was chief financial officer of the Biloxi Public Housing Authority, Representative Steven Palazzo of Mississippi had personally asked for $38.5 million.[71] Because of this, Palazzo was the most widely criticized of the dissenters, but he doubled down on his position, initially opposing the bill that would become the Disaster Relief Appropriations Act of 2013 and included $50.7 billion for Sandy relief. A week before the bill was to be voted on, however, he paid a personal visit to the wreckage of Staten Island and had a change of heart. Breaking with most of the Republican vote, Palazzo changed his vote and backed the bill, saying he realized that "Mississippians have been through much of what the Sandy victims are experiencing."[72]

Oklahoma's senators and three of its five congressmen also voted against the disaster bill. After the Sandy funding passed 241–180 in the House, freshman representative Jim Bridenstine expressed the sentiment of much of the GOP when he argued, "When we increase spending in one area, we must cut spending in another area. I hope my colleagues will consider this principle in future relief packages."[73] Bridenstine and his fellow Oklahoma lawmakers faced the music just four months later when a severe weather system swept through their state, sending one large EF-5 twister through the city of Moore. The press clamored to know if these hardline Congress members would stick to their principles and refuse extra disaster funds for their own state. Bridenstine stuck to his, guessing (correctly, as it turned out) that FEMA's $11.6 billion reserve would cover the disaster anyway. (Bridenstine then took his budgeting ideas on the offensive, introducing a bill to reallocate funds away from climate change research into improving forecasts of tornadoes and other severe weather.)[74] Another Oklahoman, Senator James Inhofe, was open to the possibility of extra funds, arguing strictly on grounds of the superior Sooner character. The Sandy bill was "totally different," he pointed out: "Everybody was getting in and exploiting the tragedy that took place. That won't happen in Oklahoma."[75] The question was simple, though, for the two members who had voted yes to the Sandy aid. Representative Tom Cole, who grew up in Moore, had pledged support from the start to officials in the Northeast. He recalled telling them, "We're always going to be there to help because we're always one tornado away from being Joplin," invoking the Missouri city cut through by a tornado two years previously. "I didn't think it was going to be quite this soon."[76]

Evidently all of these positions were acceptable to Oklahomans: all of their congressional incumbents fared well in the next election.

People pay close attention to how political leaders deal with disasters of all kinds; politicians, in turn, are very much aware that they are being watched. If climatic hazards become more frequent and intense, and if more and more citizens make the connection between the prospect of a more disaster-ridden future and the reality of greenhouse warming, will that put unbearable pressure on governments to take aggressive action toward curbing emissions? In her book *This Changes Everything: Capitalism vs. the Climate*, Naomi Klein argues that climatic chaos can deliver "a powerful message—spoken in the language of fires, floods, droughts, and extinctions—telling us that we need an entirely new economic model and a new way of sharing this planet. Telling us that we need to evolve." The result? "Rather than the ultimate expression of the shock doctrine—a frenzy of new resource grabs and repression—climate change can be a People's Shock, a blow from below. It can disperse power into the hands of the many rather than consolidating it in the hands of the few, and radically expand the commons rather than auctioning it off in pieces." In short, she writes, climate change can be "a civilizational wakeup call."[77]

Those who, like Klein, have raised the climate alarm in hopes of provoking a comfortable society into a political uprising have often been hit with spurious charges of "catastrophism." Canadian ecosocialist Ian Angus has observed three lines of attack taken against those who attempt to warn of human-induced calamity.[78] First and most obvious are the right-wing climate change deniers who characterize any such warning as no more than a ploy. Second, there are some on the left who see any argument based on looming ecological catastrophe as a variant of "economic catastrophism," a now-discredited theory according to which a sudden, dramatic economic crisis will bring the demise of capitalism. Catastrophist arguments have been widely dismissed because they ignore capitalism's extraordinary resilience and its ability to recover from almost any economic shock. Angus responds that to extend the economic logic of capitalism to the physical breakdown of ecosystems across the planet "is to equate an abstract error in economic theory with some of the strongest conclusions of modern science." But there is another, perhaps more damaging charge of rhetorical catastrophism, this one coming from the left, from some who acknowledge the threat of ecological collapse but urge that discussion of it be muted. This argument is part of a narrative strategy to avoid inducing hopelessness and apathy in the general public and avoid giving big business and the state yet another excuse to impose austerity and roll back the rights of the 99 percent.[79]

Angus takes aim at these calls for self-muzzling, citing social science research as well as the rapid growth of the climate justice movement to show
that ecological wake-up calls induce anything but political apathy. And the
possibility of co-optation, he argues, is nothing new: "Capitalists *always* try
to turn crises to their advantage no matter who gets hurt, and they *always*
try to offload the costs of their crises onto the poor and oppressed."[80] It's
well known that simply repeating scare stories of future ecological collapse
with no focus on root causes or systemic change is ineffective. But it should
be just as obvious that trying to achieve economic and ecological justice by
avoiding discussion of the calamitous, globe-spanning ecological impact of
capitalism will also fail.[81]

The wake-up call analysis is normally applied in an attempt to leverage
past events or calamities in the making; to imply that one is *hoping* for a
disaster to come along and deliver a needed shock is seen as an especially
hard-hearted brand of catastrophism. But Kristen McQueary, a member of
the *Chicago Tribune* editorial board, displayed no such concern when, with
the tenth anniversary of Katrina approaching, she wrote in an op-ed piece,
"I find myself wishing for a storm in Chicago—an unpredictable, haughty,
devastating swirl of fury. A dramatic levee break. Geysers bursting through
manhole covers. A sleeping city, forced onto the rooftops. That's what it
took to hit the reset button in New Orleans. Chaos. Tragedy. Heartbreak.
Residents overthrew a corrupt government. A new mayor slashed the city
budget, forced unpaid furloughs, cut positions, detonated labor contracts.
New Orleans' City Hall got leaner and more efficient. Dilapidated buildings were torn down. Public housing got rebuilt. Governments were consolidated."[82] McQueary was widely and rightly excoriated for trivializing
the suffering of New Orleans and summoning fresh calamity down on her
own city, all in hopes of fulfilling her personal scorched-earth austerity wish
list.[83] McQueary did not apologize, either; instead she tried to claim that
she'd simply been using "metaphor and hyperbole" and blamed readers
for misunderstanding her. She wrote, "I am horrified and sickened at how
that column was read to mean I would be gunning for actual death and
destruction."[84]

Disasters seemingly capable of transforming our political life are happening regularly anyway, without any columnist having to summon them
up. Many climate activists had counted on Superstorm Sandy to rouse the
American public and Washington to action, and for a while it seemed as
if it might. Even Barack Obama seemed to have been convinced when he
taunted climate change deniers in his second inaugural address, which came
less than three months after Sandy: "Some may still deny the overwhelming

judgment of science, but none can avoid the devastating impact of raging fires and crippling drought and more powerful storms." By the time he addressed the State Department's Conference on Global Leadership in the Arctic: Cooperation, Innovation, Engagement and Resilience (GLACIER) in Anchorage, Alaska, two and a half years later, Obama's language had strengthened further. Citing the unprecedented wildfires that had ravaged 5 million acres of Alaska that summer, he proclaimed, "Climate change is no longer some far-off problem. It is happening here. It is happening now."[85]

Sweeping statements notwithstanding, the sad reality is that in times of disaster governments seem to act even less creatively than they do in times of calm. And since it is in military matters that governments have developed the greatest control and freedom to operate, they relish the chance to put their troops to work in roles that are perceived as benign. Lifesaving rescue and relief actions can spruce up the image of any armed forces. That was the case following the 2011 earthquake and tsunami in Japan, when the country's Self-Defense Forces (SDF), long held on a tight leash under the post–World War II constitution, engaged in their largest-ever deployment. The government considered the operation a huge success. Equally positive, they said, were joint operations with the U.S. military, which stepped in to help even though treaty obligations did not require it to do so in a natural disaster. Decades-long ambivalence over the presence of American troops in the country was set aside, and an aggressive right wing hoped this would be a cure for Japan's "military allergy," wrote scholar Richard Samuels in his book *3.11*.[86] Other new allies popped up in unexpected places. Samuels observed that in 2011 "cooperation on the ground between Peace Boat, a leading pacifist organization, and the Ground Self-Defense Forces attracted special notice."[87] A month after the earthquake, the newspaper *Mainichi Shimbun* carried out a national survey and found that 95 percent of respondents supported the actions of the SDF in eastern Japan, and 88 percent agreed that it was appropriate for the forces to have worked closely with the U.S. military. This contrasted with feelings about the government in general: only 32 percent thought highly of its overall response.[88]

In its attempts to crush Maoist rebels in the late 1990s and early 2000s, Nepal's military became notorious for brutal killings and human rights abuses. But in the aftermath of the magnitude-7.8 killer earthquake that struck the country on April 25, 2015, at a time when foreign aid organizations were being accused of stinginess and Nepal's civilian government officials were bungling and sometimes abandoning relief efforts, the nation's military took decisive action, saving many lives and earning broad praise.[89] Disaster can even prompt cooperation between military rivals. In late 2015, according to the *Times of India*, "disaster diplomacy" had become "the new buzzword" in the country's peace talks with Pakistan. The two countries, along with six others in the South Asian Association for Regional Cooperation, were planning to form a common disaster response force that would function similarly to a UN peacekeeping force. India's National Disaster Response Force had performed well in response to the Nepal quake, and, reported the *Times*, Pakistan "can hardly afford to be left behind in the India-led initiative in strengthening regional bonds on a key sensitive issue."[90]

Military leaders can't take it for granted, however, that coming to the rescue in a disaster will make them more popular. Around the world, uniformed rescuers are often eyed with suspicion even as their assistance is gratefully accepted. These mixed feelings are nowhere as strong as in the United States, where, for example, there were bitter political struggles over the deployment of active-duty troops in response to Hurricane Katrina.[91] The debate continues. Police, the military, and paramilitary forces such as the National Guard have infrastructure, equipment, and skills that are useful in responding to disasters, but critics say those in uniform have acquired an outsize role. Military "mission creep" has happened partly because wrongheaded beliefs about the public's reaction to disasters—beliefs that panic spreads like a virus, that people are shocked into a helpless state, that epidemics of looting and other crimes break out—provide an additional, powerful incentive for sending in the troops.

In taking a closer look at these beliefs, Erik Auf der Heide, a medical officer with the U.S. Department of Health and Human Services, wrote, "Emergency authorities often rely on a 'command-and-control' model as the basis of their response. This model presumes that strong, central, paramilitary-like leadership can overcome the problems posed by a dysfunctional public suffering from the effects of a disaster. This type of leadership is also seen as necessary because of the belief that most counter-disaster activity will have to be carried out by authorities." Yet "authorities may develop elaborate plans outlining how they will direct disaster response, only to find that members of the public, unaware of these plans, have taken actions on their own." The overwhelming weight of evidence, concluded Auf der Heide, demonstrates that while disaster scenes are often chaotic, they don't call for heavy-handed controls; panic is rare, people not only retain their capabilities but also put them to good use, and the crime rate usually drops rather than rises.[92]

The most glaring and controversial case of militarization after an American disaster was that of New Orleans after Katrina. UCLA law professors

Cheryl Harris and Devon Carbado have described how, across the country, initial shock over the "unrelenting spectacle of black suffering bodies" that sparked debate over the question "Why were those New Orleans residents who remained trapped during Katrina largely black and poor?" dissolved inexorably into questions of law and order: "As time progressed, the social currency of the image of blacks as citizens of the state to whom a duty of care is owed diminished. It rubbed uneasily against the more familiar racial framing of poor black people as lazy, undeserving, and inherently criminal. Concern over the looting of property gradually took precedence over the humanitarian question of when people might be rescued and taken off of the highways and rooftops. Thus, while armed white men were presumed to be defending their property, black men with guns constituted gangs of violent looters who had to be contained."[93]

As the Katrina disaster was deepening, a furor over captions of two news photos arose. In one, a black man carrying a bag of goods through the floodwaters was described as "looting," while in the other, a white couple was described as "finding bread and soda" from a grocery store. The photographer who took the latter shot and wrote the caption later noted that there had been both black and white people in the water, and that he had simply "looked for the best picture." Harris and Carbado argue that this proved no conscious racism on the photographer's part but rather a cultural imperative. If he wanted to convey through a photo that there were people "finding food" during a disaster, it was more or less expected that the people in the photo would be white. That's because the majority of American readers draw a "racial association between black people and looting, particularly on the heels of a natural disaster or social upheaval," even though there is no factual reason for doing so.[94]

The racial framing that inserts looting and violence into disaster scenes whether or not they actually exist also provides a motive for militarizing disaster relief internationally. In his book on the world's response to the 2010 earthquake in Haiti, Jonathan Katz concluded that the expectation of panic and violence led to a highly centralized command-and-control structure that focused on sections of the capital, Port-au-Prince, because they were seen as safe from imagined disturbances elsewhere, and largely ignored outlying areas. And the military response to such imaginary problems had some disastrous consequences: "Expectations of desperation, near famine, and chaos led responders to expect riots at food and water distributions. To force food out while maintaining distance from crowds, the U.S. Navy threw boxes of bottled water and rations from hovering helicopters until

other responders complained that this method was itself causing panic,"
presumably by exaggerating the sense of emergency and triggering a scramble for the goods. Throughout the country, expectations of crime and chaos
triggered a flood of rumors, which in turn spurred people to take unneces-
sary protective measures that further heightened the general sense of fear.
In driving from the capital to Léogâne, the town closest to the quake's center, Katz and his partner came to "a roadblock of rocks and branches piled
across the cracked highway. About a dozen men brandished machetes and
wooden clubs. When we stopped the car they greeted us with smiles." Katz
recalls the exchange that ensued:

> I asked what the weapons were for.
> The man told me they were on guard for bands of looters they
> heard were running wild in the capital.
> "Where did you hear that?" I inquired.
> "On the news," he replied.

There were of course no such bands of looters. And the climate of fear was
not the only by-product of military intervention. When Haiti's first-ever
cholera epidemic broke out nine months after the earthquake, the source
was traced not to the direct impact of the quake but to a contingent of
United Nations troops who had been sent to maintain order in a country
that hadn't needed that kind of help in the first place.[95]

SILVER LININGS FOR THE PEOPLE

Despite all the official hand-wringing over threats to security, what Jonathan Katz actually saw in Haiti was mostly peaceful cooperation. "In fact,
while some Haitians committed crimes after the quake, far more appeared
to be doing everything possible to restore a sense of security," he wrote.[96] In
his reporting, people seemed to forget their private interests and even their
place in the social world.

> The survivors now began gathering around [President] Préval. They
> told him about the young job seekers who regularly loitered outside
> parliament. The parliamentarians called them *chimère*, gangsters.
> (The young men referred to the politicians the same way.) But when
> the columns snapped and the upper chamber fell, the survivors ad-
> mitted in hushed and grateful tones, the *chimère* were the first on top

of the pile, pulling out by the suit arm anyone they could find. They said the young men had even found and handed back wallets full of money and cell phones lost in the debris.[97]

Today, at long last, it seems that when authorities and popular culture imagine disaster, hopeful stories such as Katz's are beginning to supplant the overblown tales of panic and savagery that characterize far too much disaster coverage. In an online video address after Sandy, President Obama assured the country that he understood this hopeful vision: "When a storm hits, there are no strangers. Only neighbors helping neighbors."[98] In a sense, this emerging vision represents a breakdown of traditional authority, perhaps a state of anarchy. A world without strangers is a weird sort of world, one few of us have experienced.

Rebecca Solnit has popularized this updated vision of disaster in her book *A Paradise Built in Hell: The Extraordinary Communities That Arise in Disasters*. Reviewing a century's worth of social science and history that she saw as tearing apart all the myths of panic and crime, Solnit argued that spontaneous communities of mutual aid are the much more striking reality, time after time. These communities blossom in the absence of the state, she wrote, while command-and-control strategies fall apart and leaders preoccupy themselves with phantoms of unrest. As suggested by her use of Peter Kropotkin's anarchist concept of "mutual aid," Solnit sees a political truth revealed: by stripping away hierarchy, property, and capitalism, disasters reveal a basic utopian nature in humanity.[99]

Disaster can't necessarily take all the credit for such transformations. For example, the Occupy Sandy movement was a clear case of post-disaster mutual aid blossoming from a preexisting political assemblage. But neighbors helping neighbors is simply what happens after a disaster. In account after account, those neighbors remember the days after the disaster as a dark but enchanted time. Hard work, discomfort, loss, and fear become gilded with a profound sense of togetherness. Old divisions seem to belong to another lifetime.

Does this sort of silver lining have staying power? New Jersey governor Chris Christie hoped so, in a speech a year after Sandy.

Reverend [Joe] Carter this morning at the New Hope Baptist Church talked about something that he calls the Spirit of Sandy. . . . When the lights went out, nobody cared what color you were. He said when you were hungry, nobody cared whether it was someone who lived in a city or the suburbs, or out of state who was bringing you food to feed your family. He said when you were cold and someone offered you a warm place to sleep, no one cared if that person who offered you the warm place to sleep was a Democrat or a Republican. That was the Spirit of Sandy to him, that our state came together in a way that inspires people, and his prayer, one of his prayers this morning was to make sure that we try to make the Spirit of Sandy infect other things that we do. His last commentary—again, not mine—was he was hoping that the Spirit of Sandy from the New Hope Baptist Church today would make its way through Newark on Route 280, get itself on the New Jersey Turnpike, and, free of tolls, would head its way down to Washington, D.C., to infect some of those people down there as well because they need an infection really.[100]

The thought of a spirit that not only lives on but spreads like a contagion of goodwill is a heady one. It's safe to say that people have striven and searched for this spirit in the aftermath of every major disaster. Eight months after Yolanda, a *PBS NewsHour* headline proclaimed, "As Reconstruction Crawls, the Spirit of Tacloban Prevails."[101] Nine months earlier, Roza Sage, a representative for the Blue Mountains district in the New South Wales legislature, had connected her constituents' disaster spirit directly to ecological restoration in the aftermath of the October 2013 fires: "What we have all witnessed this past week is the spirit of the Blue Mountains. A spirit which runs deep within our community. It is a determination that while we will be tested in coming weeks, months and years, our community—like the burnt bush—will recover from this devastation and regenerate into something more beautiful than before."[102] And every now and then, the hoped-for contagion of change seems to come true. It may be an end to conflict; for Aceh, Indonesia, the great 2004 Indian Ocean tsunami is remembered as a "wave of peace" that brought an end to decades of fighting between rebel and government forces.[103] It can also be revolutionary, empowering the people to stand up against political machines. The most prominent example of this is the 1985 Mexico City earthquake, in which the efficacy of mutual aid, and the inefficacy of the government, turned volunteer brigades into a new force of civil society that challenged the ruling party.[104]

For a time at the end of 2012, radical commentators hoped that Occupy Sandy would spread and rekindle the cooling embers of Occupy Wall Street with similar passion. (This was not, in fact, the intention of many Occupy volunteers, who were simply helping their neighbors.) Across the wider political spectrum, everyone seemed to have much the same wish as Governor Christie: for the Spirit of Sandy to live on and make the post-superstorm world a little kinder than before. But as the status quo began reasserting itself in New York and New Jersey, it became difficult to say whether or not the world had changed. London's *Occupied Times* posed the question to Solnit herself:

> OT: How can the co-operation that emerges in these situations be kept going? Will we inevitably fall back to old habits?
>
> RS: Nothing is inevitable, and not everyone goes back after these extraordinary moments, but what can they go forward toward? There is a vigour in carnivals, revolutions, uprisings and disaster responses that is hard to sustain, but maybe we should not ask it to be sustained; while falling in love is great, growing from infatuation to a long-term relationship often means trading fizzy for solid.[105]

In *A Paradise Built in Hell*, Solnit wrote, "Disaster belongs to the sociologists, but carnival to the anthropologists, who talk of its liminality. . . . The anthropologist Victor Turner noted that liminal moments open up the possibility of *communitas*, the ties that are made when ordinary structures and the divides they enforce cease to matter or exist."[106] Her discussion of Turner ends there, which is a shame, because he and his wife, Edith Turner, initiated the work that is, in the academic world, the key sustained engagement with the post-disaster paradise Solnit describes. In the late 1960s, Victor Turner, an anthropologist of ritual, used the word "communitas" to describe the state of equality and unity that emerges during rites of passage, festivals, pilgrimages, and other largely sacred events. Initially drawing on his studies of African tribal cultures, he included among such events "rituals of status reversal, either placed at strategic points in the annual circle or generated by disasters conceived of as being the result of grave social sins."[107] Edith Turner's later work brought the communitas of natural disaster into the same category of "collective joy," as something that "develops in full strength in an environment sheltered by the hope and love between the members of a badly shaken community."[108]

What these events have in common is a between-ness, a "liminality" (a word Victor Turner chose from the early-twentieth-century ethnographer Arnold van Gennep's *The Rites of Passage*). In this transient space outside society, structure gives way to an "anti-structure" that turns hierarchies on their head and replaces them with a community "whose boundaries are ideally coterminous with those of the human species."[109]

Published in 1969, Victor Turner's *The Ritual Process* spread as widely through the American counterculture as it did through the field of anthropology. Its Vietnam-era adherents were quick to differentiate communitas from the camaraderie of wartime. Unlike the environments in which communitas flourishes, they pointed out, war was deeply structured. Wrote Edith Turner in 2011, during another time of war, "Communitas should be distinguished from Emile Durkheim's 'solidarity,' which is a bond between individuals who are collectively in opposition to some other group. In 'mechanical solidarity,' unity in one group depends on the opposition of the alien group for its strength of feeling." As in war, so in industrial disasters, and anywhere a human hand can be seen in the destruction. As evidenced in Indonesia in Chapter 5 and Italy in Chapter 7, Durkheim's us-versus-them passions emerge here, with "them" being the ones to blame.[110] "But in the way communitas unfolds," Edith Turner wrote, "people's sense is that it is for everybody—humanity, bar none."[111]

Disaster anthropology really got started as its own field soon after the publication of *The Ritual Process*. A formative event was the terrible Ancash earthquake and resulting landslide in the Peruvian Andes in 1970. The regional capital Huaraz was flattened, while upriver an immense flow of rock, ice, and snow fell from the side of a mountain on top of the town of Yungay. More than seventy thousand people—a quarter of the valley's population—died. Ancash was the Western Hemisphere's deadliest disaster to that point, and it held the record until the Haiti quake killed three times as many people forty years later. American researchers spent time in the emergency camps of the Andes, testing new anthropological theory against the reality of devastated communities. One of these was Barbara Bode from the American Museum of Natural History. She wrote,

> In this absence of facade, in a total lack of privacy in these over-crowded camps, where everything could be heard through paper-thin walls, and in occupations and chores suddenly shared by all, a sense of exposure—of being bereft of social structure—was experienced. The image most frequently evoked by refugees was of the doctor or lawyer and the *campesino* side by side carrying buckets of water along the camp paths, or across the pampa. In this empty space, there being no more colonnades, corners, adobe walls, all could see. On the eve of May 31, 1972, over Radio Huascaran, a local bard recited: ". . . Huaraz is naked, I ask pardon for this nakedness."[112]

This was a post-disaster image somewhat less appealing than Solnit's, especially for the doctor and lawyer. Yet it had a positive side, noted Bode, "a kind of release or liberation from the bonds of social structure, not unakin to what Victor Turner describes as communitas." [113] It was a release from the grip of property, too—the townspeople called it *desprendimiento de las cosas*, "detachment from things." [114]

Anthony Oliver-Smith was there in Ancash, too, and he went on to become the most prominent disaster anthropologist in a growing field. He also observed the blurring of class and ethnic lines and saw the concept of private property disappear, as herders donated their livestock to the common good. He then watched the spell lift, as "over time, the solidarity became strained and conflicts broke out over the use of private resources for the public good. When aid arrived, old schisms reemerged and differential access to resources was not only sanctioned, it was demanded." [115]

In 1997, anthropologist Linda Jencson saw that during the Red River Valley flood in North Dakota and Minnesota people consciously ritualized and mythologized their actions, creating an expanded sense of self and community. Like Turner's rituals, the events described by Jencson occurred in a specific bounded space: islands of defended land, "cut off from the 20th century." The ordeal led to inversions of status, with prisoner chain gangs throwing sandbags alongside mayors and stretch limousines transporting grubby relief crews. It was a time full of mystery and avowed miracles. The Red River experience also met Turner's stipulation that ritual communitas requires a simplification of life. Wrote Jencson, "Flood life in the Red River Valley became simplified in many ways. Conversation dwelled on basic sandbox topics: piling things on top of other things so they don't fall over, dirt, lots of dirt, getting dirt, good dirt: 'I'm cold.' 'When will the food get here?' 'Can I take a nap now?'" [116]

Such times of mystery and miracles are ephemeral by their very nature. Susanna Hoffman is another disaster anthropologist who wrote from very personal experience: she lost every worldly possession in the Oakland firestorm of 1991. As she struggled through the long recovery, she, too, saw all the hallmarks of communitas around her. But she relegated this to the first of three stages, which she labeled the crisis, the aftermath nexus, and the passage to closure. In the second stage, the communal spirit turned inward: "Survivors become engulfed in practices related to their circumstances. They blanket themselves in a coating of new alliances, in essence initiating a nascent subsociety and culture within their larger traditions. New devotions and new aims overrun their lives, alienating them from the commonality they held with kin, ally, and neighbor before the calamity." [117]

In this kind of environment, conflict arises. In Oakland the great enemy became the insurance companies, but everyone outside the community of loss—the "unsinged"—were also regarded as strangers. Correspondingly, Hoffman wrote, "The community surrounding disaster survivors, once sympathetic, begins to react rigidly against the new class among them. Disaster victims threaten change, and in the face of that potential, the larger community retrenches. . . . We in Oakland were told we didn't deserve new houses. We were robbing insurance companies and committing fraud. At the same time we were frequently met with such comments as: 'I wish my house would burn down so I could have a new one.'" [118]

To wish to sustain the "paradise" part of disaster after the chaos and suffering have passed is understandable. Yet this wish denies the nature of the liminal. For Victor Turner, it was a "moment in and out of time," a gap in the social fabric that ultimately makes that fabric stronger. "There is a dialectic here, for the immediacy of communitas gives way to the mediacy of structure, while, in *rites de passage*, men are released from structure into communitas only to return to structure revitalized. . . . Communitas cannot stand alone if the material and organizational needs of human beings are to be adequately met," he wrote. [119]

So what *do* we do with disaster communitas while it lasts? In a book he co-edited with Hoffman in 1999, Oliver-Smith, by then a respected professor at the University of Florida, explored possibilities for "operationalizing communitas":

> In practice, the question deals with how social policy responds to communities afflicted by a crisis such as disaster or involuntary resettlement. The usual response has been to assuage the situation with waves of material assistance of varying kinds at various stages of the crisis. As we saw in the case of Yungay [in Peru, after the 1970 quake], the arrival of aid tended to dissolve the bonds of solidarity which had been created in the immediate aftermath of impact. It was not until a further challenge appeared that those bonds were reawakened, this time to confront and reject a regional reconstruction policy and to impose their own intentions. The Yungay case suggests that, while certain forms of aid in disasters are absolutely necessary, the continuation of other kinds of aid may have very deleterious effects. The Yungay data further suggest that much aid also is characterized by a fairly consistent underestimation of the internal resources and capacities of communities to respond to crisis. Perhaps there needs to be less consideration toward the delivery of more aid and more attention devoted to devising culturally appropriate ways to nurture the potentials represented by post-disaster solidarity. [120]

This is part of the recognition that is gradually steering disaster policy (whether Oliver-Smith intended it or not) from a focus on human vulnerability and aid and toward a focus on resilience and community-based solutions.[121] From its semimystical beginnings, communitas is emerging as a phenomenon that resilience-minded disaster planners rely on. With governments and international agencies depending more and more upon vulnerable individuals and communities to deal with their own catastrophes, it's a good time to demystify disaster communitas as well as the broader embrace of self-reliance, mutual aid, and resilience as totems of disaster response. The heroic selflessness shown in disaster accounts and in Solnit's paradise is inspiring, but let's admit how very normal it is—if we can do so without diminishing that inspiration.

David Graeber, an anthropologist who is as familiar as Solnit with the anarchist tradition, outlined another interpretation in his world history *Debt: The First 5,000 Years*. Graeber identified three principles of human economic relation, which he provocatively called communism, hierarchy, and exchange. Rather than describing different types of societies, all three types of relations exist in different contexts in every society. Graeber's communism, like all communism, follows from the principle of "from each according to their abilities, to each according to their needs." But he differs from most interpreters of communism by seeing it everywhere.

> Almost everyone follows this principle if they are collaborating on some common project. If someone fixing a broken water pipe says, "Hand me the wrench," his co-worker will not, generally speaking, say, "And what do I get for it?"—even if they are working for Exxon-Mobil, Burger King, or Goldman Sachs. The reason is simple efficiency (ironically enough, considering the conventional wisdom that "communism just doesn't work"): if you really care about getting something done, the most efficient way to go about it is obviously to allocate tasks by ability and give people whatever they need to do them. . . .
>
> This is presumably also why in the immediate wake of great disasters—a flood, a blackout, or an economic collapse—people tend to behave the same way, reverting to a rough-and-ready communism. However briefly, hierarchies and markets and the like become luxuries that no one can afford.[122]

While using loaded words, Graeber does well to bring post-disaster relations back to the practical level of efficient allocation toward shared goals.

Even without the Turnerian ritual atmosphere of crisis, neighbors helping neighbors just makes more sense than waiting around for someone else to help. To remain in awe of disaster communitas means to expect less than that of our own neighbors.

Demystification of post-disaster selflessness is important for another reason: it allows us to talk about labor. It seems so callous, so lacking in reverence, to speak of the spirited support and mutual aid of disaster-affected communities as labor that we almost never do so. No economist adds up these person-hours or works them into calculations of the cost of disaster. This taboo is the most ritualistic quality of disaster: labor is not only voluntary but also sacred. Communitas makes burdens weightless. Yet there comes a point at which we can't ignore the hard work of defense and care, the mucking out.

Now the world has arrived at precisely that point—when international authorities have begun taking Oliver-Smith's advice to cultivate "the internal resources and capacities of communities to respond to crisis," and when the global economy is producing more and more crises for those communities to absorb. If communities really are going to have the resources and capacity to navigate that more disastrous future, the ephemeral nature of all silver linings and the unnatural causes of disasters must be recognized and acted upon, not only in books and scholarly papers but also throughout the world's economy, politics, and culture—and across all the grim, gray landscapes ravaged by disasters that didn't have to happen.

5

GRAY GOO

East Java, Indonesia

The mudflow disaster shows the price we have to pay for allowing politics to blur efforts to mitigate disaster, which is a matter of humanity.
—Editorial, *Jakarta Post,* April 2, 2014[1]

We the victims of Lapindo mud thank the government, in this case Mr. Joko Widodo. We feel no need to be grateful to Lapindo.
—Residents displaced by the Lusi mud volcano, celebrating the end of their nine-year struggle to obtain compensation from the gas-drilling company PT Lapindo Brantas, October 11, 2015[2]

Every disaster, from the sinking of a cargo ship to the destruction of a city by a tornado or earthquake, has multiple entangled causes. For legal and political reasons, attempts are usually (but not always) mounted to unweave the factors behind a disaster and determine their relative contributions. We can clearly elucidate the web of events leading up to a disaster, and yet this may still be rejected by those who have a large economic or political stake in some other explanation that works more to their own benefit. Political struggles are then inevitable. Deborah Stone, a professor of government at Dartmouth College, has written, "Causal stories need to be fought for, defended, and sustained. There is always someone to tell a competing story, and getting a causal story believed is not an easy task." The struggle may be carried out in scientific journals, legislatures, the courts, the media, the streets, or all of those arenas at once. Some strands in the causal web inevitably become more visible than others, as Stone explains: "Assertions of a causal mechanism are more likely to be successful—that is, become the dominant belief and guiding assumption for policymakers—if the proponents have visibility, access to media, and prominent positions; if the theory holds with widespread and deeply help cultural values; if it somehow captures or responds to a 'national mood'; and if its implicit description implies no radical redistribution of power or wealth."[3] If the wrong story prevails, recovery from the disaster can be hobbled and vulnerability to the next disaster deepened. And in a world dominated by the extraction and burning of fossil fuels, struggles over causes are becoming more contentious and more important than ever.

The following story of Indonesia's Lusi mud volcano is just one of many accounts in this book that reveal the human hand in creating hazards and disasters. But it shows, more sharply than any other, how doubt and indifference over the causes of a disaster can have grave consequences for the communities that have been devastated. It's a good moment to heed the words of epidemiologist Nancy Krieger, who looked beyond the web of causation to ask, "Has anyone seen the spider?"[4]

LUSI

At 5:00 a.m. on May 29, 2006, a mix of water, steam, and gray clay erupted from a rice paddy in Porong subdistrict, just south of the city of Sidoarjo in the eastern part of the island of Java in Indonesia. Nearly a decade later, the eruption had slowed but showed no sign of stopping. It could continue for another decade or for several decades. Meanwhile, more than forty thousand people have lost their homes, and property losses have approached a billion U.S. dollars.

The day before the mud first sprouted from Porong's soil, there had been an accidental blowout in a natural gas exploration well less than 160 yards from the eruption. The well, called Banjar-Panji 1 (BJP-1 for short), had experienced a series of mishaps over the previous two days, and its operators were attempting to extract the drill from deep in the earth when the blowout occurred. The prevailing explanation for the origin of the mud eruption is that the drill fractured a deep rock formation, triggering a fault slip and allowing boiling-hot groundwater to rise through the borehole, pass through a thick layer of clay, and bring a steaming mud slurry to the surface. But there's an alternative causal story: that the eruption was triggered by a magnitude-6.3 earthquake that struck two days previously near the city of Yogyakarta, about 170 miles from the eruption. The gas exploration company that owned the well, PT Lapindo Brantas, has insisted on the earthquake hypothesis, and therefore its own blamelessness, from the beginning. This creation of doubt over the cause of the eruption has had serious consequences for those whose homes, communities, and livelihoods were lost forever.

Political scientist Richard Stuart Olson has written, "When disasters were perceived as acts of God or otherwise outside the realm of possible human control or modification, political accountability was low and blame

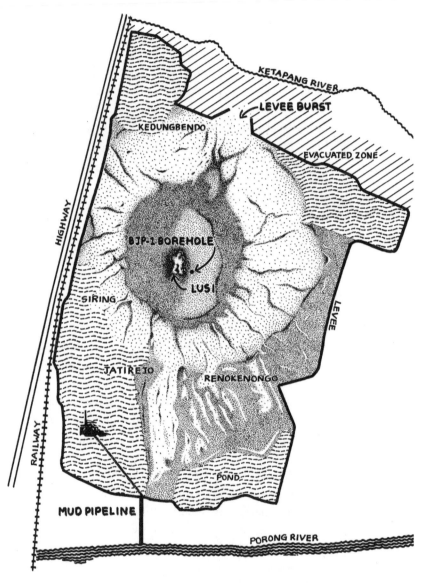

Figure 5. Area buried by the Lusi mud volcano, Porong subdistrict, East Java, Indonesia. Names of former villages now under up to sixty feet of mud are shown. The eruption site is labeled "Lusi," and "BJP-1" indicates the location of the gas exploration well that is believed to have triggered the eruption.

management a minor concern," but that in more recent times communities and societies have expected the cause (or causes) of disaster to be positively identified as a first step in assigning responsibility for response, recovery, and reconstruction, as well as in preventing another such disaster.[5] Recognizing those facts only too well, Indonesia's biggest industrial conglomerate, the Bakrie Group, of which Lapindo is a part, has worked tirelessly to convince the world that the mud volcano was purely an act of God or nature, in hopes of dodging responsibility for dealing with the mudflow and the victims' recovery. In that, the company has been only partially successful.

The eruption of the Porong mud volcano—which became widely known by the nickname "Lusi," a contraction of "Lumpur Sidoarjo" (Sidoarjo's Mud, named for the city nearby)[6]—isn't over. Even in the chaotic early months the disaster was an excruciatingly slow-motion affair. The burst of steaming slurry that emerged in 2006 flowed out into a densely populated, economically active suburban area. Residents of neighborhood after neighborhood watched and waited as the gray tide crept toward their homes and eventually buried them. The area was occupied by several townships that locals refer to as "villages" in English. A portion of Renokonongo village went under in August 2006, two and a half months after the eruption began. On November 21, a major toll highway passing through the center of the mudflow zone was closed forever and was buried soon thereafter. Fingers of the mudflow reached the villages of Jatirejo in September, Perumtas in November, and Ketapang in December. By the end of 2006, three more villages—Jatirejo, Siring, and Kedungbendo—had disappeared under the mud.[7] The volcano would eventually bury thirteen thousand households, thirty-three schools, sixty-five mosques, and thirty factories (a diverse array of facilities that included a watch factory, a brewery, a snack-food processor, and a rattan furniture firm, among others) in mud thirty to sixty feet deep. Some of the companies simply went out of business, while others moved to nearby Surabaya, Indonesia's second-largest city, leaving their former workers behind. With the loss of each village, a new wave of residents would evacuate to a very unpleasant temporary refugee camp in the town of Porong, outside the mud zone. Nine years later, many people had yet to be permanently resettled. Over an expanse of almost three square miles, there was no visible evidence of any of the former communities; only a single, mangled, mostly buried steel frame of a factory protruded from the mud.

Mud volcanoes are common in several seismically active zones around the world, but Lusi was the first to affect a populous, built-up area. Sidoarjo is Indonesia's second-largest industrial center, and Porong subdistrict plays a big role in the regional economy. A video documentary made by former residents of the now-lost villages contains extensive scenes of what no lon-ger exists: a well-established cluster of working- and middle-class neighbor-hoods interspersed with businesses and rice paddies. It was in the midst of this bustling scene that Lapindo drilled its exploratory well in March 2006. The video moves on to shots of the eruption's first few days, before anyone could have known how much destruction was to come, and then to much later scenes in which only rooftops are visible, and then still later, when there's nothing but mud. It is heartbreaking to see pre-2006 images of lively communities followed by scenes of their slow-motion interment.[8]

MAN AGAINST MUD

Early on, a "National Team" assigned by the government to manage the mudflow—with the effort paid for by the drilling company Lapindo—built emergency levees to shield villages, but the mudflow overtopped or burst through them one by one. Lagoons and a canal were dug in an attempt to dilute and divert the volcano's outflow to the Porong River, about a mile away. The canal failed at several points and inundated more land. The day after that accident, as Pertamina, the national utility company, was prepar-ing to relocate a gas pipeline threatened by the mudflow, the line exploded, killing fourteen people. An attempt was even made to plug the mud vent with giant steel balls strung together with chains, but to no avail. Through 2007, some of the villages that had not yet been buried were plagued by sudden scattered eruptions of water, natural gas, and hydrogen sulfide. As a result, some neighborhoods outside the mud zone were declared uninhab-itable. Many buildings and a rail line outside the levee were damaged as a result of land subsidence caused by the weight of the mud buildup.[9]

Eventually the National Team managed to completely corral the mud-flow into a roughly rectangular area with a thick thirty-five-foot-high earthen levee. But flow from the Lusi vents was unstoppable and threat-ened a catastrophic overtopping of the levee. The Sidoarjo Mud Mitiga-tion Agency (abbreviated BPLS in the Indonesian language Bahasa), which in 2007 had taken over management of the disaster, finally succeeded in pumping excess mud and water out of the levee-bounded area through a large pipeline out to the Porong River. Despite such heroic efforts, numer-ous small overflows, leaks, and breaches occurred. The area eventually took on the appearance of a gigantic, heaped-up bowl of extra-thick porridge. By 2014, the mud was about sixty feet deep toward the center and thirty or so feet deep around the edges. Lusi continued spewing water and mud at the center of the bowl, and a large portion of the central area had become a shimmering, watery quagmire of gray goo.[10] An unidentifiable, unpleasant smell lingered over the entire area. The incidence of respiratory disease in the surrounding area has been double the pre-eruption rate, but the precise source of that problem has not yet been identified.[11]

In the north and east, much of the surface had become crusty enough that vehicles could be driven on it. But even such dry areas were hazardous, because flowing in every direction from the central vents were meandering two- to four-foot-wide streams carrying hot, muddy water downhill toward the levees, cutting deep channels along the way. The entire surface was in constant flux, with the streams often changing course dramatically in the course of a single day. Unseen at the bottom of the big bowl lay all of the old villages and industrial sites.[12]

On a warm, hazy January morning in 2014, out on the gently sloping surface of Lusi's northern sector, we met BPLS geologist Hardi Prasetyo, who since 2007 has been in charge of managing the mudflow and preventing further destruction. Prasetyo sees that work as just the beginning of a much bigger effort to turn the disaster to his country's advantage. He exudes optimism, saying his philosophy is to "forget the past, look to the future." As the nature of the disaster has evolved, his team has had to improvise. He views the entire BPLS effort as a journey into the unknown. Because this disaster was unprecedented, he said, "we had no references to follow." Typically, noted Prasetyo, disasters unfold in a certain sequence. There's the "driving force"—the hazard—and after that, in order, come the rescue, recovery, and rebuilding phases. But here, rescue and recovery have had to proceed even while the driving force continues erupting. And BPLS has had to run to keep up.

Prasetyo and BPLS can monitor the action at Lusi's central vents only from helicopter overflights, via satellite images, or through binoculars. From the mud flats, we watched as a continuous flow of steam and occasional ejections of dark mud leapt high out of the vents about a mile to the south. We asked how far one could walk toward Lusi's center before sinking. Prasetyo suggested that we find out by engaging in one of his regular pastimes, a trek out across the mud. He said, "The first time I walked out there, people panicked—they thought I was crazy. They were forgetting I am a professional geologist. I know what I'm doing. Let's go!" Joined by his son Dwinanto Prasetyo, who handles media relations duties for BPLS, and three other employees, we set out in the direction of Lusi's central vent, which Hardi referred to as the "summit." The footing was firm at first, as long as one avoided the damp bubbling spots dotting the mudscape where water percolated up from the wetter mass below. Hardi took the lead, and we stepped carefully in his tracks. On such walks, he always probes the surface ahead with a cane to ensure that he and those following don't plunge a leg hip-deep into the mud. At one point, Dwinanto pulled from the mud a half-buried Indonesian flag attached to a short pole. It had been used to mark the location of the previous deepest foray across this sector of Lusi. We were able to walk perhaps fifty yards farther before the firm surface began to give way to an expanse of gooey grayness. Hardi planted the flag at the limit of the dry mud, declaring with tongue in cheek that this "ascent" marked the closest approach yet to Lusi's summit via its "north slope."

KICKED, NOT SHAKEN

Almost a decade after the initial eruption, most of Lusi's victims are still waiting to be fully compensated. The dispute over whether the volcano's origins could be found in a drilling accident or an earthquake has been a major impediment to them in their quest to achieve justice. By creating doubt about Lusi's trigger, Lapindo has been able to skirt legal responsibility even while agreeing to provide partial compensation to victims. In this, the company got a boost from a February 2008 resolution by Indonesia's House of Representatives declaring that the mud eruption was a "natural disaster," followed by a Supreme Court ruling to the same effect. Those decisions met with widespread dissent. The disaster was attributed instead to "human error" by the country's minister for energy and mineral resources, the Indonesian police force, Medco Energi (a company with a 32 percent stake in Lapindo), the government-appointed Supreme Audit Agency, and the majority of geologists and engineers worldwide who have studied the evidence.[13]

Mark Tingay, an expert in petroleum geomechanics at the University of Adelaide in Australia, says that in the academic debate over the mud volcano's genesis, "the two arguments are not actually that far apart. Both sides of the debate agree that Lusi was initiated following the reactivation of a major fault zone, the Watukosek Fault, underneath the point of eruption. Hence, the real argument is over what reactivated this fault. Again, both camps agree that something caused the stresses acting on the fault to change, triggering the fault to slip and allowing high-pressure water trapped underground to escape to the surface." Where the camps disagree is on the source of that water and the nature of the force that caused the fault to slip.[14]

In Tingay's view, the "something" that caused fracturing was the drill blowout, a type of accident that's supposed to be prevented by enclosing most of the drill's length in a concrete-and-steel casing. But as the drill reached a point a mile and three-quarters into the earth, its bottom mile or so had been left uncased. On the way down, the drill tip had passed through a thick clay layer called the Kalibeng Formation and then penetrated a much deeper stratum of porous limestone (the zone where the drillers hoped to find gas). With the drill unprotected, the operation experienced a series of small accidents and then a big one. Mark Tingay explains what happened next: "This culminated in a major 'kick' in which large volumes of high-pressure fluid rushed into the well and then out of the well head. The drillers then did standard procedure of 'shutting in' the well—closing big valves at the wellhead to seal the well off and stop fluids rushing out. However, this does not stop fluids rushing into the hole underground, and as fluid cannot escape, the pressure in the hole spikes. They recorded a big spike in pressure in the well, which our analysis shows was large enough to have caused the fault to reactivate."[15] On its way to the surface, the highly pressurized, superheated water pushed up through the Kalibeng Formation, accumulating clay along the way to form the mud slurry. Lusi burst into daylight less than two hundred yards from the well.

Tingay and colleagues have published drilling data indicating that pore pressures in the earth deep under BJP-1 and nearby wells just before, during, and after the BJP-1 blowout were indeed sufficiently strong to have set off the eruption; no distant quake was required. Pressure changes under the well were consistent with the timing of mudflow at the surface (at the center of what is now Lusi) and provided evidence of a direct connection between the well and Lusi through which water and mud flowed. The data also suggested that the source of water was the deep limestone layer, not the Kalibeng Formation, lending further support to the drilling trigger hypothesis.[16]

Those who argue that the May 27 Yogyakarta earthquake triggered the eruption have offered a very different explanation: that seismic waves from the distant quake shook the Watukosek Fault to life, causing "liquefaction" and eruption of the Kalibeng clay layer; this scenario requires that the source of the water erupting from Lusi be the Kalibeng Formation itself, not the deeper limestone.[17] Tingay and others in the drilling trigger camp maintain that the site of the May 29, 2006, mudflow was too far from the May 27 quake to have experienced any more than a small nudge. Tingay notes that if the quake had prompted the eruption, "it would be the smallest stress change ever to have triggered any sort of fault reactivation." He calculates that the stress or pressure change caused by the drilling kick was twenty-four to forty-eight times bigger than that caused by the earthquake. Tingay's colleague Michael Manga has shown that the seismic waves hitting below Lusi on that May 27 were not strong enough to trigger a mud volcano. Furthermore, he has identified twelve earthquakes that since 1973 have hit the site of BJP-1 with stronger seismic waves than did the Yogyakarta quake and triggered no eruptions.[18]

Papers supporting the earthquake trigger and published early in the dispute were written by researchers working with or for Lapindo or receiving support funds from groups associated with the Bakrie Group and are generally dismissed for that reason.[19] But in some subsequent studies, scientists with no such conflicts of interest have also concluded that the earthquake could have created Lusi. One of them, Adriano Mazzini of the University of Oslo, has generated some of the more important research results supporting the involvement of the Yogyakarta earthquake in the Lusi disaster. First of all, he points out, the geological formation under the BJP-1 well was under stress and primed for eruption, if given a jolt. He argues that Lusi "was meant to happen at some point, whether naturally or artificially triggered." He is convinced that "a man-made scenario alone cannot explain the observations collected on the field," because "certain things cannot be explained without invoking a natural trigger or at least a concurrence of a natural trigger" (meaning concurrence of the earthquake with the drilling accident). Only on-the-ground research can fully resolve the mystery of Lusi, he says: "Interpretation and debate of drilling data is more of a drilling engineer's exercise rather than research. So the drilling data have been and can be easily debated and interpreted in different ways depending on the side you stand and are therefore not a conclusive argument."[20]

Richard Davies, a petroleum geologist now serving as a pro-vice-chancellor at the University of Newcastle, and co-author with Tingay and others on several Lusi studies, bristles at Mazzini's contention that analysis of drilling data is not real research. In an e-mail, Davies wrote, "Thousands of papers have been published by drilling engineers and related well operations experts. To say that it's not research and dismiss it betrays a lack of appreciation of how critical these data are." On Mazzini's other comments, Tingay e-mailed, "I can easily explain ALL observations we have of Lusi to date solely invoking a man-made trigger. As such, I think that this disaster was not 'primed,' nor would it 'have happened anyway.' The disaster was . . . entirely avoidable had proper well planning, execution and management been conducted."[21]

A last nail appeared to have been driven into the coffin of the earthquake trigger theory when, in June 2015, Tingay, Davies, Manga, and others published a paper in *Nature Geoscience* that, in Tingay's words, provided "definitive evidence that the earthquake did not trigger any response in the

Kalibeng clays, and, thus, that all existing earthquake trigger models are now debunked." Their key finding was based on gas concentration data collected directly from the BJP-1 borehole by the drillers' own instruments. If, as the earthquake trigger argument requires, the Yogyakarta quake had caused liquefaction of the Kalibeng clay stratum—which contains high concentrations of hydrocarbon gases such as methane, ethane, and propane—large quantities of those gases would have been released, and large increases in their concentrations would have been detected in the borehole, which ran through the clay. The seismic waves coming from Yogyakarta would have reached the site of the BJP-1 well between five and fifteen minutes after the quake, and that is when gas would have been detected had the quake triggered the fault.[22]

But, as the paper showed, there was no increase at all in gas concentrations in the twenty-four hours that followed the earthquake. Gases finally did spike at the time of the drilling accident the next day, far too late for the quake to have been responsible, and presumably released by the kick that followed the accident. Concentrations of another gas, hydrogen sulfide, also leapt, going from zero between the quake and the drilling accident to five hundred parts per million immediately following the accident. This is important, because water from the deep limestone layer into which the drill bit is thought to have penetrated is rich in hydrogen sulfide, which is not found in the Kalibeng clay. This hydrogen sulfide spike would not have occurred under the earthquake hypothesis, which says that the Kalibeng clay was the source of the water that gushed from Lusi, but it's entirely consistent with the drilling trigger hypothesis, under which the deep limestone was the source of the water that rose through and mobilized the clay to create the eruption.[23]

At least some of the scientists who had been advocating for the earthquake trigger stuck to their guns despite the new findings. Stephen Miller of the University of Neuchâtel in Switzerland was perhaps the most emphatic, somewhat inexplicably telling Rachel Nuwer of the New York Times, "All science screams that Lusi is natural."[24] Mazzini also continues to insist that the earthquake reactivated the already stressed Watukosek fault, mobilizing gas- and water-soaked mud that moved up through the fault to create Lusi and cause gas releases along a 5,000-foot stretch of the fault. Since Mazzini published work on this scenario in 2009, Tingay and colleagues have not disputed the reactivation of the fault or its role in providing a passageway for water and mud to reach the surface. But they insist that it was the jolt from the drilling kick (vastly stronger than the jolt from the earthquake at the site of Lusi) that reactivated the fault; the fact that gases were released only after the drilling kick, they say, rules out the earthquake as the cause of reactivation.

BPLS, the agency that is still struggling to restrain the mudflow, takes no official position on the disaster's origin. Dwinanto Prasetyo claimed to be unswayed by the Tingay group's definitive 2015 study, attempting to keep intact BPLS's veneer of neutrality on the trigger issue when he told Nuwer, "If the majority of scientists said it's Lapindo's fault, then maybe the government could base a decision off of that. . . . But every year there is a new scientific paper, and it's always 50-50 about whether this was a natural or man-made disaster."[25] (Any examination of published studies shows that it's hardly 50-50; the drilling trigger has by far the greater support.) For his part, Hardi Prasetyo insists that BPLS is interested only in the future and that the cause of the eruption is irrelevant to his mission. But when Hardi discussed with us the growing body of research on the trigger issue—something he often does—he focused almost exclusively on the few studies supporting the earthquake hypothesis. That would be consistent with the position taken by the national government of which BPLS is a part. After relieving Lapindo of a good portion of its responsibilities for compensation and mitigation, failing to fully enforce presidential decrees targeting the company, and then shouldering the burden itself, the government could hardly justify any conclusion other than that the disaster was "natural."

FOSSIL EMPIRE

Hundreds of millions of dollars' worth of damage had been done and tens of thousands of people's lives had been disrupted, and still, by the time the Tingay gas paper was published, Lapindo and the Bakrie Group had managed to drag out the controversy over the mud volcano's trigger for more than nine years. By doing so, they had managed to dodge their responsibilities to the disaster's victims for nine years as well. Like the trigger battle, the battle over victims' compensation would also reach closure of sorts in the summer of 2015.

The struggle over responsibility in the aftermath of Lusi has been a three-way scramble among the government, Lapindo, and the citizens of Porong subdistrict who were displaced by the disaster. The forces are not symmetrical. In the aftermath of the eruption, Lapindo, with its wealthy owners and longtime close friends in high political places, was much more powerful than the former inhabitants of what was now a mudbowl. The sprawling Bakrie Group has many interests in addition to fossil fuels: mining of metals, oil palm and rubber plantations, construction, real estate, media, telecommunications, and finance. In addition to being a dominant force in the Indonesian economy, the Bakrie Group wields economic and political clout worldwide through its holding company Bakrie and Brothers and another investment arm, Bakrie Global. Bakrie had some shady dealings with the predatory Phoenix-based resource giant Freeport McMoRan, Inc., in the 1990s.[26] Bakrie Metals' website counts among its "major clients" the North American companies Exxon-Mobil, Chevron, Hess, and Unocal.[27] The family's influence is pervasive; for example, in 2010 the Washington, D.C.–based Carnegie Endowment for International Peace established a new Southeast Asian studies department with funding from the Bakrie Center Foundation, the conglomerate's philanthropic arm.

In a September 7, 2006, diplomatic cable made public by Wikileaks in 2010, the U.S. ambassador to Indonesia, B. Lynn Pascoe, noted that in the rush to prepare for a visit by President Susilo Bambang Yudhoyono to the disaster site, Lapindo had hired the U.S. petroleum services giant Halliburton as a "geotechnical consultant" to provide "much needed expertise" in improving the containment levee.[28] Yudhoyono had been reassured by what he saw and heard at the disaster site, and, wrote Pascoe, an "apparent result of the President's visit is the slowing of the criminal cases against several executives" of Lapindo and Bakrie. By the time of the cable, three months into what Pascoe himself called a "clearly man-made disaster," the East Java police chief had been claiming to have "sufficient evidence to make arrests of corporate executives involved with the well due to gross negligence and willful misconduct directly leading to the accident"; however, according to consular officials, the police "were ordered to hold off on any arrests related to Lapindo for the time being by the 'highest levels of the central government.'" That led to fear "that Lapindo executives would be able to wiggle out of their financial responsibilities to the East Javanese who lost their homes, land and businesses." Law enforcement efforts would be directed elsewhere, Pascoe reported: "Provincial and local government officials claim they are financially unable to provide any kind of meaningful assistance to their displaced residents, other than police and military security to keep out looters and outside provocateurs from organizing dissent among those displaced."

Lapindo and the Bakrie Group have always contended that they mounted a more than generous response to what they consider a huge "natural" disaster not of their own making. In 2014, Aburizal Bakrie, one of the sons of conglomerate founder Achmad Bakrie and a former chairman of the Bakrie Group, mounted an ultimately unsuccessful election campaign to succeed his good friend Yudhoyono as president of Indonesia. When asked early in the race if he was worried that the still-roiling mudflow crisis could hurt his chances, Bakrie said, "I have no problem with that, because we are not the ones to blame. The Supreme Court ruled that we did not have to pay for the damage. But my mother said to me: 'God has given everything to you. I don't care whether you win or lose in court. Help them [the victims].' Then I paid them, but not as a compensation, because we are not guilty. . . . That is why I am no longer in the list of 40 Indonesia's richest people!'"[29] The Bakries have complained from the beginning about how much they have had to fork over as compensation to the mudflow victims while failing to acknowledge that the government has spent more than twice as much as they have in order to deal with a disaster for which almost everyone believes the Bakries' company Lapindo was responsible.

At the time of the initial eruption in 2006, Aburizal Bakrie was not only Indonesia's richest person but also the senior minister for welfare in the Yudhoyono government. Nevertheless, with virtually everyone in the vicinity of Lusi convinced by what they were certain they had seen with their own eyes—that Lapindo's drilling accident had caused the eruption—Yudhoyono ordered the company's executives to begin paying for remediation efforts and restitution. It has been suggested that in doing this the president may have actually been attempting to provide his friends in the Bakrie Group some political cover.[30] Having Lapindo pay compensation right away without formally blaming the company may have been the government's way of cooling off the crisis and helping inoculate both itself and the Bakries against criticism.

The plan to have Lapindo compensate mudflow refugees was outlined in three presidential decrees in 2007–9. Temporary payments were made for food, water, shelter, health care, transportation, and other living expenses and for lost crops and wages. But the thorniest issue was compensation for land and property buried by the mud. The basic terms said the company would pay each household 20 percent of the value of its lost real estate immediately upon qualification, and then pay the remaining 80 percent by May 2008, two years after the initial eruption. Lapindo was to pay on a per-square-meter basis, with buildings, urban land, and agricultural land valued at different rates. That left renters out of the money altogether.

LUSI'S CASTAWAYS

Lapindo shelled out the compensation at a glacial pace, and that prompted Lusi's refugees to launch frequent, often massive, public protests. Working from a news database, Lynette McDonald and Wina Widanigrum of the

University of Queensland have listed a small sample of events that occurred during that period:

> Thousands of mud-flow victims from Siring village staged a demonstration outside Sidoarjo district administration hall on August 22, 2006. A violent protest by Sidoarjo residents on September 1, 2006 damaged Lapindo Brantas' property. Protesters dumped a truckload of mud on September 27, 2006 outside Minister Bakrie's welfare ministry. On April 26, 2007 thousands of mudflow victims occupied Surabaya airport to raise international awareness of their plight. On June 26, 2007, protesters blockaded a road, one of the few remaining access points to the mud volcano, stopping trucks loaded with sand and rocks for the construction of dams.... On January 30, 2008, more than 4,000 homeless residents marched to the government office in Sidoarjo, urging Lapindo Brantas to pay the remaining 80-percent compensation.[31]

At many of these and other events, groups of protesters covered themselves head to foot in mud from the volcano. The protests, however, did not shake loose much compensation. The two-year deadline for completing all compensation for property loss came and went, and Lapindo, having barely completed the initial 20 percent payments, was hardly even getting started on the remaining 80 percent. Then came the House of Representatives' resolution calling the mud eruption a "natural disaster," followed a few months later by the worldwide economic meltdown—a crisis in which the Bakrie empire, including Lapindo, incurred heavy losses. Company owners saw these events as providing retroactive justification for reneging on their commitments to the mudflow victims, arguing that (1) as they had claimed all along, they bore no responsibility for the disaster since it was "natural"; and (2) they could not continue paying compensation anyway, because the economic crisis had choked off their cash flow. Relief and compensation payments were suspended, and the government took no decisive action to force Lapindo to resume them.

In September 2009, a new presidential decree provided that all costs of physically managing the mudflow and protecting infrastructure would thenceforth be borne by the government. Thus Lapindo didn't merely wriggle out of its original legal commitments—the company was now relieved of any responsibility for dealing with the still-evolving mud emergency or for helping the growing number of new victims. It did agree to restart compensation payments, but only in small monthly installments rather than lump sums—a process that could go on for years, making it difficult or impossible for families to buy or build their new homes. (Most residential construction in Indonesia is a pay-as-you-go affair; mortgages aren't an option.) Lapindo also offered an alternative: the company could provide housing in a Sidarjo housing development called Kahuripan Nirwana Village (KNV), named for Aburizal Bakrie's brother Nirwana and owned by the parent company's real estate development subsidiary Bakrieland. The value of a refugee family's new house would be subtracted from the cash compensation they were to be paid.

At this time, most former residents of one of the buried villages, Renokenongo, were still living in refugee camps, and they split on the question of whether or not to accept the offer of housing in KNV. In the end, one faction yielded to heavy pressure from Lapindo and decided to move to KNV, while about five hundred families, seeking to keep as much of their community together as possible, decided to move as a group and build their own subdivision in a former sugarcane field about four miles from their former home. To make that possible, they needed their legally mandated cash payments from Lapindo, so they stepped up pressure on the company. As soon as they received enough to get started, they began construction of the new community, which they called Renojoyo. By 2014, about six hundred houses had been completed.

The quiet new village has long, straight streets lined with neat houses. It has the nondescript look of brand-new suburbs everywhere, but it seems a comfortable enough place to live. One such house belongs to a former rice farmer named Gunardi and his family. Almost six years after they should have been compensated the full amount that they were owed by Lapindo, they said they had received only 25 percent. In the meantime, they had settled for building a house smaller than the one they had had in Renokenongo. Gunardi told us, "I liked the old house better, but this is what we could afford." If he ever gets the other 75 percent, he said, he would allocate it among his children so they can build their own small houses. But whatever he received, he planned to hold a little back to buy a small piece of land for growing rice. He's a farmer, and with his land now buried in Kalibeng clay, he has no way of making a living. He was glad to have a house and be surrounded by his old neighbors, he said, but what he really needed was a way to earn an income. Most of Gunardi's neighbors were in similar predicaments, now separated from their former occupations. Along with the endless wrangling over compensation, one of the biggest concerns was still the difficulty of reestablishing livelihoods.[32]

Gunardi's neighbor down the street had his own business, which he had operated out of his home back in Renokenongo and was able to keep going in the new neighborhood. His house was larger and fancier than Gunardi's, but he, too, said it was much smaller than his former one. He had acted as a de facto community leader in the struggle for compensation. He had also done better personally than Gunardi, getting 75 percent of the compensation he was owed; nevertheless, he was not satisfied with Lapindo's progress in paying restitution to victims in general. "Either the company should speed things up or the government should take over," he said. "I don't want to hear about political parties and maneuvering. We just want the compensation."[33]

FINALLY, JUSTICE—OF A SORT

The dispute over compensation, like the mudflow, dragged on year after year. It was not only a painfully slow process; it was also opaque. Powerful Lapindo negotiated directly with individual families and transferred money to them only a little at a time. (The company also settled directly, and confidentially, with owners of industrial properties.)[34] Meanwhile, the government, through BPLS, was compensating residents still living in designated uninhabitable zones outside the levee. Yet many other households—those of people who had owned no land and had rented their housing or who owned ruined property not encompassed by the sometimes arbitrary boundaries set by presidential decrees—got nothing beyond some emergency assistance. They had been rendered homeless, and many of them jobless, with their formerly strong social ties disrupted. In rebuilding their lives, they were on their own. And the protests against Lapindo and the government rolled on. In all, Lapindo had owed an estimated 3 trillion rupiah in compensation for residential property (amounting to $235 million, or approximately 5 percent of Aburizal Bakrie's personal net worth before the 2008 crash); by early 2013, according to President Yudhoyono, they had paid about 73 percent of that amount.[35] But that didn't square very well with the stories of Gunardi and others who say they have received far less than three-quarters of what they were owed.

Bosman Batubara is an activist who worked for several years on behalf of the mud-inundated communities. At the time of the well blowout and eruption, he was working for Bumi Resources, Indonesia's biggest coal company, which is also owned by the Bakrie family. He began seeing a stream of internal e-mails in which management attempted to convince employees that the volcano was a "natural" disaster. The claims in the e-mails seemed to him both desperate and dishonest. Disillusioned with his employer, he quit his job and moved to the city of Yogyakarta, on Java's south coast—coincidentally, the city that had suffered badly in the earthquake Lapindo claimed was the cause of the mudflow. There he met Paring Waluyo Utomo, a leader in the Lumpur Lapindo resistance movement, and joined the struggle. In 2014, with the compensation struggle limping along with no end in sight, Batubara said, "nobody really knows how much Lapindo has paid. They make a series of partial transfers directly into people's bank accounts, when the people have already signed for the whole amount." He believed that the high percentages of the payout being cited at the time by the government and media referred to the total amount that households have signed for even though they were receiving that money only a little at a time.[36]

As desperately needed as the compensation payments were, it will take more than cash to heal the damage done by the mud. For one thing, the formula that reimbursed households in proportion to the value of the property they lost served to cement in place the imbalances of wealth and power that prevailed before the eruption. Batubara gives the residents of Renokenongo who stuck together and built the new village Renojoyo plenty of credit for their persistence but says their victory was incomplete: "They held out for their rights and stayed in the refugee camp longer than most, and they managed their move to the new land collectively. From the outside they seem strongly organized. But if you go deep there are a lot of problems. Better-connected people got better houses closer to the main road." And the better-connected were for the most part those who had owned more property back in Renokenongo, had been assigned bigger compensation packages, and had been able to extract from Lapindo a bigger share of what was due them.

The London-based International Institute for Environment and Development, which has worked to help the mudflow victims, explains how strong community cohesion has not been enough to ensure justice for all:

> Before Lusi, these communities would not have been considered poor, given their relatively high levels of social and economic welfare based on a thriving marine economy (fish and prawns) and great natural wealth, including fertile lands and oil and gas. They are organised according to traditional patron-client patterns, and rely on the links that community leaders (who are mostly religious heads) have with government. These are villages that rely on the collective. Thus, when the disaster hit, their co-operation was very strong, and it was easy for them to unite. But the fact that the company was

working hand-in-hand with the government made it very difficult for the people to work collectively. The struggle began strongly, with significant support from the media, NGOs, CBOs and international organisations and advocacy from local, provincial, national and international organisations. But as the media coverage fell away and many NGOs failed to develop any long-term solutions and also moved on, any hope for finding a just and agreed solution to the situation began to fade.[37]

Batubara warned us not to be too hard on the refugees: "You sometimes hear that the victims are all just in it for themselves—that they are greedy, that there's no solidarity. But see it from their point of view: it's a new chance at a new life. Several former residents are now even running for office. They are taking advantage of their experience, using their networks. They feel if they get into the legislature, they can speak for their people."[38]

In July 2014, former Jakarta mayor Joko Widodo, a populist universally known as Jokowi, was elected president of Indonesia. Aburizal Bakrie no longer had a close friend in the presidential palace.[39] Soon after taking power, the Widodo government restarted negotiations with Lapindo aimed at obtaining full compensation for those who lost their property to Lusi. A year later, in July 2015, the parties signed an agreement under which all victims would, at long last, receive full payment. But once again it was the government and the people of Indonesia, not Lapindo and the Bakries, that would come to their rescue. Through BPLS, the government would pay all remaining compensation claims, which amounted to $59 million. Lapindo, in turn, would be expected to reimburse the full amount within four years; to secure the deal, the company put up more than $200 million worth of assets as collateral.[40] Some claimants would have to wait to get their money until the government worked its way through verifying a large backlog of property documents. Yet finally, after more than nine agonizing years, full compensation would be paid.

"A WORLD WONDER"

Whatever his position on the trigger issue was, Hardi Prasetyo of BPLS has worked tirelessly to limit the damage from the mud volcano. It was a grim battle in the early days, with some geologists forecasting that the eruption would continue for another thirty to eighty years. He therefore took great encouragement from a study published in 2013 that predicted a tapering off of the eruption to a relative trickle within the coming decade.[41] If that projection holds (and there is no guarantee that it will), Prasetyo and BPLS will be able to shift from remediation to redevelopment.[42] This is where Prasetyo sees a bright silver lining in Lusi's gray goo; in his words, "We are standing right now between reality and a dream." In addition to establishing the Center of Excellence for Mud Volcano Studies, he plans to develop the mud zone as a tourist attraction—a "world wonder," he calls it—with water skiing on the lagoons, a mud spa (he regularly slips on a bathing suit and jumps into one of the hot streams to show visitors he's serious), and a domed arena.

One obstacle standing in the way of Prasetyo's plans was the fact that Lapindo owned the land buried by Lusi; the company technically had been "buying" it from the former residents as it delivered compensation. But the collateral Lapindo put up in the final compensation deal included the Lusi mudbowl itself. And because Lapindo remained in shaky financial shape, there were doubts that the company would ever manage to pay back the $59 million. Were it to default, ownership of the land buried by Lusi would revert to the government of Indonesia, and then, if someday the eruption finally settles down, the way could be cleared for Hardi Prasetyo to pursue his dream of a combination research center, theme park, and mud spa—his "world wonder."

But until the day Lusi finally sputters into silence, BPLS will have to stay focused on the race to keep up with the unceasing flow of mud and water. If you stand on the western portion of the levee, look at the mud that reaches almost to the top, and then look to the other side, toward the major highway and railway line far below, it would seem there is still plenty of reason to worry. But BPLS's struggle is especially intense in the northern sector, where occasional breaches continued to occur into 2015. Dwinanto Prasetyo told us, "Flows to the north are uncontrollable—it's not a stable area." And subsidence of levees on all sides makes continuous monitoring and maintenance essential. (For structural reasons, the levees cannot be built any higher.)

Concerns have also been raised that BPLS's pumping of millions of tons of mud into the Porong River has caused environmental damage. For example, both its water and fish pulled from it now contain elevated levels of cadmium and lead.[43] Prasetyo believes there's no reason to worry and argues that even before Lusi's mud managers encased its embankments in concrete and dredged its channel, the Porong was not so much a river as a managed floodway designed and constructed long ago by Java's Dutch colonial rulers to rid the surrounding landscape of excess monsoon rainfall. He pointed out to us that the Porong is broad and straight with a current strong enough to carry the excess sediment from the mud volcano all the way to the sea without the flora or fauna along the route suffering any ill effects. He stressed, "The river is not our target but simply a transfer medium." Looking downstream from above the outlet of the mud-transfer pipe outlet, he described how the Porong was carrying clay that originated deep in the nearby earth to join similar clays on the seafloor just beyond the river mouth. He is fond of summing up that situation as "Lusi, you have come home!" But some of Lusi's mud never made it home. As more and more sank to the river bottom, the Porong's flow began losing the strength needed to carry sediments all the way to the sea. In a massive operation, BPLS began dredging up the mixture of Lusi mud and other sediments from the river bottom and hauling it away. At the mouth of the Porong, they heaped up the sediments and smoothed them out to create a 237-acre island. There BPLS has struggled to establish new mangroves and fish hatcheries in canals and around the shoreline. Prasetyo has big plans to turn what he calls "Mud Island" into an ecotourism getaway, but so far the mangrove trees that BPLS has transplanted onto Mud Island's shore remain spindly and pale green, compared with the natural stands from which they came.

Some former residents, meanwhile, have continued their efforts to earn at least a partial living back on their home ground. Having demanded and gotten free access to their former land, they have created a combination tourist attraction–political protest site along the west levee, overlooking the highway. They have built stairs of wood and bamboo leading from the roadside up the outer wall of the levee, where they charge visitors admission for climbing up and taking in panoramic views of Lusi. On the dirt road that runs along the top of the levee, in the great tradition of local disaster media, some are selling homemade Lusi documentaries on DVD. Meanwhile, motorcyclists pick up tourists and take them for a ride across the firm southwest slope of the mud zone. Unofficial tour guides move in and out in a steady stream, taking visitors a hundred yards or so up the mudslope, to the point where solid mud fades into impassable slurry, and telling them the story of Lusi.

In mid-2014, protests over compensation escalated, with activists blocking BPLS from carrying out its Lusi management duties. Mud within the levee began building to dangerous heights, and at several spots on the west and north sides it overtopped and breached the levee. Those threats were controlled, the highway and rail line were not damaged, and BPLS announced plans to resume pumping mud out to the river. A new breach of the north levee wall required construction of another barrier beyond. And Lusi's mud keeps coming.

The writer Wendell Berry once wrote, "I am against climate change, but even more I am against the things that cause climate change."[44] Among the issues Berry had in mind were the ecological crimes humanity is committing in order to acquire fossil fuels, crimes such as mountaintop removal in the Appalachians, the ravaging of Alberta's tar-sand-rich landscape, and fracking all across North America and the world—all clearly industrial disasters. The burial of a large swath of Porong subdistrict under the Lusi mudflow fits well in that list. It differs, however, in that it had the outward appearance of a "natural disaster," and the perpetrator has therefore been able to crouch behind a cloud of doubt over the mud volcano's trigger. Having a universally recognized corporate villain to target does not guarantee justice and restoration after a disaster (the handling of the 1984 Union Carbide gas leak disaster in Bhopal, India, will suffice as evidence of that); however, it is easier for radical opposition to build against a human perpetrator than it is against "natural" forces. The ordeal suffered by Lusi's castaways may hold within it a preview of the future for all of us as we face more and more manufactured hazards that look perfectly natural.

6

HOW TO BOOBY-TRAP A PLANET

In the most general sense of progressive thought, the Enlightenment has always aimed at liberating men from fear and establishing their sovereignty. Yet the full enlightened earth radiates disaster triumphant.

—Max Horkheimer and Theodor Adorno,
Dialectic of Enlightenment, 1972[1]

Late in his life, Paul Cézanne sat down with the author Joachim Gasquet to explain to the younger man the process by which he painted some of the most important landscapes in the history of art: his more than sixty post-Impressionist views of the Sainte-Victoire mountain ridge in Provence.

"In order to paint a landscape correctly," Cézanne said, "first I have to discover the geographic strata. Imagine that the history of the world dates from the day when two atoms met, when two whirlwinds, two chemicals joined together. I can see rising these rainbows, these cosmic prisms, this dawn of ourselves above nothingness." He feels himself melt into the iridescence, a "rainbow of chaos," until night falls.[2] The next morning he begins to sketch out the rocky skeleton of the mountain. Then, as he tries to impose form, a new chaos seems to tear the land out from under him. "The geographical strata, the preparatory work, the world of drawing all cave in, collapse as in a catastrophe. A cataclysm has carried it all away, regenerated it." He said, "These boulders were made of fire. There is still fire in them." All that remains is color, "smoke of existence above the universal fire," rising toward the sun. "Genius would be to capture this ascension in a delicate equilibrium," but he doesn't consider himself a genius, no matter how many times he tries. "My canvas is heavy, a heaviness weighs down my brushes. Everything drops."[3]

Cézanne saw all of these forces—rainbows, whirlwinds, cave-ins, cataclysm, smoke, and fire—emanating from the stones of Sainte-Victoire in his attempts to master the mountain's form. If this much catastrophe lies in the act of painting a landscape, how much more can happen when we shape and live on the landscape itself?

In the drought summer of 1989, a huge forest fire swept across the carefully maintained south face of Sainte-Victoire, consuming houses and farms.

In subsequent years the bare slopes released landslides, one destroying the nearly thousand-year-old chapel of Saint-Ser.[4] A television reporter, looking out on "lunar landscapes, with white-gray rocks and pine trees standing like blackened matches," mused, "Cézanne has cause to turn in his grave."[5]

We have now made the entire planet our canvas, and the potential for creating fresh geoclimatic chaos has grown as a result. How we weigh the risks posed by such hazards is strongly influenced not only by the sizes of the risks and costs of dealing with them but also by what we know or believe to be true about the origins of the hazards. If we believe that human activities don't affect the probability or magnitude of a given event—if it is as inevitable and unpredictable as it is disruptive—then we can only adapt to its reality. As we have seen, confrontation with that sort of common adversity is often an occasion for cooperation, mutual aid, and heroism, even if no longer-term improvement comes out of it. If instead there is doubt or conflict over the origin of a destructive hazard, as there was in the Lusi debacle, all parties may see an opportunity to shirk responsibility for recovery and rebuilding. Yet if there is consensus that a hazard arose at least partly because of human actions, then the collective can-do spirit that emerges during and immediately after a disaster might be channeled into ensuring that those who contributed to the hazard are held responsible for the resulting damages and suffering, and seeing to it that no further such hazards are created. Now we will examine the many ways in which we are blowing out, cranking out, and pumping out all sorts of disasters—in cases when we suspect, and in cases when we know, that a climatic or geological event would not have happened without a helping human hand.

DIY DISASTER

Attitudes toward geoclimatic hazards, and the disasters associated with them, have evolved dramatically over the centuries. Once treated exclusively as acts of angry or capricious deities (usually assumed to have been provoked by humans' sinful acts), disasters came to be viewed during the Enlightenment as the inevitable consequences of natural processes; with the Industrial Revolution, disasters continued to be regarded as natural events, but ones that could be at least partly tamed with technology. As time went on, it became possible to redirect the flow of floodwaters, fortify unstable slopes, and maybe even knock an asteroid off course if it was to threaten us. Eventually, however, it became clear that fending off disaster can be a very tricky business, and that technological solutions often misfire or even trigger fresh calamities.

Our public policies are now falling even further behind as they fail to account for another, potentially even more jarring prospect: that the world economy is not only fostering disasters but also actually producing some of the geoclimatic hazards from which disasters emerge. There are greenhouse gases, of course, with their potential role in spawning heat waves, storms, floods, droughts, and, indirectly, landslides and wildfires.[6] Non-greenhouse-gas air pollutants such as black carbon and sulfate aerosols play their own wild-card role in many of the new, disturbing weather patterns we are seeing—heating some places at some times, cooling others, and altering precipitation patterns in unpredictable, sometimes dangerous ways.[7] And climate disruption is only one of several links between everyday human activities and hazards that look "natural." The Lusi mud volcano is just one of many recent examples. Extensive manipulation of rivers and groundwater causes coastal lands to subside, increasing the threat of flooding at least as much as rising sea levels. That, along with destruction of coastal ecosystems, results in storm surge hazards where none existed before. Thousands of landslides are triggered every year by damming rivers, road construction, mining, quarrying, irrigation, burning, and logging. And in recent years, as we will see, the dirty and green energy sectors both have been setting off earthquakes.

Nowhere do political or business leaders appear to be very interested in halting the production of new hazards and disasters. That is disturbing but not surprising, since doing so would mean curbing profitable activities and leaving valuable resources in the ground or under the sea. In preceding chapters, we have seen only a small sampling of humanity's many recipes for cooking up natural-looking hazards. There are many more. In the arduous struggle to prevent, prepare for, and recover from disasters, communities and nations can no longer avoid the even more difficult question of deeper causes. In the words of the historian Ted Steinberg, "The next time the wind kicks up and the earth starts to roar, what will we tell ourselves? Will we rise up in indignation at what nature has done to us? Or will we reflect on our own roles as architects of destruction? It is how we answer these questions that will determine the future of calamity."[8]

"WEATHER ON STEROIDS"

Tropical cyclones are among the most destructive of geoclimatic hazards, and it now seems they are being supercharged by climate change. Excess energy trapped by greenhouse gases can result not only in rising average global temperatures and locally extreme heat; it can also allow regions of the atmosphere to hold larger quantities of water vapor or generate stronger winds. The possibility that such energy could be dissipated by more powerful storms was once controversial but is now widely acknowledged. By the time Hurricane Katrina struck the Caribbean and U.S. Gulf Coast in 2005, researchers were already predicting that greenhouse warming could strengthen tropical cyclones. But in the broader public discussion, claims that Katrina was a symptom of anthropogenic climate change were loudly and effectively shouted down. A mere seven years later, similar views on Superstorm Sandy were widely accepted. People began talking openly about the possibility that natural disasters not only are not wholly natural, but that the underlying hazards themselves are increasingly created or aggravated by the normal operation of the human economy. Twelve months after Sandy, the record-shattering Typhoon Yolanda struck the Philippines; even as the storm was still raging over Southeast Asia, climate scientists were linking it to the warming global climate.[9]

Now whenever disasters strike, greenhouse emissions are widely, but not universally, viewed as suspects. Some qualifications remain necessary. While the *frequency* of hurricanes, for example, is not thought to be affected yet by greenhouse emissions, it has been all but confirmed that their *intensity* can be strengthened by the warmer surface waters in the North Atlantic.[10] Among several lines of evidence are Atlantic storm surge data going back to 1923 showing that Katrina-size storms are twice as frequent in warm years as in cold ones.[11] A 2015 study including Superstorm Sandy and Super Typhoon Yolanda showed that both were influenced by high sea surface temperatures that had a discernible human component.[12] When, in 2015, Hurricane Patricia exceeded even Yolanda's intensity, the new record was attributed to unusually warm sea surface temperatures in the region of the Pacific where Patricia originated. That warming was attributed to a major El Niño event, but the implication is that greenhouse warming in the Pacific will similarly fuel stronger tropical cyclones.[13] Record warm years are now routine, while record cool years no longer occur. To quote an Environmental Defense Fund vice president, "We can't say that steroids caused any one home run by Barry Bonds, but steroids sure helped him hit more and hit them farther. Now we have weather on steroids."

If climate science is making progress in figuring out the trend in tropical cyclones, it is still struggling to understand that quick-hitting scourge of the inland plains: the tornado. Individual twisters are notoriously difficult to study, but the kinds of weather fronts that spawn them are well understood. The towering storm clouds that are characteristic of such fronts feed on high convective available potential energy (CAPE) in the atmosphere.

Climate experts project that a warming atmosphere over North America in coming decades will generate higher CAPE levels and increase low-level wind shear (that is, wind currents passing each other in opposite directions, creating a potentially dangerous rotation) in the spring and fall across much of the continent east of the Rockies, with a very likely increase in the number of days per year in which the kind of severe thunderstorms that spin off tornadoes will occur.[14] In any case, a longer storm season could itself constitute a disaster—in 2011, losses caused by the most severe thunderstorms across the United States were a staggering $47 billion.[15]

Out-of-season tornadoes became a grim reality in the eastern and central United States in the 2015 Christmas season, when, under freakishly warm temperatures of fifteen to twenty-five degrees above normal and an atmosphere saturated with water vapor, a storm system grew to cover almost half the country at once. The giant storm was attributed to a strong El Niño in the Pacific being superimposed on a background of greenhouse warming. It incubated sixty-eight tornadoes across fifteen states, ranging in strength up to EF-4. Bob Henson and Jeff Masters of Weather Underground noted that "2015 is the first year in records going back to 1875 that has seen more confirmed tornado-related deaths in December than in the rest of the year combined." As the storm front churned east, it also produced severe flooding and mudslides and brought blizzards behind it. On some lands flanking the Mississippi River, waters grew even deeper than in 1993's record summer floods (and this in the dead of winter, not normally flood season on the river). Just before the first tornadoes struck on December 23, the *New York Times*, noting that the extreme warmth wave was expected to bring severe storms, twisters, and floods, had already labeled that anomalous Christmas week "a fitting end to the hottest year on record."[16]

The same year also fanned a continuing intensification of the country's fire season. In August 2015, as numerous large, intense wildfires burned throughout the heat- and drought-stricken western United States, the U.S. Forest Service released a report projecting that the agency would spend half of its entire budget on firefighting that year, compared with only 16 percent twenty years earlier; furthermore, the report predicted, the expense would rise to two-thirds or more of the Forest Service budget by 2025. The reasons were clear, according to the report: "With a warming climate, fire seasons are now on average 78 days longer than in 1970. The U.S. burns twice as many acres as three decades ago and Forest Service scientists believe the acreage burned may double again by mid-century." The Forest Service is part of the U.S. Department of Agriculture. Commenting on the report, agriculture secretary Tom Vilsack told the *Washington Post* "that to get a handle on the problem, extremely large or intense fires—which are only 1 to 2 percent of the total, but chew up 30 percent of firefighting costs—should be treated as natural disasters, much like hurricanes and floods are, and funded accordingly."[17] That proposal may have made budgetary sense to Vilsack, but it flew in the face of fire science. A growing body of research, some done in his own department, shows that the disastrous trend is not natural, but that the potential for wildfire is increased by the worsening heat and drought stress that comes with human-induced climate change. Based on historical data and tree-ring data going back a thousand years, a study published in *Nature Climate Change* in 2013 concluded that "drought has been, and remains, a primary driver of widespread wildfires" in the U.S. Southwest.[18] The epic droughts that the western states have experienced in these early years of the greenhouse era, and with them the wildfires, are expected to worsen and extend across the country. In a 2012 paper published in the journal *Forest Ecology and Management*, Forest Service scientists, using future climate models, projected that "fire potential is expected to increase in the Southwest, Rocky Mountains, northern Great Plains, Southeast, and Pacific coast, mainly caused by future warming trends. Most pronounced increases occur in summer and autumn. Fire seasons will become longer in many regions."[19] The association among human-induced warming, drought, and wildfires is only one reason that Vilsack erred in calling the more frequent extreme fires "natural disasters." Another is that U.S. forest lands all have long been under some degree of human management. Yet another, as acknowledged in the Forest Service's own 2015 fiscal report, is that more and more houses and other buildings are being constructed in areas that are almost certainly doomed to burn at some point.

It's a cruel irony on our greenhouse planet that an increased threat of drought in many regions coexists with a higher moisture content in the atmosphere overall. In recent years, that warmer, wetter atmosphere, the melting of Arctic sea ice, and unusual air circulation patterns—all associated with rising concentrations of greenhouse gases—have been bringing highly destructive winter storms and paralyzing quantities of ice and snow to parts of the United States, Europe, and China.[20] By the winter of 2014–15, the repeated onslaughts were sorely testing the endurance of the U.S. Northeast's residents. Research is increasingly finding what climate scientists call "teleconnections": links among extreme climatic events across vast distances. In early 2014, the low-pressure system that persisted over Southeast Asia for weeks, which brought disastrous rains, caused the deadly landslide that hit Barangay Andap in the Philippines (see pages 42–3), and eventually grew into Tropical Depression Agaton, is also suspected of having played a teleconnected hand in the extraordinary onslaught of cold air, snow, and ice that plagued the United States and Europe that winter.[21] The catastrophic deluges that struck Pakistan in 2010 were part of the same greenhouse-charged weather system that brought deadly heat, drought, and wildfires to Russia.[22] Climate models project large increases in the number and size of floods in India, Bangladesh, and Pakistan in the coming century; by destroying crops, such flooding could halt or reverse the progress that is being made in South Asia to alleviate hunger and poverty.[23]

A global increase in the annual number of catastrophic floods had become evident as early as 2002, when a group of National Oceanic and Atmospheric Administration (NOAA) scientists writing in *Nature* declared, "We find that the frequency of great floods increased substantially during the [late] twentieth century . . . and the model suggests that the trend will continue."[24] A more recent study, finding that prospects remain dire, concluded that in the twenty-first century severe flooding will be far more frequent and human exposure to flooding will be four to fourteen times as great as it was in the twentieth century, even when the effect of population increase is removed.[25]

RISE AND FALL

In a 2013 article, Wolfgang Kron of Munich Reinsurance Company made an extraordinary claim: "There are few natural catastrophes that are not somehow related to coasts. While not all of them occur right on the borderline between land and sea, their causes can be found either in meteorological events produced over the water or in geological events that happen at the crustal plate boundaries along the continents or mid-ocean ridges."[26] In the hazardous century ahead, it is indeed coastal areas that are expected to be imperiled most severely (but not exclusively; some other places, ones distant from any coast, can expect a heightened probability of geoclimatic disasters such as wildfires, tornadoes and landslides). The obvious aggravating factors affecting coasts will be the melting of polar ice caps and glaciers and the more rapid swelling of shallow coastal waters as they warm. The greater the relative sea level rise, the bigger the threat from heavy precipitation, storm surges, and tsunamis. The terrible flooding that struck Bangkok in 2011 could become the norm if sea levels rise as projected. By 2050, moderate flooding of Thailand's Chayo Praya River at a level not even reaching that experienced in 2011 could cause far more damage, inundating almost half of the sprawling metropolitan region; in 2100, floodwaters could cover two-thirds of the area.[27] In Venice, Italy, "alarm level" floods, which now occur about four times a year, would happen 250 times every year in the event of a sea level rise of less than two feet.[28] All told, with a worst-case scenario like the global six-foot sea rise that could happen if warming goes unaddressed, we could see 10 million people on the U.S. Atlantic coast and 18 million Europeans evacuating their homes, and 120 million people forced out of their homes across Asia.[29]

More than 100 million people live at elevations within one meter of sea level, and, as if increasing sea levels weren't bad enough, the land under many of the world's great coastal cities is sinking to meet the sea. Much of the subsiding urban land is in river deltas. A delta is not a fixed geographical feature; a satellite image of any delta is just a snapshot of soil and water on the move. Over the past two thousand years, delta growth was accelerated by upstream cutting of forests and tilling of agricultural lands, both of which expose soil to rainfall and allow it to be washed into rivers and their tributaries. But in recent decades, most deltas have stopped expanding, and some are shrinking. That's because the constant flux of sediments that is required to replenish them is being interrupted, primarily by damming of rivers. One-quarter to one-third of sediment flux through the world's rivers is being trapped in reservoirs behind an estimated 45,000 large dams. And dam building is gaining momentum worldwide. Even as they are being robbed of sediment, many coastal lands are sinking as well, and that is putting a lot of people and property at risk in many of the world's largest cities. (The world's forty most important deltas account for only 0.4 percent of the Earth's land area but are inhabited by 4.6 percent of the human population.) Pumping of fresh groundwater for irrigation, residential, and industrial uses is allowing the underlying sands and clays that held the water to undergo slow-motion collapse. Removal or degradation of dunes, wetlands, and mangroves has made matters worse. Other local factors, such as extensive petroleum extraction in the Niger River delta and industrial dredging in Italy's Po River, are accelerating subsidence. In many places, the land is subsiding as fast as or faster than the adjacent sea is rising. In fact, a 2006 survey of the world's deltas led by a group at the University of New Hampshire concluded that climate-induced sea level rise has so far been a "relatively minor influence" on the condition of delta regions when compared with human manipulation of landscapes and water resources.[30]

Of the ten world cities projected to have the largest populations exposed to coastal flooding by 2070, nine are in Asia. (The one non-Asian city, number nine on the list, is Miami.)[31] Where Bangkok's outskirts meet the South China Sea, twelve hundred acres of native mangrove forest have been cut, leaving the city even more vulnerable to storm surges. In the period

1951–80, China's east coast was hit with up to fifty storm surge disasters. Surges are becoming more dangerous with sea level rise, but now the impacts of rising seas are expected to be eclipsed by other human-induced changes that continue to lower coastal lands in some areas of China: urban development, inadequate enforcement of groundwater protection laws, and the decreasing ability of the country's great rivers to carry sediment all the way to their deltas—largely a result of river course alterations and sediment capture by numerous upstream dams. Seven thousand miles of coastline already have some sort of engineered protection against flooding from the sea. The Yangtze River delta and the Jiangsu coastal plain upstream from it currently enjoy the highest level of engineered protection; without it, twenty thousand square miles in that region would be flooded by only a one-foot sea level rise. But if there is a three-foot global ocean rise and China's barriers are not raised, current defenses won't be enough: those twenty thousand square miles will go under.[32]

DENYING GRAVITY

Year in and year out, one of the world's most common geoclimatic hazards is the sudden, destructive descent of large masses of rock, soil, and/or ice. In just the eight-year period ending in 2011, the world saw 3,059 fatal landslides—excluding ones caused by earthquakes—causing 35,287 deaths.[33] Catastrophic slides can bring devastation in many forms: mudslides that bury whole villages or towns; rock avalanches that crush all in their path; masses of water-soaked volcanic ash called lahars that can move at incredible speeds up to fifty miles per hour; landslides that temporarily block the flow of rivers, creating the threat of outburst floods; or earth and rock crashing into a body of water or sliding down an undersea mountain slope to trigger a tsunami. Human activities such as deforestation; mining; building of roads, tunnels, and dams; house construction; and even crop irrigation can create conditions that favor mass earth movements and set communities up for disaster.

In the coming era of intensified warming, heavier rainfall is expected to trigger even more landslides.[34] And we could face an even wider array of destructive earth movements as the planet's cold places thaw. The wall of water, ice, mud, and rock from the Himalayan glacial dam described in this book's introduction, which almost killed Ramala Khumriyal and did kill tens of thousands of others, is a stark example (see also Chapter 14). Melting of glaciers and permafrost is very likely to trigger more, and more devastating, rock flows and rock avalanches. Canadian researchers have found that in the largely mountainous province of British Columbia, the frequency of large landslides has risen in recent decades to more than two per year. They conclude, "The causes and triggers are numerous, but climate warming in recent decades has probably increased the incidence of catastrophic slope failure in northern British Columbia."[35]

A spectacular 2002 event in an obscure corner of Russia provides a glimpse of the future of cold mountain regions. On the evening of September 20, an enormous avalanche of ice, water, and rock hurtled down the Genaldon Valley in the Russian republic of North Ossetia, scouring it of trees and soil to a height of more than three hundred feet above the valley floor and leaving a trail of devastation more than ten miles long. Eyewitnesses noted that "the next morning, the view of the path of destruction was stunning. The whole floor of the Karmadon depression was buried under a mass of scree [2.5 miles] long and up to [426 feet] deep. . . . The village of Nizhnii Karmadon was completely destroyed and several rest homes along the Genaldon river below the gorge were devastated. Over 100 people were killed."[36] It seems that the huge Kolka glacier on a mountain at the head of the valley had "left its bed completely, revealing the bare rock bed" and had "virtually flown out" onto its path of destruction. This happened because an extraordinary quantity of liquid water had accumulated under the glacier, thanks to an unusual melt-off of ice and snow at high altitudes across the Caucasus in the four years leading up to the disaster. Temperatures in the region had been much warmer than normal over that period and, to make matters worse, the summer of 2002 had been especially rainy.[37]

A rockslide and debris avalanche along the Yigong River in Tibet in 2000 also featured a projectile of rock, ice, and water. The 300-million-cubic-foot mass, which traveled six-tenths of a mile horizontally and 1,600 feet downward, was airborne for an incredible thirteen seconds. The sudden movement came as a result of snowline retreat on the warming Tibetan Plateau and consequent saturation of the underlying rock.[38] Research by Christian Huggel of the University of Zurich finds that climate disruption will lead to more such hazards, but it's hard to say exactly where and when they will strike.[39]

The graceful, nine-hundred-foot-high Vaiont Dam, which stands in the picturesque mountains of far northeastern Italy, has not held water behind it for a very long time. On October 9, 1963, not long after the then-new dam had finished filling the Vaiont Reservoir, a three-quarter-square-mile section of a steep slope above the reservoir gave way, hurling an eight-hundred-foot-thick mass of solid rock into the water. Pushing the bulk of the lake's contents ahead of it, the rock mass triggered a towering tsunami.

The dam, amazingly, held firm, but the wave overtopped it by 460 feet. A wall of water more than two hundred feet high plunged down through the gorge below, destroying five villages.[40] According to observers, "hundreds of houses, factories, fifteen bars and cafes, three bridges, several miles of railroad tracks and the state highway, a large wooded area, etc., were leveled and obliterated. . . . For a radius of almost two miles across and about four miles up and down the valley, the destruction was all but total." Two thousand people were killed.[41] Studies of the Vaiont disaster found that filling of the reservoir had triggered the rockslide through two mechanisms: by saturating the clay layer that underlay the slope above, thereby reducing the friction that had held the rock mass in place, and by putting increased pressure on the slope. And, of course, the reservoir supplied the water that formed the tsunami that devastated the valley below.[42]

In 2015, there was widespread concern over the severe damage done to Nepal's hydroelectric industry by landslides that followed that year's big earthquake.[43] But causation is rarely unidirectional. The Vaiont tragedy revealed the power of dams to trigger devastating landslides, and, as we will show in Chapter 14, dam construction led to astonishing dislocations of earth in the northern India flood of 2013.

Scars of prehistoric landslides are visible on mountains and hills around the world, and their silent warnings often go unheeded by developers. Kathleen Nicoll of the University of Utah's Geography Department has examined development on and around the Traverse Mountains south of Salt Lake City. She wrote, "In the Traverse Mountains, the 'development footprint' since 2001 is emerging as one of the key factors that compromise slope stability" and raise the risk of landslides. All along the range there is visible evidence, obvious only to the trained eye of the geologist, of repeated landslides in the era before human habitation. Such old deposits are being restrained from sliding today only by the geologic pause button; even modest disturbance could be sufficient to prompt renewed movement. That nudge could well be supplied by twenty-first-century development of "master-planned communities" in the area.[44]

America's most landslide-prone large city is Seattle. As in most such badly exposed places, residential development leads to disaster through two routes: the tendency to build houses and roads in the path of potential slides and the triggering of slides through construction and earthmoving. For a century, area contractors have expended great effort to prevent outward movement of sloping land by halting "slope-toe erosion"—that is, by grading the bottoms of steep hillsides. But William Schulz of the U.S. Geological Survey has concluded that "more than about 16,400 years is required to naturally achieve slope stability in Seattle, given past climatic conditions. It does not appear that human-constructed, slope-toe erosion protection has made significant impact on landslide activity to date." Meanwhile, other construction activity has continued to encourage slope failures. Examining more than thirteen hundred landslides that have occurred in and around Seattle, Schulz found that human-induced events outnumbered natural ones by 7.4 to 1. Chief culprits included placing bodies of water above or below slopes, saturating portions of slopes with groundwater, and landfilling and construction on or above slopes.

In El Salvador in 2001, earthquake-triggered landslides struck across the entire southern half of the country on January 13 and February 13. Many occurred in the vicinity of road construction and quarrying, which had created large zones of instability. In these areas, noted one damage report, "In many cases, it was impossible to tell where one landslide ended and another began." The most devastating slide struck the neighborhood of Las Colinas, which had been built in an area previously off-limits to development at the foot of a volcanic slope south of the capital, San Salvador. The slide left a half-mile-long path of destruction and killed as many as five hundred people. It has been blamed on a buildup of water under the lower part of the hillslope, behind a retaining wall that had been constructed at the foot. Two hundred Las Colinas survivors brought a case in the Supreme Court of Justice, charging that the government had failed to prevent a preventable tragedy. They lost.[45]

WHERE THE QUAKES COME SWEEPING DOWN THE PLAINS

The Lusi mud volcano disaster is one of many cases in which the quest for fossil energy created a geoclimatic hazard. The number of earthquakes occurring under the soil of middle America has increased sharply since 2009, especially in states such as Oklahoma that have active oil and gas drilling industries. Earthquakes don't come naturally to Oklahoma. From 1978 through 2008, the state averaged two quakes per year of magnitude 3 or greater.[46] After that, the number rose dramatically year by year, reaching 109 in 2013, then leaping to 584 in 2014 and 907 in 2015. The Sooner State now far surpassed all of the other lower forty-eight in its earthquake frequency.[47]

The strange phenomenon first gained nationwide attention in 2011 when a severe magnitude-5.7 quake struck near the town of Prague, east of Oklahoma City. Damage occurred to 174 homes, with six destroyed and twenty suffering major damage, and shook the ground as far away as Chicago.[48] It was the biggest quake yet recorded in a state where threats normally come from the sky, not the earth. A local resident told a reporter, "We're in tornado country, man. These earthquakes, it just scares the hell out of everybody here."[49]

For decades, researchers had been studying the induction of small earthquakes by deep-earth injection of wastewater from the oil and gas industries. Such induced quakes had been detected in Colorado, New Mexico, Texas, Arkansas, Ohio, and other states, but the outbreak in Oklahoma was extraordinary.[50] In the summer of 2013, a paper in the journal *Geology* found a link between wastewater injection and the Prague earthquake. Katie Keranen at the University of Oklahoma and colleagues at Columbia University and the U.S. Geological Survey showed that injection had raised subterranean fluid pressures to the point at which they set off a cascade of small tremors that eventually induced the record 5.7-magnitude quake.[51]

Fluids are injected deep into oil and gas fields for three different purposes. The most widely discussed use of injection is in hydraulic fracturing, or "fracking," of shale formations to release natural gas deposits. Water is also injected for the purpose of "enhanced oil recovery," that is, to help push the last barrels of oil out of largely depleted formations. But the recent huge increase in earthquake activity in the central United States is associated with the third type of injection, known as "salt water disposal." Many oil-producing formations contain large quantities of water as well as oil; in some, four barrels of water may be pumped to the surface for every barrel of oil. To get rid of this dirty, salty waste, companies reinject it deep into old, fully depleted oilfields. Saltwater injection rates began rising dramatically in several parts of north-central Oklahoma in 2005–9 and have remained high. Near many wells, the increase in salt water injection was followed by an increase in earthquake activity.[52] This connection extends throughout the central and eastern United States; between 2011 and 2014, seven out of every eight earthquakes of magnitude 3.0 or larger occurring in those states were associated with injection wells. Across the region, more than eighteen thousand injection wells—about 10 percent of all such wells—have had earthquakes occur nearby. Of the forty-five most active wells, each injecting more than a million barrels of oil per month, thirty-four were associated with earthquakes.[53]

In 2014, Keranen, who by then had left Oklahoma for Cornell University, was the lead author on a study laying out strong evidence that intensifying earthquake swarms east and north of Oklahoma City were associated with handful of the highest-volume wastewater injection fields in that area.[54] She and others have worried about a situation in which injection induces, say, a magnitude-4 quake along a small fault, and its shock waves set off a quake in a larger, already stressed fault nearby, one that happens to be connected to a fault that's primed to produce an even larger earthquake. For example, such a chain reaction starting near the town of Jones, an active drilling area that has seen a steep escalation in seismic activity, could conceivably end up at the Nemaha Fault, which runs through the heart of Oklahoma City. The Oklahoma City section of the Nemaha could theoretically experience a quake of magnitude 7 or greater, a thousand or more times as strong as the magnitude-4 quakes that have become common in the state. Keranen and her coauthors concluded, "The increasing proximity of the earthquake swarm to the Nemaha fault presents a potential hazard for the Oklahoma City metropolitan area." [55] But industry leaders continued to deny responsibility for the increased seismic activity and refused even to take small steps; they were backed up by the Oklahoma Geologic Survey (OGS), a state government agency associated with the University of Oklahoma.[56] When, in 2014, some OGS employees began expressing doubts that the escalation in quake activity was all "natural," petroleum tycoon (and major donor to the university) Harold Hamm urged OGS to fire those employees and allow him to help choose a new director for the agency.[57] He was ignored.

By then, a far higher percentage of Oklahoman than Californian homeowners had earthquake insurance policies. But policyholders found themselves stranded in a shaky no-man's-land. Companies were denying more than 90 percent of their claims, largely on the grounds that the quakes were human-made and not "acts of God" as specified in the policies.[58] But those stiffed by their insurers would also find it exceedingly difficult to recover their losses from the petroleum companies through lawsuits. If a propane truck were to blow up in an Oklahoma resident's driveway, the gas company or its insurance company would have to pay for the damage; however, when injection-induced earthquakes damaged Oklahoma houses, there was no way to prove that a specific quake was caused by a specific company's injection well and sue for damages.

In 2015, the link between disposal wells and earthquakes became impossible to ignore. The OGS finally began conceding that the link was real, and the state started taking action. Through the spring and summer, the Oklahoma Corporation Commission (OCC) identified hundreds of problem wells in the state's oil-and-gas belt and ordered companies to limit the quantities of wastewater being injected into them. Then, on August 3, the commission turned its attention to twenty-three wells in the seismically active area north and east of the capital, directing that disposal volumes at the wells be cut 38 percent within two months. Unlike most other troublemaking wells, these were not especially deep or high-volume. It was possible that the quakes were being prompted by more distant wells, which would be consistent with Keranen and her colleagues' findings that waves of pore pressure could migrate more than twenty miles from sites of injection and induce seismicity. But no one could say immediately why the dramatic spike had occurred.[59] For geologists and seismologists, each new quake swarm helped reveal previously unknown faults, and concern grew that a future event could occur along a small fault connected to the Nemaha system, leading to a quake much stronger than those that hit Prague in 2011, and in a much more built-up, densely populated urban area.[60] Around that time, Governor Mary Fallin made headlines when she provided the first official admission that the earthquakes were linked to injection wells, stating, "I think we all know now that there is a direct correlation between the increase of earthquakes that we've seen in Oklahoma with disposal wells."[61]

In September, the OCC ordered two more disposal wells completely shut down, and the crackdown would continue in the months to come. Oklahomans waited and watched for signs that stepped-up regulatory activity was starting to calm the seismic activity, but 2016 did not begin auspiciously.[62] On January 6, in a rural area about 120 miles northwest of Oklahoma City, twenty earthquakes occurred in a single ten-hour period; two of them, with magnitudes 4.8 and 4.7, struck thirty seconds apart. They ranked as the state's fourth- and seventh-strongest quakes on record—but only for a month. On Valentine's Day, a magnitude-5.1 whopper, the third-biggest ever, shook the state.[63]

CLIMATE-FRIENDLY QUAKES

The problem of earthquake creation is not confined to the fossil fuel industry. Renewable energy production, too, is capable of inducing seismic events. An attempt to develop a geothermal energy system beneath Basel, Switzerland, which involved high-pressure water injection, triggered four small quakes in 2006–7. The project was abandoned and the entire investment lost.[64] The magnitude-7.2 El Mayor–Cucapah (EMC) earthquake that struck the Mexican peninsula of Baja California in April 2010 was the strongest in the region in more than a century. It originated in the rupture of a fault system that had been considered dormant. Four people died, hundreds were injured, and many landslides were triggered. Across the border in the United States, the entire city of Calexico was shifted two and a half feet to the south. A study published in the journal *Geophysical Research Letters* in 2014 revealed that fluid extraction at the Cerro Prieto Geothermal Field (CPGF), which hosts the world's second-largest geothermal power plant, had been creating extremely strong stresses in the spot where the quake occurred, potentially strong enough to have caused the rupture. The authors concluded, "Although we cannot definitively conclude that production at the CPGF triggered the EMC earthquake, its influence on the local stressfield is substantial and should not be neglected in local seismic hazard assessments."[65]

On May 11, 2011, a magnitude-4.7 quake hit the town of Lorca in Spain, followed by a much stronger 5.2 shock that caused heavier than expected damage, given its size. An hour and a half after the second quake, *Earthquake Report* noted, "Given the relative abnormality of such large earthquakes in this area and the fact that two descended upon Lorca in short time, there is a feeling of chaos in Lorca, with much damage reported." In the end, there were nine fatalities, three hundred injuries, seven thousand families made homeless, and fifty buildings, including a seven-year-old high-rise, that would have to be demolished. Total damage was estimated at €70 million.[66] The disproportionate damage and loss of life caused by the Lorca quake was eventually attributed to the fact that its focal point, a slip in a fault approximately ten thousand feet below the surface, was unusually shallow. The most likely culprit has turned out to be groundwater extraction, largely for irrigation of artichokes, broccoli, olives, lemons, and other crops in a region known as the "kitchen garden of Europe."[67] The water table on one side of the fault had dropped a whopping eight hundred feet since 1960, greatly lightening the gravitational load on the underlying rock and causing a "stress perturbation" on the Alhama de Murcia fault system. The ruptures in the fault occurred where they would be expected to occur if the dropping water table was the source of stress.[68] In a *Nature Geoscience* commentary on the study revealing the possible role of water extraction in causing the tragic quake, Jean-Philippe Avouac, a geology professor at Cal Tech, managed to find a silver lining, writing that "if ever the effect of human-induced stress perturbations on seismicity is fully understood . . . we might dream of one day being able to tame natural faults with geo-engineering." But for the present, he argued, we should remain wary of imposing our own stresses on the Earth's crust; he noted a new such concern over proposals for pumping carbon dioxide into subterranean repositories to keep it out of the atmosphere, which could create seismic stresses over vast areas.[69]

It has long been known that damming of rivers to form reservoirs in seismically active areas can produce not only Vaiont-style landslides but earthquakes as well. As of 2002, reservoir-triggered quakes of magnitude 4.0 or greater had struck forty-two sites around the world, and many more have occurred since.[70] In 2008, the horrific magnitude-8 Wenchuan earthquake in Sichuan province, China, the strongest to strike the country since 1950, killed more than eighty thousand people and destroyed many buildings and roads. Although still officially classified as a tectonically triggered quake, Wenchuan may actually have been not that natural in origin. Since the day of the quake, there has been a running argument among seismologists regarding the possible role played by the nearby Zipingpu Reservoir in the disaster. The reservoir began filling in 2004, reached its greatest depth in January 2008, and was drawn down relatively quickly over the next four months, until the day of the quake. That timing prompted considerable suspicion, which was supported in a 2009 paper by Christian Klose, a Columbia University researcher in geophysical hazards. Klose showed how the pressure changes from the huge mass of the reservoir could have created conditions that hastened the onset of the quake.[71] Fan Xiao, chief engineer on the Regional Geological Survey Team of the Sichuan Geology and Mineral Bureau, is convinced that the reservoir induced the earthquake and has concluded that "widespread and largely unchecked dam-building" in China's seismically active southwest provinces could lead to more catastrophes.[72] Other seismologists dismissed the reservoir trigger idea, pointing out that all large quakes previously associated with reservoirs had been in the magnitude 6.0 to 7.0 range, only one-tenth to one-hundredth as strong as Wenchuan.[73] But Leonardo Seeber of the Lamont-Doherty Earth Observatory at Columbia, citing Klose, told the *New York Times* that with the reservoir having filled, the quake could well have occurred "a few hundred years" sooner than it would have without the reservoir. "It would have occurred anyway," Seeber told the *Times*, adding drily, "But of course the people who were affected might think the timing is an important difference."[74]

Eleven months after Wenchuan, the people of L'Aquila, Italy, learned that when it comes to earthquakes, timing is indeed everything. No drilling, dam building, or other human action was suspected of causing the quake that struck L'Aquila, so when it left three hundred people dead, the bitter dispute that followed was not over the production of a hazard but over how terrible death tolls are made.

7

FORESHOCK, SHOCK, AFTERSHOCK
L'Aquila, Italy

We all know that the earthquake could not be predicted, and that evacuation was not an option. All we wanted was clearer information on risks in order to make our choices.

—Vincenzo Vittorini, survivor of the L'Aquila earthquake, 2011[1]

As the experience across many countries demonstrates, stringent building codes and seismic retrofitting regulations are the most effective measures communities can adopt to ensure seismic safety.

—International Commission on Earthquake
Forecasting for Civil Protection, 2011[2]

In the wee hours of April 6, 2009, a magnitude-6.3 earthquake struck L'Aquila, a city seventy-five miles northeast of Rome, Italy. More than three hundred people died, and tens of thousands of buildings were destroyed or damaged. There was no controversy over the source of the quake; L'Aquila is located in a seismically active area where earthquakes occur intermittently and unpredictably over decades and centuries. It is the dispute over where to place blame for the disaster—one that crippled a historic city and produced an unusually high death toll—that has been wrenching. Nothing complicates disaster quite like blame.

L'Aquila's experience holds lessons for public officials everywhere who are eager to prove that they are the people's protectors. The chief issue setting L'Aquila apart from other seismic calamities was an official accusation that scientists bore responsibility for the deaths and injuries that occurred. On May 25, 2011, after two years of public debate over whether L'Aquila should have been evacuated before the quake, the government hauled into court six scientists and an official of the national government. The so-called L'Aquila Seven were charged with giving false reassurances that led to the deaths of twenty-six residents who, it was claimed, would not have stayed in their homes had they not believed that there would be no quake that night.

On October 22, 2012, the seven men were convicted of manslaughter;

the presiding judge sentenced each to six years' imprisonment. The sentence was two years longer than had been requested by the prosecution. The unprecedented legal action against the scientists drew attention away from an important issue: had officials ensured that L'Aquila had more seismically sound housing before 2009, that could have saved far more lives than any earthquake warning. Another strange outcome of the disaster was that the centuries-old city center was saved from destruction but remains a ghost town frozen in time.

The battle over blame for the lives lost in the quake has been fed by a public misapprehension that experts can always provide accurate warnings of imminent disaster. Such preventive actions are possible in the case of some other types of hazards, including tornadoes and tropical cyclones and (with a little luck) wildfires and volcanic eruptions. But with earthquakes, there is never definitive evidence to support precise decisions on when and where an evacuation ought to take place. Nevertheless, many residents of L'Aquila came to view the deaths in their city on that April night as preventable. Had people been urged to leave their homes and stay outside that night, fewer would have died—that much is true. The problem was that there was no way of knowing in advance when to give such a warning (or, just as important, when to announce that it was safe to go back inside). With earthquakes even more than with major storms, the most lives are saved and the most property protected by actions that are taken not when the hazard strikes but long before, with the proper location, arrangement, and construction of the places where people will be living and working on the day that the unthinkable happens.

LOST IN 2009

The city of L'Aquila lies in a valley of the Apennine Mountains, which form the spine of the Italian peninsula and are prone to seismic activity. The 2009 quake was the eighth-largest shock to hit the city since the 1300s.[3] But the last quake this large had been in 1915, and none had done as much damage to the centuries-old heart of the city as the 2009 shock did. In the five and a half months after the 2009 quake, residents endured eleven aftershocks of magnitude 4 or greater.

The disproportionately severe destruction of life and property in L'Aquila has been attributed to the energy transmission properties of the soil in the area (which sits on layers of limestone and clay over an ancient lake bed) and the fragility of buildings, many of which had been constructed before stronger seismic codes went into effect. The picturesque center of

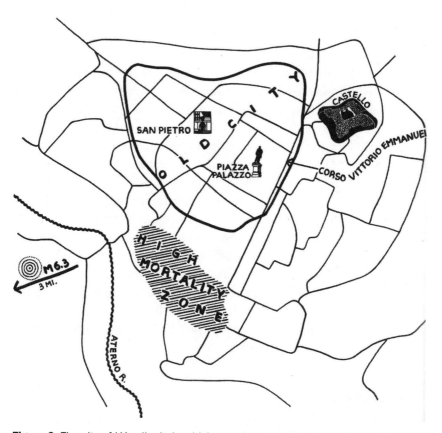

Figure 6. The city of L'Aquila, Italy, which was devastated by a magnitude-6.3 earthquake that struck approximately three miles to its southwest on April 6, 2009. Heavy casualties were suffered not only in the area labeled as the high-mortality zone but also in the old city and elsewhere.

L'Aquila, the "old city," which dates back to medieval times and has long been a tourist destination, was rendered uninhabitable. The village of Onna, a few miles east of L'Aquila, was completely destroyed.[4]

By August 2013, the rubble was long gone, and rebuilding had proceeded to the point that the casual visitor could pass through most sections of the city or its suburbs without noticing any evidence that a major earthquake had recently occurred. Although the interior of the city's imposing 440-year-old fortress, locally called *il Castello*, had been damaged and was closed for repairs, its basic structure was intact, and it appeared from the outside to be unaffected. But the quarter-square-mile old city, the original L'Aquila, was now a ghost town on life support. Untold quantities of debris had been removed and two large sinkholes had been filled in, making it permissible to walk, but not drive, through most of the area. Only a few sections remained enclosed within chain-link fences with signs reading "Zona Rossa" (Red Zone). On weekend nights, tourists and L'Aquila University students brought life to the eastern edge of the old city, along the pedestrians-only section of Corso Vittorio Emanuele where some shops, bars, and restaurants had reopened. But the rest of the old city was silent. Block after block was lined with two-, three-, and four-story buildings, their facades crisscrossed with various types of bracing: metal pipes, wide steel beams, cables, and straps. Superficially, some of the shored-up buildings appeared unscathed; many others displayed long vertical or diagonal cracks or were missing large hunks of stucco and stone. Doors were chained and padlocked, but some gaped a few inches, emitting a musty smell. The glass in most windows was intact, but here and there curtains fluttered from broken panes, and through some windows could be seen entire rooms occupied by forests of supporting scaffolds. In the thresholds of doorways along Via Antinori, weeds had gone to seed and dried up, while in a couple of second-floor planter boxes above Via Garibaldi, long-dead plants shared space with still-colorful artificial flowers. In one house, the bathroom appeared to have been left undisturbed since the quake four years earlier. A bar of soap remained on the edge of the lavatory, and a washcloth lay on the floor of the shower. Inside no. 6 Via Camarda, a calendar hung on an otherwise bare wall, open to March 2009.

A block away from the Corso Vittorio Emanuele is the parklike Piazza Palazzo, which was once a social hub in the old city. The streetlights still burned, but the public toilets were locked up and the benches were empty. Some of the governmental buildings surrounding the plaza were hidden behind tarps as they underwent repairs. A marble plaque on the face of one edifice still honors those who died in the city (whose name means "The Eagle") in June 1944: "L'Aquila liberata dall'occupazione nazista donava nuove ali alla missione affidatole dalle comunità fondatrici quale artefice di giustizia pace e libertà" (L'Aquila, liberated from Nazi occupation, gave new wings to the mission entrusted to it by the community as a founding architect of peace, justice, and freedom). After sunset, light came from just one doorway in the plaza, that of a tiny bar called Farfarello. It was established more than twenty years ago but was shuttered for years after the quake while the building above it was being secured and the city dragged its feet on issuance of a new business permit. Farfarello finally reopened in 2013, and a banner stretching over the threshold bore another inspiring message, translated as "If a dream meets so many obstacles, it means that it's right." The bartender that evening, Daryoush Shoyajee, told us that the old city had been home to many of the area's senior citizens, and some university students had rented rooms there as well. They no longer frequented the park, and tourists had stayed away for the first two years after the quake for fear that an aftershock or fresh quake could bring the fragile buildings tumbling down. Before the quake, he said, the plaza was the place where residents of the old city and other people from all over L'Aquila, young and old, would come. "Now they don't. And that"—he waved toward the tourist foot traffic out on Corso Vittorio Emanuele—"is not the same."[5]

The propping up of the old city's otherwise neglected buildings turns out to be a legal necessity. Mauro Dolce is a civil engineer and professor of structural engineering, and at the time of the quake was director of seismic risk prevention and mitigation for Italy's Department of Civil Protection (DPC). Dolce oversaw much of the reconstruction effort. In a 2013 e-mail, he explained that the preservation of L'Aquila's old city "is the current and usual policy of post-earthquake reconstruction in Italy, adopted after relatively recent earthquakes, at least since 1997's Umbria-Marche earthquake. In Italy not only the cultural heritage sites (churches, palaces, etc.) but also the historical centers of cities and villages are considered worthy of being preserved. Demolition of old buildings, unless they have been almost totally ruined by the earthquake, is undertaken very rarely." But that policy, Dolce noted, made his job that much harder, because the costs of restoration are always far higher than the costs of demolition and reconstruction. The result, of course, is that restoration of historic districts takes a very long time. "This is not just the case of L'Aquila, as you can find that reconstruction after previous earthquakes has needed some decades." Finally, he added, "there are often administrative questions to be solved, because the property of flats in old/ancient buildings is often very much fragmented, and sometimes it is difficult even to find the owners." Thanks to all of those factors, the old city remains stuck in the spring of 2009, deserted.[6]

But Mauro Dolce's struggle to rebuild the new city and preserve the old one, difficult as it may have been, was not the worst ordeal he had to face in the years following the quake. That came when he was convicted of manslaughter, as one of the group that came to be known as the L'Aquila Seven.[7]

FROM FALSE ALARM TO FALSE ACCUSAL

Like most L'Aquila residents, Daryoush Shoyajee supported the prosecution of the scientists. He said that in the days before the quake, "the officials told us there is no problem, stay in your homes." Then, with hundreds killed, he said, "nobody has had to pay the price for it, and that is not right!" He was not satisfied that the committee members were then facing six years each in prison: "Pffft. Maybe someone will do a month in jail, no more. This is Italy! It will be like Berlusconi," he said, referring to Italy's often-accused, never-convicted former prime minister. "You have heard of Berlusconi, no?"[8]

But how did the L'Aquila Seven end up in jail for not warning of an imminent earthquake, an event whose precise location, timing, and destructive power cannot be predicted by any known method? The whole affair emerged out of an unfortunate series of events involving bad timing, miscommunication, tragic coincidences, and a fateful comment to the media. Starting in January 2009, swarms of seismic tremors were felt in the region around L'Aquila. They caused no damage and were not viewed as evidence of an imminent major quake, but they did raise anxiety levels in the valley. Confusion arose when a government laboratory technician named Gioacchino Giampaolo Giuliani, who practiced amateur seismology, began making predictions based on levels of radon emissions. Initially, in a March 24 interview, Giuliani had said he believed that the swarm of small quakes was a "normal phenomenon" in that area and would probably be over within a week. Five days later, Giuliani changed his mind when a radon detector that he had placed in a basement in L'Aquila registered elevated radon levels. He alerted public safety authorities that a strong earthquake could strike "within a week and probably centered upon Sulmona," a town about forty miles from L'Aquila.[9] His warning was leaked, and fear spread through the region. When a magnitude-4 tremor further heightened tensions, officials decided to address that fear by convening the National Commission on Major Risks on March 31. The group included several scientists associated with the National Institute of Geophysics and Volcanology (INGV), along with public safety officials. The scientists' task was not to decide whether evacuations should be ordered—that was the job of government figures—but rather, in light of Giuliani's highly publicized projections, to explain the current state of science's capacity to predict earthquakes from seismic sequences and how that applied to the current situation. But critics later claimed that the meeting was a public relations stunt to calm the region's residents down.[10]

There are no recordings or transcripts of the meeting, only minutes that were written up after the quake. It is known that, not surprisingly, no heated debate occurred; the scientists involved in the meeting all knew that (1) the risk of a large quake had increased a hundredfold or more in recent weeks; (2) the probability of a major quake striking any particular place within any narrow time window remained small; and (3) there was nothing contradictory between those two facts. In a departure from usual practice, the meeting was attended by public officials who were not members of the Commission on Major Risks, and there is evidence that they did not find the scientists' discussion reassuring at all. The mayor of L'Aquila later stated publicly that he was much more worried after the meeting than he had been before. As a result, he ordered closure of the city's schools in the days after the meeting. (In the end, the quake hit at night, so the children who died were mostly asleep in their homes.) As was usual practice, the commission did not issue a formal public statement after the March 31 meeting. Two members of the group, Franco Barberi and Bernardo De Bernardinis, did speak with the press after the meeting, but the recording was destroyed in the quake, and the judge did not appear to have considered their comments in finding the group guilty. Among pre-quake public statements that were recorded and preserved, two of the most highly publicized were made by De Bernardinis, who participated in the commission meeting in his capacity as vice director of DPC; he was not a seismic expert. Asked in a television interview about the recent series of tremors, he told the reporter, "The scientific community continues to assure me that, to the contrary, it's a favorable situation because of the continuous discharge of energy," erroneously giving the impression that the shocks felt up to that point were relieving stress and lessening the threat. Then, when asked whether, despite the recent strong tremor, viewers should not worry but rather relax and have a glass of wine, De Bernardinis said, "Absolutely," adding, "Montepulciano d'Abruzzo, of course," a reference to a locally produced wine. But what often goes unmentioned is that both comments were made *before* the commission meeting. The meeting's attendees were not even aware of either statement. But the comments, especially the Montepulciano quip, were highlighted in the local evening news without making clear to viewers that they were not official statements from the Commission on Major Risks and in fact were made before that body had even met.[11]

The L'Aquila Seven were convicted of manslaughter two years later, not by a jury but by the presiding judge, Marco Billi. In a 950-page explanation of the verdict, Billi accused the defendants of participating in a "media operation" to convince the public that they were in no danger. He convicted them, he wrote, for improperly analyzing seismic data in the days before the quake and then providing inadequate explanations of the threat. "The deficient risk analysis was not limited to the omission of a single factor," he wrote, "but to the underestimation of many risk indicators and the correlations between those indicators."

The scientists who participated in the March 31 meeting were not being asked for recommendations on whether people should evacuate buildings, and they did not themselves issue public announcements of any kind after the meeting. Nevertheless, they were accused by prosecutors of allowing their deliberations to be misused by government officials who were more interested in calming the residents of the L'Aquila area than in protecting them. The seven were eventually convicted "for having given out falsely reassuring information to members of the public" and thereby causing the deaths of twenty-six citizens who, the prosecutors were able to convince the court, had responded to those reassurances by staying in their homes when they otherwise would have gone to a safer open area.

Discussion of the disaster often frames it as a case of complex, flawed scientific expertise undermining traditional wisdom. In an article published by the journal *Nature* in 2011 just before the trial began, surgeon Vincenzo Vittorini, whose wife and daughter died when the family's house in L'Aquila's old city collapsed, is quoted as saying, "My father was afraid of earthquakes, so whenever the ground shook, even a little, he would gather us and take us out of the house. We would walk to a little piazza nearby, and the children—we were four brothers—and my mother would sleep in the car." That is the just kind of action some people took when a magnitude-3.9 tremor occurred late that Sunday night, and that action saved their lives when the main quake struck at 3:33 a.m. Monday. Vittorini claimed, "That night, all the old people in L'Aquila, after the first shock, went outside and stayed outside for the rest of the night," But, he said, "Those of us who are used to using the Internet, television, science—we stayed inside." [12]

A draft version of the risk commission meeting minutes, but not the publicly released version, has a DPC official (not a defendant herself) thanking the commission members "for your statements that allow me to reassure the population through the media which we will meet at the press conference." An opinion piece published by the Australian newsletter *Risk Frontiers* argued that this official's statement implies "that the scientists were well aware that the information they had provided was to be used to play down the risk," and that they were rightly prosecuted for failing to correct the false impression that the public was being given. The draft minutes also cite a comment by commission member Franco Barberi, a professor of volcanology at the University of Rome III and a former DPC director, to the effect that "there is no reason for saying that a sequence of low magnitude shocks could be considered the precursor of a strong event."[13] But, say the scientists' defenders, there exists no recording or transcript of the meeting, and therefore no way to judge precisely what Barberi said in the context of the full discussion—one that the officials in attendance apparently found not reassuring but alarming. And, they say, there is no reason to believe that the scientists endorsed statements made by officials either before or after the meeting; indeed, there is no evidence that any of them, other than De Bernardinis himself, were aware of the pre-meeting "Montepulciano" interview.[14]

THE NEEDLE AND THE DAMAGE DONE

When the case was filed against the L'Aquila Seven, outrage swept through the world's scientific community. The International Commission on Earthquake Forecasting for Civil Protection, which the Italian government convened soon after the quake to report on the state of the art in quake forecasting, concluded, "Given the current state of scientific knowledge, individual large earthquakes cannot be reliably predicted in future intervals of years or less. In other words, reliable and skillful deterministic earthquake prediction is not yet possible."[15] The fact that large quakes are often preceded by swarms of much smaller tremors does not imply the converse: that swarms are predictors of large quakes. Warner Marzocchi, an INGV seismologist (who had not been on the March 2009 Commission on Major Risks and therefore was not prosecuted), notes that if action were taken every time seismograph needles recorded signatures like those seen before the L'Aquila quake, authorities would be ordering frequent, needless evacuations of cities all over Italy. Like wildfire modelers in Tasmania, seismologists everywhere know that sounding such false alarms not only would be disruptive but also could severely erode overall preparedness when the public comes to view scientists as no more than wolf criers.[16]

Judge Billi repeatedly stressed that commission members were not accused of drawing incorrect scientific conclusions and that it was not science that was on trial. Yet during the trial, Billi and prosecution witnesses held extensive discussions of scientific aspects of the incident—often erroneously, according to the defense. An appeal filed in 2013 says that the judge used "brief parts of scientific papers or single tables ... to draw general conclusions," while "many scientific statements made in the verdict are not supported by the vast literature available."[17]

The biggest problem lies in confusion between the size of a risk and the size of the *increase* in a risk that comes with an uptick in seismic activity. For L'Aquila, the probability of a quake strong enough to cause a hundred or more fatalities in any given twenty-four-hour period was about one in a million in the months before the tremor swarm began in January 2009. With increased activity in early February, the probability had risen to one in thirty thousand. The day before the main quake struck, the probability was about one in ten thousand. At that point the risk of a major deadly quake had increased a hundredfold—cause for some concern. But the risk was still tiny, a long way from a threshold that would prompt an evacuation. In fact, on that day there was better than a 99 percent chance that no quake capable of killing even a single person would occur in the next twenty-four hours.[18] But quakes do happen, and this time a much deadlier one did occur, even at a probability of much less than 1 percent. From that point onward it became very difficult for the public or the government to look back and ask what should have been done during the first five days of April 2009 without being influenced by what actually did happen on April 6.

Marzocchi knew all of the defendants, and he has played a prominent role in the debate over their conviction. Sitting in his office at the institute, in a quiet suburb on Rome's southern fringe, he told us that his colleagues' estimates of the risk were correct, and that he or any other scientist with expertise in the discipline would have assessed the risk similarly. He says, "People tell us, 'You said it was a low-probability event. But it happened! You were wrong!' Well, no, we were not wrong. Before the event happened, it really was very unlikely to happen at that time." Marzocchi believes that low-probability events are the hardest ones to discuss in public without caus- ing confusion: "People are reluctant to accept our explanations of unpredict- ability. I ask, 'But why do you accept the reality of the odds in the lottery?' It's because, of course, it's only a game and the unpredictability is the beauty of it." By contrast, in matters of life and death, "people are driven by hope, which overcomes rationality. But Nature is not interested in our hopes!"[19]

The University of Southern California's Thomas Jordan, who headed the international commission that examined the L'Aquila affair, told *Nature* in 2011, "The role of science is to present information about hazards. But it's the role of the decision-makers to take that information, and a lot of other information, in order to make decisions about public welfare."[20] Da- vid Alexander, a professor in the Institute for Risk and Disaster Reduction at University College London, isn't buying any such arguments. In a 2014 article on the L'Aquila affair, he asserted that the scientists were indeed criminally negligent, that their trial "was an attempt to bring some sense of morality, responsibility and accountability into Italian public life," and that "many observers now believe that the trial did have a solid basis of reasoning."[21] Marzocchi says he was "astonished" by Alexander's article, stressing, "The committee met only because of Giuliani's radon predictions. Yes, there was a seismic sequence that raised the probability of a quake, but we see ten to fifteen sequences like that one or even more intense every year in Italy. What if we ordered evacuations every time? Normally, the Major Risks Commission doesn't even meet under such circumstances."[22] He points out that before the terrible 2011 earthquake and tsunami in eastern Japan, there was a very intense seismic sequence, but no action was taken and no one was charged with negligence. In public and private, Marzocchi stresses the basic fact that while science can tell us with some confidence the probability of an earthquake greater than magnitude x happening in region y within twenty or a hundred or five hundred years, the odds of an earthquake striking tomorrow or next week are always so small that they are useless as a basis for urging immediate action—of no more use than the odds of winning that are printed on the back of lottery tickets.

Confusion was created, Marzocchi argues, not by the scientists on the commission but days earlier, by the amateur seismologist Giuliani's unfounded attempts at prediction. He described Giuliani as one of many amateurs who have tried to forecast earthquakes using radon data despite the fact that such readings are far less useful predictors of near-term events even than seismic signatures. "And by the way," he adds, "Giuliani *didn't* win the lottery. His prediction was for a quake in a spot seventy kilometers away from where the quake actually happened a few days later." Had Giuliani's advice been followed at that time, points out Marzocchi, the most likely destination for evacuees would have been the closest large town: "People would have fled to L'Aquila! There might have been even more deaths."

The Giuliani incident is part of a long tradition of failure in short-term earthquake forecasting. One of the more notorious episodes came in 1977, when Brian Brady, a research physicist in the U.S. Bureau of Mines, predicted that a powerful magnitude-8.4 earthquake would strike Lima, Peru, in late 1980; he later pushed the date back to July 1981 and the magnitude up to an almost inconceivable 9.8. For a long four years, Peruvian society was thrown into anxiety and tumult, while U.S.-Peruvian relations were severely disrupted. After a populist, antimilitary government was swept into office in the May 1980 elections, some of its leaders ridiculed the military's top brass as "earthquake believers" and labeled Brady, the author of the prediction, as an "international terrorist." Brady was no terrorist, but he was a man obsessed with vindicating his prediction and the wildly erroneous theory of rock failure on which he had based it. Needless to say, the big quake never happened.[23] In contrast to the Giuliani-L'Aquila affair, there was no loss of life and no criminal proceeding. But the Brady affair shows that even an event that never happens can have dire consequences.

Despite the long, troubled history of forecasting specific earthquakes, Marzocchi is especially concerned that the L'Aquila verdict will encourage seismologists and geologists everywhere to practice the equivalent of "defensive medicine," repeatedly overprescribing emergency measures in order to keep prosecutors at bay, but also, as in the Brady affair, creating confusion and disarray with repeated false alarms. He further worries that after one too many needless evacuations, people will start ignoring the basics of earthquake safety. Furthermore, a too-tight focus on warnings and evacuation draws attention away from longer-term, more effective safety measures such as adherence to strong seismic building codes.[24]

Marzocchi and other experts are by no means arguing that scientists should shy away from making public statements; instead, they should communicate more explicitly. Five years after the quake, members of the international commission that had studied it recommended future use of a system called operational earthquake forecasting (OEF), which involves communication by scientists of quantitative information on short-term earthquake probabilities and changes in those probabilities. Such information would remain descriptive only. Scientists themselves would not dispense advice on whether and when to evacuate; those questions would remain the responsibility of public safety officials. In the 2014 paper, Jordan, Marzocchi, and two co-authors mounted a point-by-point defense of OEF against critics who had warned of dangers created by any kind of public announcement of highly uncertain short-term probabilities, writing, "Though communicating OEF and its uncertainties is a difficult issue, not communicating is hardly an option."[25] (The OEF idea grew out of knowledge gained after the 2009 quake, so none of its conclusions could be considered applicable to the case of the Seven.)[26]

CODES OVER ALARMS

In the larger picture, communication should not be the first line of defense in preventing earthquake disasters. The report of the international commission was clear in stressing that establishment and enforcement of stringent building codes and mandatory retrofitting of older buildings for seismic safety are the most effective means of preventing death and injury.[27] Those rules are guided by seismic hazard maps. Based on long-term probabilities of earthquake activity, such maps are much more robust than attempts to issue warnings of short-term quake risk. L'Aquila had been highly vulnerable to earthquake damage. Although reinforced-concrete buildings predominated in the area immediately south of the old city, where the strongest seismic shaking and the majority of deaths occurred, most of those structures dated to the 1960s and 1970s, before stricter seismic codes had gone into force. And some reinforced buildings and bridges turned out to have had badly rusted reinforcing bars. Many of those who died in this high-mortality zone were students staying in cheaply built reinforced-concrete apartment or dormitory buildings.[28]

A serious attempt to build back safer was launched. The European Union Solidarity Fund (EUSF) sent $675 million in aid for temporary housing; that, along with about $2 billion in funds from the national treasury, opened an opportunity not only to provide shelter but also to begin the necessary conversion of L'Aquila into a less seismically vulnerable city. The flagship project, regarded by some as a silver lining coming in the form of seismic safety, would provide housing to fifteen thousand residents in groups of new apartment buildings under a program called CASE—an Italian acronym for "seismically isolated and environmentally sustainable housing."[29]

The CASE blocks were built atop giant shock absorbers for earthquake protection and included many green features. Seventy percent of the EUSF funds (along with a larger sum from the Italian treasury) was used to construct CASE, so in 2013 the European Union's Committee on Budgetary Control assigned a team led by rapporteur Søren Bo Søndergaard to examine how the funds were being used. In its report, the team concluded that much of the aid money had been spent inappropriately. The entire CASE green building project, according to Søndergaard, was in violation of agency rules because it consisted of new, permanent buildings rather than the much cheaper types of temporary housing that the EU was willing to pay for. Reacting to the report, Mauro Dolce told us it must be remembered that with funds to restore much of the old city center remaining far out of reach for the foreseeable future, the CASE buildings needed to be durable enough to be occupied for ten years or more, and that DPC has "demonstrated that the actual costs were consistent with the Italian market and with the quality of the construction." The European Union eventually approved the expenditures.[30]

The problem of the city center and its very old, unsafe buildings also remains in DPC's hands. To the casual observer, it would appear that no expense was spared in the reinforcement effort. Not only have vast quantities of steel and wood been affixed to outer walls, but the interiors of many buildings are packed with dense support networks of black two-inch pipes linked by large, seemingly brass-plated joints; these impressive-looking scaffoldings also adorn the exteriors of some commercial buildings. At the Farfarello bar, Daryoush Shoyajee said he believed the massive shoring-up was one reason why the old city remained a ghost town more than four years after the quake. Just above his head, over the bar, largely obscuring the bottles of whiskey and grappa on a high shelf, was one such network of black pipes. It was supposed to be helping hold up the three-story building above. Pointing to one of the countless shiny joints that held it all together, he said, "See these? Five euros each one! I tell them, for what you are spending to hold up all these buildings, you could simply build new ones!" So why this great expense? "There is only one person in Italy who can be responsible for this!" he declared, then winked. "You know. Mafia."[31]

Asked about Shoyajee's conclusion, Mauro Dolce disagreed—sort of. "I would not say that propping works are a question of 'Mafia,'" he told us, "at least not more than it is for reconstruction works in general."[32] The EU report had indeed found that at least in housing construction, some portion of the reconstruction funds provided to the city by the Solidarity Fund "were paid to companies with direct or indirect ties to organized crime." The report noted that as the rebuilding effort was booming in 2010, a prosecutor with Italy's National Anti-Mafia Department, Olga Capasso, had said, "It seems to me that among the problems related to combating organized crime, Aquila is one of the biggest problems at the national level."[33]

In the end, Shoyajee's other prediction—that none of the L'Aquila Seven would ever serve serious prison time—came close to being fulfilled. On November 10, 2014, six of the convictions were overturned by an appeals court. The remaining one, that of former civil protection head Bernardo De Bernardinis—who had made the infamous "Montepulciano" comment—was upheld, but his prison term was set at two years, not six. A wave of relief swept the world scientific community, but people in and around L'Aquila were not so happy. In the courtroom, cries of "Shame! Shame!" greeted the reading of the decision. A spokesperson for quake victim Vittorini told reporters, "Today we have an earthquake after the earthquake."[34] Three months later, in an official statement on their ruling, the three appeals court judges wrote that official reassurances given by De Bernardinis were decisive in causing some of the quake victims to stay indoors, but the other six were blameless because they never discussed the false idea put out by De Bernardinis that the series of small tremors, by discharging energy, was reducing the likelihood or strength of any large quake. But the scientists were not fully out of the woods; under the Italian system, the prosecutors still had the option of appealing the verdict.[35]

A TIME TO HEAL AND A TIME TO BLAME

Whether or not the question of blame for the L'Aquila disaster has finally been resolved in the public mind, much harm has been done by the controversy. Just as those responsible for the Lusi mud eruption refused to accept blame for the destruction of a community, the attempts by Italian authorities to blame the L'Aquila Seven for the high death toll rather than accept their own failure to enforce safe construction has hampered the recovery and left a city in limbo.

Both of these events illustrate how blame complicates disasters. If it's still possible to define a category of "natural" disasters, it consists of those destructive events for which nobody *can* be blamed. Clearly, any such disaster has social and economic roots; it's just that there's no specific entity that can be held responsible. The insurance industry recognizes this in every clause invoking "acts of God," but it's not just a technicality—it's something felt inside. Anger reigns when there's someone to prosecute, fine, or pillory; when there's not, all that's left to do is heal.

Yet sometimes blame *is* the right response. The more we see behind the natural-looking facade of disasters, the more culpability comes to light. For guilty parties—and especially for corporate entities whose only moral compass is self-preservation—the category of natural disasters is a necessary shield. The greatest monument to this form of damage control is, of course, the bulwark of climate change denial that the fossil fuel companies and allied industries have spent decades and millions maintaining. If the people of Earth are having a hard time accepting the fact that we are not only victims but also producers of disasters, it's because there's so much wealth and power still being protected by the wall of denial.

But where climate peril is already pounding on the roof or lapping at the doorstep, that bulwark can start crumbling quickly. That's why, at least among those who are looking to the greenhouse future with deep apprehension, all eyes are on Miami.

8

ATLANTIS OF THE AMERICAS

Miami, Florida

Have you ever seen a rainbow so beautiful? I took this shot last week. Each end of the rainbow landed on Biscayne Boulevard.

—Lucas Lechuga, photograph caption,
MiamiCondoInvestments.com, 2010[1]

WE LIVE IN A RAINBOW OF CHAOS
—Mural on a building at NE First Avenue and NE Seventh Street, Miami,
by artists five and Kemo, 2014, after Paul Cézanne[2]

"When I started this job, people kept asking me, 'Why do we have so much flooding now?' and I said, 'Well, there's just one problem: the whole city's four feet too low—that's all!'" Bruce Mowry, Miami Beach's city engineer, was driving us around the island, explaining his department's strategy to keep it dry as sea levels rise. "If we get the four feet of rise that's predicted, all of this area will be two and a half feet underwater."[3] He showed us parts of the city where the streets already can flood during twice-a-day high tides for weeks at a time, rain or shine. Clearly, Mowry had a lot of work ahead of him.

The city of Miami Beach consists of a long, low barrier island accompanied by a scattering of man-made islets. It's one of the lowest-lying municipalities in Miami-Dade County, and its residents are leading the way into the region's wetter future. By 2013, people living and working on the low western side along Biscayne Bay had come to dread full-moon high tides, when salt water would seep into street drain outlets and the porous limestone that provides the island's foundation, forcing water from the ground and drains up and out into the streets and sidewalks, putting them deep underwater and threatening buildings and infrastructure. It's a phenomenon that the island's sixty miles of seawalls could do nothing to stop; the water was slipping in quietly from below.[4] The October "king tide," when most flooding occurred, was not only a danger and a nuisance but also a national media event.

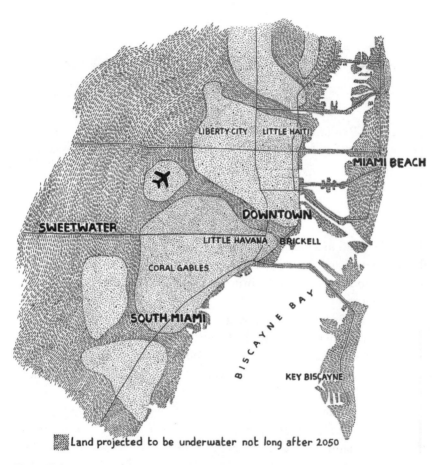

Land projected to be underwater not long after 2050

Figure 7. A portion of Miami-Dade County, Florida. Stippled areas are projected to be either underwater or frequently flooded if the sea level offshore rises five feet above its current state—an event that some expect to occur by the end of this century.

Miami Beach is just a small part of a region that's in big trouble. If sea levels rise as projected, no major U.S. metropolitan area stands to rack up bigger losses than Miami-Dade County. Sixty percent of the county is less than six feet above sea level. Even before swelling of the seas is factored in, Miami has the greatest total value of assets exposed to flooding of any city in the world: more than $400 billion. Its projected average annual losses, taking into account vulnerability to flooding and existing protection measures, are equaled only by those of New York City and Guangzhou, China. But with future sea level rise and continued economic growth, the exposure of property in Miami and Guangzhou will far outstrip that of any other urban area, reaching almost $3.5 trillion in each city by the 2070s.[5] The sea level around the South Florida coast has risen nine inches over the past century. Experts expect the sea level to edge up another three to seven inches in the next fifteen years and nine inches to three feet in the next forty-five years. Even the very gradual rise of recent decades will make necessary extensive infrastructure reengineering; however, Florida's own Department of Transportation forecasts that it will become difficult, expensive, and maybe impossible for road- and property-raising efforts to keep up with the accelerated sea level rise that's expected.[6]

SAVE OUR STREETS

Mowry assumed the city engineer's position in Miami Beach in 2013, and it has fallen to him to carry out an ambitious plan to pump water out of the city's storm drainage system and into the bay whenever the drains get too full and threaten to flood the streets. Eventually it will require fifty to sixty pumps to keep the city dry. He had gotten a few pumps running by the time of the 2014 king tide, and they prevented some of the worst flooding. Three months later, as we stood at the west end of Tenth Street where it meets the bay, Mowry pointed to the two big Swiss pumps that had kept the notoriously flood-prone Alton Road area of the city relatively flood-free during the king tide. To protect the pumping station, he had put up a seawall rising 5.7 feet above mean sea level, which is an ample 4.5 feet above high tide; clearly, Mowry had built the wall with climate change in mind. Stretching north and south from the new wall, privately owned seawalls rose less than two feet above high tide, and that, said Mowry, will be a problem because "everybody in the world agrees we'll be getting at least two more feet of sea level rise." And while higher seawalls could keep high tides or small storm surges from pouring in over the top, they alone cannot prevent

Miami Beach's regular "sunshine flooding," which comes up from below. Pumps are required to keep that water out.

Unfortunately, if the bay continues rising, at some point too much effort and expense will be required to keep pumping the island's low spots dry. So Mowry has made even more ambitious plans to raise the lowest-lying streets throughout the west side of the island. He showed us block after block that he planned to raise a full two feet: on West Twentieth Street, Tenth Street, Sixth Street, Purdy Avenue (down which residents had been known to kayak during high tide), West Avenue, and others. He even intended to take one block of Sixth Street over a hump, elevating it two feet at the ends and a full six feet in the middle, to allow level street access from new, well-elevated buildings that flood-conscious developers had planned for either side of the street. There's a problem with all of this road raising, of course. Once they're two feet higher, streets and sidewalks in many places would loom above the doorways of the buildings alongside, pouring water into them when it rains. In such spots, Mowry says, it might be necessary to leave the sidewalk low and place a short wall between it and the street. Many of the newer luxury condo complexes had been built on high mounds of fill, so their owners generally welcomed the road-raising plans. But so much elevation activity would require staggering quantities of fill soil. There are no sources on the island, and dredging Biscayne Bay for fill is now prohibited, so most would have to be imported from the mainland.

Some of the artificial dunes that overlook the beaches and the Atlantic on the island's east side and adorn the fairways of Miami Beach Golf Club are among the highest elevations on the island. But Mowry has them beat. He showed us an artificial plateau rising fifteen feet above sea level that he'd built with recycled construction fill in a secluded spot behind the golf course, created to store the city's critical machinery and the ultra-expensive pumps waiting to be installed—just in case there's a storm surge. His most precious stuff will probably be safe there, he told us, but "I just have to hope there's not a twenty-foot surge."

1926

In Miami, the possibility of a twenty-foot storm surge is not necessarily a science fiction nightmare. And looking back at the area's geologic history, the slow version—twenty feet of gradual sea level rise—would be a routine occurrence. In recent times, Miami has enjoyed much longer interludes between monster tropical cyclones and floods than has the central Philippines. But over the long term, southern Florida has had a pretty eventful run.

Miami is perched on the southeast corner of a formation jutting from the southeast corner of North America called the Florida Plateau. A geologic time-lapse video of the plateau would show it bobbing in and out of the Atlantic: for a while it's dry land; for a while it's seafloor; and in between it's part one, part the other. A mere 125,000 years ago, at the peak of the last interglacial period, Florida was a narrow, ragged stump reaching not much farther south than present-day St. Petersburg. But by the time of the most recent Ice Age, which reached its zenith eighteen to twenty thousand years ago, sea levels had dropped and the peninsula had swelled to almost twice today's width. The world's oceans have been creeping up again since the Ice Age, but when humans first made their way into the region more than ten thousand years ago, Florida's landmass still stretched much farther west into the Gulf of Mexico than it does today. Along the west coast where some of those early arrivals settled, archaeologists who study their villages have to wear scuba gear. In recent times, the continuing rise of the seas surrounding Florida has been accelerated by human-induced warming of the world's oceans and melting of ice at high latitudes.[7]

The construction of Miami began on the Atlantic Coastal Ridge, a long, low backbone of limestone that runs along the coast, right through today's downtown area. The ridge is oriented more or less north and south, and it's cut through at regular intervals by east-west channels that were eroded by tides in earlier geologic times, when the ridge was a sandbar. Having been partially filled in by soil, the channels became small, fertile, but flood-prone valleys known locally as transverse glades. Through most of Miami's history, the glades were used for grazing and agriculture, but in the 1970s they began filling with urban development. The barrier islands just offshore from Miami, including Key Biscayne and Miami Beach, were formed when ocean currents coming down the coast from the north deposited sediments on top of a submerged limestone ridge, the remains of an ancient coral reef.[8]

Miami was incorporated in 1896; three decades later it was suddenly a bona fide boom town. The year 1925 saw investment in new construction leap to more than $103 million (more than $1.4 billion in today's dollars), from just $11 million in 1924. The city more than tripled its land area through annexation. Eight years earlier, developers had started stripping the long barrier island offshore of its mangroves and filling in low, wet areas to create the new city of Miami Beach. By 1925, the Miami area was home to more cars (105,000) than people (70,000) and had been swept up in a phenomenal real estate bubble.[9]

The boom was short-lived. By March 1926 the good times were already winding down, and speculators were departing for greener pastures. Six months later came the Great Miami Hurricane.[10] In his history of Miami in 1926, Frank Sessa observed, "Miami, in truth, did need a period of stabilization, a chance to catch up with itself. That period did not materialize. Whatever chance the city might have had to recover some measure of its boom-time economy was wiped out by the hurricane that struck in September, 1926."[11]

Hurricane-force winds started being felt late on Friday night, September 17. Many motorists were trapped overnight as a storm surge flooded the causeway between the mainland and Miami Beach. The winds roared and the sea poured in all night, but the calm that suddenly descended around dawn turned out to hold the greatest danger. According to the *New York Herald-Tribune*, when the eye arrived, many believed the storm had passed and decided to head out for work or start cleaning up debris; however, "none went far. Great waves, tangled wreckage and houses blown across their paths sent them scurrying to safety." Once the storm had passed for good, Miami's "ornate skyline was twisted into a wild medley of cocked roofs, crushed towers and suspended beams. . . . Of the estimated 55,000 homes in the greater Miami district about 40 per cent had been damaged."[12]

Miami's overall economy recovered from the 1926 hurricane, but the great storm brought with it no rainbow. Many who suffered large losses never received adequate relief, because providing full compensation would have been a tacit acknowledgment of the true extent of the damage, and that, it was believed, would hurt the tourism industry. By the end of the 1920s construction activity remained far below pre-hurricane levels, and it plummeted further after the Wall Street crash of late 1929. Sessa characterized 1926 as the year of "recovery from the boom," not necessarily from the hurricane.[13] In ecological impact as well, the boom was far more destructive than the storm. In his book *The Swamp*, Michael Grunwald characterized the architects of Miami's rapid westward expansion as "declaring war" on the Everglades. They ended up almost destroying the great swamp, but the war wasn't won easily:

> The Everglades turned out to be a resilient enemy, resisting man's drainage schemes for decades, taking revenge in the form of brutal droughts and catastrophic floods, converting the Florida swampland into an enduring real estate punchline. In 1928, a hurricane blasted Lake Okeechobee through its flimsy mud dike and drowned 2,500 people in the Everglades, a ghastly foreshadowing of Hurricane Katrina's assault on New Orleans. Mother Nature did not take kindly to man's attempts to subjugate her.[14]

THREAT NUMBER ONE: STORM SURGE

Accounts of the 1926 hurricane tended to dwell on the terrible damage done by its 145-mile-per-hour winds, but even greater destruction came with the storm surge. It completely submerged Miami Beach and reportedly carried ships all the way across Key Biscayne without their hulls touching the ground. In height, it eclipsed all surges striking the U.S. Atlantic or Gulf coasts over the next eighty-six years, until Sandy came along. The surge had been set up by east and northeast winds that blew constantly for more than a day before the storm's arrival and raised the sea level on the east side of Miami Beach to six and a half feet above normal, even before waves began pushing across the island. Miami Beach was fully submerged under two to four feet of seawater.[15] The U.S. Weather Service later estimated that the 1926 surge hitting the east shore of Miami Beach reached as high as ten feet. As southerly winds coming behind the storm's eye forced water up through Biscayne Bay, the surge reached almost twelve feet along mainland Miami's southern shoreline.[16] Had a storm with the path and power of the 1926 hurricane struck the far wealthier, far more built-up Miami of 2005, it would have caused more than $150 billion worth of destruction. That would make it the most expensive hurricane in U.S. history, eclipsing twice over the damage wrought by Hurricane Katrina that year.[17]

Today when Miamians think of hurricanes, they think of Andrew, which slammed into Florida's coast south of Miami in 1992, causing about $40 billion in damage (in inflation-adjusted dollars). Although, like the storm of 1926, it ranked as category 5, Andrew was "drier" and faster-moving, did not make a direct hit on the city, and did not produce a large storm surge. If an Andrew-type storm struck Miami head-on today, the damage bill would be almost $70 billion—enormous, but still less than half of the destruction that a monster like the Great Miami Hurricane would cause today.[18]

The city recovered quickly from Andrew. Two decades later, rebound from the Great Recession brought another building boom to Miami Beach and low-lying parts of Miami's coastal mainland. It was the biggest that the city had seen in the eight-plus decades since the original boom of 1925. Miami was more exposed than ever and had no intention of backing down.

THREAT NUMBER TWO: THE WATER TRAP

Two of Miami's chief north-south thoroughfares, Biscayne Boulevard and Interstate 95, follow the Atlantic Coastal Ridge, which rises to dizzying heights of eight to fifteen feet above sea level along much of its length, dipping a little at each glade.[19] This strip of land will be among the last to be submerged as the seas rise. But routine flooding has come to plague other parts of the county. Miami Beach has gotten most of the headlines, but in some mainland communities flooding has become a chronic problem. The metropolitan area that runs north-south through Miami-Dade and Broward counties is sixty miles long, narrow, and hemmed in by water from the left, from the right, and from below—trapped between the soggy, low-lying Everglades to the west and the Atlantic to the east, while sitting atop the porous, water-laden limestone that forms the Biscayne Aquifer. The aquifer is the area's source of potable freshwater, but as the seas rise more and more saltwater will force its way in, not only threatening the water supply but also raising the water table; in inland neighborhoods, saltwater is creeping noticeably toward the surface.

Harold Wanless, a professor of geology at the University of Miami, is well known around southeast Florida as the indefatigable Paul Revere of the climate crisis; he is to Miami's future floods what David Bowman is to Tasmania's future wildfires. According to Wanless, what many see as Miami's worst-case scenario is actually a conservative estimate. Using the latest U.S. government projections, which include ice melt, he is confident that the waters offshore will rise two more feet by 2048, three feet by 2064, and four to six and a half feet by the end of the century. At five feet, more than half of Miami-Dade County will be submerged. Miami Beach will be gone. For the rest of the region, the worst will still be yet to come, Wanless says: "When we're at five feet, it will mean we're on an accelerating curve because of ice melt. The sea will be rising at a foot per decade. Now if that's happening, try to make any grand scheme work in a place like this." But problems will begin arising much sooner, even after a foot or two of rise, he adds. "Inland areas will see more and more days of flooding after big rains, because drainage will become more and more sluggish. And we'll be more and more prone to storm-surge damage from a hurricane."[20] Another geologist, Peter Harlem of Florida International University, told us that severe disruptions can result even from the kind of "nuisance flooding" that's already occurring in some places and will spread to many more with even small increases in sea level. He said, "People don't understand six inches. Six more inches can make life miserable."[21]

In dealing with the Everglades, Miami is faced with a delicate balancing act. Freshwater flows out of the giant Lake Okeechobee in central Florida through the Everglades, and from there it pushes eastward to keep the Biscayne Aquifer filled. Were the aquifer to be ruined by saltwater intrusion, Miami would be doomed. So water managers adjust the flow from

Okeechobee and the Everglades to apply back pressure against westward intrusion from the rising waters of Biscayne Bay. But that vital water flow also helps complicate Miami's future flood problems. Henry Briceño, a research scholar at the Southeast Environmental Research Center at Florida International University, puts it this way: "Around here, the flooding's not just coming from the sea. It's coming from behind us. We're gonna get our asses wet—with water coming from the Everglades!"[22]

For decades, Wanless, Harlem, Briceño, and other scientists met with frustration when trying to draw attention to the sea level problem. Then, once enough Miamians found themselves wading through the fallout of global warming, it seemed everyone was questioning the city's future. A *Rolling Stone* headline waved "Goodbye, Miami," while *The Guardian* announced, "Miami, the Great World City, Is Drowning."[23] The immersion that everyone knew would repeat someday in the course of geologic time seemed to be happening right now. And it was just about the worst possible news for anyone connected to the pillars of the area's economy: tourism and real estate. Miami Beach, where those industries converge in a very big way, had its effort going to pump the waters out of the storm drains, but more was needed. So in February 2014, the city commission voted to create a beefed-up flood prevention infrastructure capable of handling peak tides of 2.1 feet, plus future sea level rise of seven inches. That new plan, under which Bruce Mowry was installing all those new pumping stations, backflow prevention valves, and other features, would cost $400 million, doubling the price tag of the city's existing flood plan.[24] The massive costs of the many other planned flood prevention projects, including extensive road raising, would not be covered by those appropriations; a lot more money would be needed for that.

The pumps installed in 2014 achieved a large reduction in flooding during the October king tide, and a surge of relief swept over the island.[25] Nicole Hernandez Hammer, a climate researcher with the Union of Concerned Scientists, told us, "[U.S.] Senators Whitehouse and Nelson and the EPA director, they had all come down to see the October flooding, and there wasn't any. It was like, 'Ah, OK, I guess they took care of it,' and they went home. But look at the hundreds of millions of dollars spent to prevent just this little bit of flooding, which is only a small fraction of what we're going to see in thirty years."[26]

Hammer was referring to the fact that Miami-Dade faces not one but four related flood hazards. There is Miami Beach's tidal flooding, which affects parts of the mainland's coastal areas as well. There is the rainy-day flooding that increasingly plagues low-lying mainland communities out west. There is the possibility of a big storm surge, a threat that looms larger with every inch of sea level rise. And then, as the Atlantic continues to warm and swell, there is the long-term prospect of seawater pushing up from Biscayne Bay into mainland glades and canals, flooding more and more of Miami's coastal areas while backing up the whole hydrological system and causing the western suburbs of the county to be permanently inundated from the Everglades side. That last form of flooding will occur gradually over an extended period, but Harlem and his colleagues have modeled the endgame: with a two-degree-Celsius rise in global mean temperature, the Florida Keys, the entire Everglades, and all of Miami-Dade County will be submerged. That, they predict, will be the situation sometime after the year 2100—it's hard to say exactly when.[27]

What can Miamians expect in the meantime? Harlem, who passed away in 2016, thought in terms of a five-stage timeline, while taking care not to attach precise time periods to the stages. In stage one, only the lowest-lying areas, mostly out-of-sight, out-of-mind natural landscapes, flood frequently. In stage two, more private property is being affected. Harlem maintained that Miami-Dade County is now passing from stage one to stage two. In stage three, the majority of people become affected; at that point, sea level becomes a political issue and collective action will replace individual responses. Impacts become increasingly dire in stage four, until the region arrives at stage five, when the only exposed land in Dade and Broward Counties will be a string of islands inhabited by a relatively small population of easygoing but hardy hurricane veterans, a place Harlem has nicknamed "Margaritaville."[28]

THE BIG "SHHHH"

As 2015 rolled around, the construction and real estate booms rolled on, most visibly in Miami Beach and other low-lying coastal areas. Home prices had surged by more than 10 percent in the previous year. With the rise in real estate values slowing nationwide, Miami was the only major metropolitan area in the nation to have maintained double-digit growth.[29] Much of the increase was at the high end of the market, and it didn't seem to matter whether properties were in badly exposed waterfront locations or on higher ground. In mid-2014, throughout eastern Miami-Dade County, forty condominium projects comprising almost 6,800 units were under construction, and plans for almost 11,000 more units were in the works. In the trendy, mostly low-lying Brickell area south of downtown, prices per square foot were up 50 percent over 2012.[30] The market was projected to remain strong for another decade or two, according to local real estate experts.[31] Local geologists and climate experts weren't so sure.

What was sustaining the boom? About half of the investment was coming in from Europe, Russia, and Latin America. This was normal. Miami has long been viewed as a thoroughly international city. Within a single 2001 *Time* article titled "Miami: The Capital of Latin America," the city was also referred to as "the capital of Hispanic TV and music," "the cruise capital of the world," and "tomorrow's business capital of the Americas."[32] With economic and political chaos gripping much of the planet, Miami was looking like the world capital of lucrative, safe real estate investment, too—as long as you had an exit strategy. *Miami Herald* reporter Karen Rundlet put the situation bluntly: "No one wants to kill the market with negative talk about one of the region's primary economic drivers: growth. . . . When it comes to real estate, the market—meaning buyers and sellers—neither is likely to respond to the crisis of rising sea levels until the final hour."[33] And of course such a situation can persist so long as no one knows precisely when that final hour will come.

Philip Stoddard is a biology professor at Florida International University and mayor of South Miami, a small city situated on what (in South Florida anyway) is considered high ground. Although South Miami itself faces no imminent flood crisis, Stoddard is the most climate-active mayor in the region. Miami, as he sees it, bears a grave responsibility; he likes to tell his fellow residents, "Everyone is watching us, so we have to do something that matters. We'll go under, but other cities will learn from us." In 2015, Stoddard watched with frustration as the real estate frenzy continued to grip the coastline and islands just to his north and east. He had heard plenty of American and international investors talking about a ten-year window to invest, make money, and get out before it's too late. He calls that "a dubious calculation," but almost every person we met in Miami did talk about a ten-year horizon. And like all horizons, it appears to recede as you approach it. In Miami's economy, catastrophe is always ten years away—or rather it will be until it isn't. Stoddard told us, "There's going to be someone out there thinking, 'They're saying ten, so I'll get out in nine.'"[34] Of course, he said, someone else starts thinking eight, or maybe seven, and then one day someone's left holding the bag of sand.

The developers have their own calculations; for them, Stoddard says, the worst that can happen is to lose money on the last project before things go bust. So they have already gotten their investment and profit out of the previous ones: "If you lose badly on your last one or two projects out of seven, you're OK. It's a Ponzi scheme mentality. Mayor [Philip] Levine over on Miami Beach said, 'The best and brightest are buying. Sounds like it must be a good idea!'" Stoddard believes his fellow mayor is bringing in development as fast as he can, because "he wants a city that's too big not to save."

Like all Ponzi schemes, Miami's depends on no one pointing out that the emperor is wearing no robes, or even a swimsuit. As Stoddard says, "There's a big 'shhhh' philosophy." Among Miami's working majority, most of whom aren't involved in the real estate boom, unawareness of the sea level threat is a matter of information access, according to Nicole Hammer. When they see flooding, she says, many in Miami mistakenly believe that it's simply the result of poor engineering or maintenance. Sea level rise is discussed in academia, in policy forums, on public radio, and the like—that is, in places where many of the people most under threat never encounter it.

WHEN WATER SURGES IN, DOES MONEY SURGE OUT?

Hurricane Andrew is given a lot of credit for prompting much-needed changes in building codes and zoning laws in Miami-Dade and around the country, but that storm hit very early in the greenhouse-aware era, and Florida spent much of the next quarter century acting as if there were no climate crisis looming. In fact, after Rick Scott took office as governor in 2011, officials of the Florida Department of Environmental Protection were ordered not to use the term "climate change" or "global warming" in reports or communications, according to the Florida Center for Investigative Reporting and the *Miami Herald*. The paper reported in March 2015, "The policy goes beyond semantics and has affected reports, educational efforts and public policy in a department with about 3,200 employees and $1.4 billion budget." [35]

Living as they do in the most badly exposed region of one of the states most threatened by sea level rise, many Miamians have long been frustrated by the attitude of fellow Floridians who oppose taking any action on climate. To keep future flooding scenarios from growing even more dire, many in Miami have been urging their city to thumb its nose at the legislators up in Tallahassee and start slashing southeast Florida's own emissions. (As American cities go, Miami has a typical, meaning hefty, greenhouse gas output.) [36] A Miami-Dade Climate Change Advisory Task Force was appointed in 2006, and its 2010 report contained many recommendations for emissions reduction, as well as identification of "adaptation action areas" threatened by sea level rise on a fifty-year rolling time scale and of engineering solutions to deal with immediate flooding problems. Then in 2013 the county created a Sea Level Rise Task Force, which urged that the recommendations of the Climate Change Advisory Task Force, among others, be put into practice. Finally, on January 21, 2015, the Miami-Dade County Board of Commissioners met to vote on a set of resolutions based on the task force's recommendations. The commissioners that day faced a forty-six-page agenda covering a vast array of issues from affordable housing for the elderly to controls on assault weapons. But when the meeting kicked off with an open mike for citizens' comments, almost all attention focused on the six resolutions related to sea level and climate change. Miamian after Miamian stepped to the microphone in support of the motions, which directed local officials finally to begin acting on the climate and sea level measures after so many years of talk. As the comments continued for an hour and a half, not a single person rose in opposition to the resolutions. Even representatives of the Miami-Dade Chamber of Commerce and the Builders' Association of South Florida gave their full support. All six resolutions passed. Clerk of Court Harvey Ruvin, who had chaired the task force, saw that day's action as much more than a formality, telling reporters, "People are going to look back at this day as a turning point." He said the effort Miami needed would be on the scale of New York's post-Sandy initiatives; however, this time the effort would come before, not after, disaster.[37] Celeste De Palma, now of the National Audubon Society in Miami, expressed strong support for the resolutions in the meeting's comment session. Once they'd been passed, she told us, "My overall sense of these resolutions is that they are a baby step we should have taken years ago. However, it is a welcome step forward, since we were getting nowhere. The next step would be to address the sources of carbon dioxide—that was left out of the resolutions on purpose to, in the words of Commissioner Rebecca Sosa, 'avoid making this a partisan issue.'"[38]

Sosa, who as mayor had written and sponsored the resolutions, indeed acknowledged that she had tried to ensure broad support from her fellow commissioners and the public by avoiding references to the human hand in climate change. Nevertheless, a new day may have dawned in Miami, at least as one looked up from the grassroots. When the new mayor, Carlos Gimenez, unveiled his budget in September 2105 and it contained only a single mention of sea level rise—in a list of "unfunded capital projects" buried on page 265—his negligence made headline news. A few years earlier, no one would have noticed, but now an uproar was raised across the county. As a League of Women Voters representative told the county commissioners, "This has to be a joke. Given that we're Ground Zero for climate change."[39]

Grassroots activists like De Palma are noting that the climate resolutions were not far-reaching, certainly not in proportion to the predictions of climate modelers. If implemented, they will reduce Miami's emissions footprint and provide near-term flood protection, but the city still has not awakened to the prospect of the wholly new land- and waterscape that will become local reality later in the century. We asked everyone we met about this, and most, while they certainly did not wish for a storm surge from a major hurricane, seemed to believe that an event like that would sound a loud enough wake-up call. Stoddard was one: "A big storm would do it. A lot of water washing over people makes an impression. And as you raise the sea level, the likelihood of that happening goes up." Hammer was another, saying, "A storm surge will give the biggest shock." Even though the floods from a storm surge don't stick around long, Hammer believes the sheer terror created by a surge like that of 1926 would have a bigger impact than creeping inundation, despite the fact that the latter is permanent. Maybe, by providing a tangible preview of the future, a surge would do much more to jolt Miamians out of complacency than, say, previous awareness campaigns that included painting high-water lines on the city's streets.[40] Wanless said, "We don't appreciate how powerful storm surges are. Andrew passed south of the city, it was a fast-moving storm, it didn't have waves and currents attached to it, and it was gone. It didn't go on for hours. It was not a good representative of what a storm surge could be." Very few people are alive who remember the enormous surge from the hurricane of 1926, but with higher seas (even today, but especially in the future), a wall of water that size or worse could be produced by a much less powerful storm. No one knows what an even stronger storm might do. On the other hand, Wanless told us, "I thought Sandy would prompt some change here—and it did for a little while, bringing some greater awareness. But then in six months property values went right back up." All of this is not to say that the gradual but inexorable rise of the Atlantic wouldn't also get Miamians' attention. Hammer noted, "Now we are having three to six of those scary tidal flooding events each year. When we start having two hundred a year, this is going to be a very different place."

One often hears in Miami that the city's steep decline will be triggered by neither storm surge nor irreversible inundation but instead by economic panic. Many believe the crisis will come when the big insurance companies decide they can't afford Miami anymore. Wanless says, "Certainly there's going to come a time when the insurance companies are going to walk. They'll say, 'Look, we're not a charitable organization.'" But more and more, there is worry that the collapse of the mortgage market will come first. Philip Stoddard told us, "Investors who buy packages of thirty-year mortgages are wagering big that the property will be worth as much in thirty years as it is today. At some point that confidence will disappear. It has to." He believes companies will start bundling mortgages on coastal and other low-lying properties separately from those that are safe from sea level rise. "Someone will start selling mortgage-backed securities that are free of climate-risky investments. They don't want to do it yet, because it would mean losing a big chunk of their market. But as soon as they do that, boom! The market bifurcates and no one will buy the riskier thirty-year mortgages."

AN UNLEVEL SEA

Income inequality in Miami-Dade is higher than in any other large urban U.S. county other than Manhattan.[41] Extremes of wealth and income are strongly related to race, and both race and wealth are associated, in a complex way, with flood risk. As in most coastal cities, properties close to the bay or beach are the most desirable and expensive. Recent research indeed showed that 59 percent of people living in areas at high risk for coastal flooding and storm surge are white, while 30 percent are Latino and only 8 percent are non-Latino black. (The three groups make up 40, 38, and 19 percent of the county's total population, respectively, so whites are disproportionately exposed along the coast.) For flooding in inland areas, however, the numbers are very different: 41 percent of those at high risk are Latino and 20 percent are non-Latino black.[42]

The parts of the city that are relatively safe from flooding span the economic and ethnic range, from Little Havana to upscale Coral Gables to some predominantly black communities, including Liberty City and Little Haiti (leading a few to speculate that gentrification, not inundation, may be the chief threat to lower-income neighborhoods).[43] Other predominantly black neighborhoods are down in the Glades, places that will be submerged by a couple of feet of sea level rise. But black Miamians also live in some of the most economically and environmentally harsh places in the county. For many, flooding remains low on the list of concerns.

Flood prevention typically focuses on the most affluent areas, and Miami Beach isn't the only lucky coastal community to attract substantial investment for protection and resilience. In 2012, the town of Sewall's Point, which occupies a narrow peninsula about a hundred miles up the coast from

Miami, announced that it would receive a $3.2 million grant from FEMA to elevate ten houses and one office building. The median house value in Sewall's Point is $750,000, so the announcement also elevated more than a few eyebrows around Florida. FEMA's goal in providing the funds was to avoid having to make big payouts when, as is inevitable, those properties are flooded someday; elevation would also hold down the homeowners' flood insurance premiums. Of course, raising the houses will also raise resale values, another benefit to the community.[44] "The people in Sewall's Point do get flooding, but not that bad, and they have a lot more access to resources already," said Hammer. "Then you have places like along Sistrunk Boulevard near Fort Lauderdale, which floods a lot; you see people putting kids in shopping carts to go across the street so they don't have wade through water to go to the store." Sistrunk runs through a very low-income area, where more than 95 percent of residents are African American. They have gotten no help in building up their flood resilience, Hammer says.

This kind of discrimination has a history in South Florida. In a 2007 review of research on the roles of race, class, and ethnicity in disaster vulnerability, Arizona State University professor Bob Bolin wrote, "Throughout the Hurricane Andrew case studies, the authors highlight how race, ethnicity, and class inequalities shaped people's experiences, from impact related losses to access to assistance, inequities in insurance settlements, the effects of pre- and post-disaster racial segregation, and the calamitous effects of disaster on an already marginalized and impoverished black community. Each of these studies documents how already existing social conditions in greater Miami shaped the contours of disaster and the ways that marginalized populations variously endured continuing or increased disadvantages in the recovery process."[45]

SWEETWATER

As the years wear on, municipalities with the highest property values and the most solid tax bases will follow Miami Beach's lead—pumping, raising, and armoring. Other places may be considered expendable. Here is how University of Miami architecture professor Elizabeth Plater-Zyberk has recommended that state and county funding for dealing with rising waters be allocated: "It's going to have to be a political-economic question, and you're going to say, 'What is the most meaningful to us in terms of economic development and the GDP of the region?' And everyone would say, well, the airports, the port, the tourist industry, that means the islands, our white-collar downtowns, and you'll start thinking about what to do for those. . . . The shopping centers or the houses that are, let's say, out west—water will be coming up there as well . . . generally speaking, affecting the economy minimally. Those are the places that you might decide to give up." [46]

One such place "out west" is Sweetwater, a small suburban town that sits just north of Florida International University, at three feet above sea level. Less than nine miles farther west the Everglades begin. On maps, the landscape north of Sweetwater is largely blue, crowded with vast rectangular ponds up to a square mile each in surface area, in which the water management district attempts to corral the area's excess. Under Sweetwater's modest streets and lawns runs the constant eastward subterranean flow of freshwater from the Everglades, sustaining the aquifer under Miami. Sweetwater floods with every hard rain, and the reason is obvious. Walk along one of its residential streets, look through one of the many metal grates embedded flat in the asphalt, and there is the water table not far down, partially filling the system's lateral conduits even during the dry season. With any significant rainfall, the drains fill instantly and the streets flood. Sweetwater's population is 96 percent Latino, and its median household income is $30,000; its small slab-on-grade houses are a world away from the pastel-trimmed condo towers overlooking Biscayne Bay. Plater-Zyberk is almost certainly right: it will be the condo towers, not Sweetwater homes, that will be protected to the bitter end.

Many are arguing that the Army Corps of Engineers, which controls the level of freshwater in the Everglades, should raise the water level, thereby increasing the flow eastward through the limestone under Dade County and pushing back more forcefully against the seawater that's trying to intrude from the Atlantic. But that would quickly flood low-lying West Dade towns such as Sweetwater. For now, those towns are protected by law from being intentionally flooded. But Phil Stoddard sees the law as a threat to other communities, such as South Miami. Although he thinks Dade County's future should be guided by the motto "Shrink carefully, depopulate thoughtfully," he's ready to consider sterner action if it will protect Miami's groundwater: "One foot of sea level rise reverses everything." After that, "if they were to raise the water up to the level they'd need to, Sweetwater floods. And the Corps is not allowed to flood out a suburban area . . . This one little community is going to cost us our freshwater supply many years earlier than it would otherwise disappear. . . . No one will do anything about it because of the politics."

So far, no one has proposed a technological fix that can work for southeastern Florida's low-lying inland communities. Are Plater-Zyberk and Stoddard right? If the fragility that is Greater Miami is to be conserved so that resilience (at least for a while longer) can be achieved, will there be no choice but to surrender Sweetwater and other communities to the Everglades? And if so, will there be any Staten Island–style provision made for compensation and relocation? Will communities be able to move together to a safer place and somehow keep themselves intact, as the mud volcano refugees of Renojoyo did?

To many who live in places like Sweetwater, the prospect of having to abandon their homes someday is a tragic one. When she was young, Hammer came with her family to the Miami area from Guatemala. Then, she told us, "We lost our home in Hurricane Andrew, and we left, moved two hours north and lived up there for several years. Now my family—my mom, my uncles—they are all moving back down to Miami. The best thing about Miami for the Latino community is that you're among people who speak your language, literally and figuratively! It's a very comfortable place."

Hammer points out that if there is another monster hurricane, or when parts of Dade County simply become unlivable because of routine flooding and seawater intrusion, the economic impacts will of course run along class and color lines but may adhere even more closely to lines of home ownership. There will be something like a three-way split. Renters who find themselves in the position that Hammer's family was in after Andrew can take their remaining possessions, leave, and find housing and employment somewhere else; they will suffer, but they won't necessarily be wiped out. Affluent owners of properties in Miami Beach or Brickell will also come out fine if they can absorb their uninsured losses or if their losses are only on second homes or investment properties. It's middle-class homeowners who will take a big proportional hit, according to Hammer, De Palma, Stoddard, Wanless, and others we spoke with. Hammer says their sad fate will be shared by low-income families, often immigrants, who have saved, scraped together a down payment, and bought a small place of their own. If they lose that, she says, they are broke and in deep trouble.

In some Dade communities, chronic flooding has provided cover for actions that border on the shocking. In 2014, for example, the Sweetwater police obtained two armored military vehicles, two military helicopters, twenty-four military assault rifles, and a military grenade launcher. The source of the hardware was the same Pentagon surplus-equipment program that had supplied the armaments used by the Ferguson, Missouri, police against protesters that year. What could be the reason for police officials in Sweetwater (and adjacent Florida International University, which also took advantage of the program) for buying the armored vehicles, choppers, and weaponry? Sweetwater's police chief said they would need them in flood and hurricane emergencies and to deal with assault-rifle-toting criminals. Asked when was the last time he'd had anyone shooting an assault rifle, the chief admitted, "In Sweetwater, none that I can recall." In discussing flood disasters and assault rifles, law enforcement officials made no mention of the intense struggles over police violence that were then under way in several U.S. cities. The city of Sweetwater does have its flooding problems, but Howard Simon of the American Civil Liberties Union told a local TV reporter that, as a basis for purchase of heavy military equipment, disaster preparedness in this case was obviously "an after-the-fact concocted rationale."[47]

SETTLING INTO MARGARITAVILLE

At the end of September 2015, a climate change workshop held at the Hyatt Regency Hotel in downtown Miami and hosted by former Vice President Al Gore coincided with an early king tide. But this was no ordinary full moon; it was a so-called supermoon, which not only lit the sky in its closest possible approach to Earth but also simultaneously swept through Earth's shadow, resulting in a front-page-grabbing lunar eclipse. The close orbital approach also resulted in three days of extra-high tides that severely flooded several parts of Miami Beach, Fort Lauderdale, Hollywood, Hillsboro, and Deerfield Beach. In Fort Lauderdale, fish followed the tide; a resident observed nonchalantly, "Look at that. A mullet in the street." Miami Beach resident Robert Wolfarth was less pleasantly surprised: "We still have street flooding, even in a neighborhood where we have pumps that are online and working. It's not just this neighborhood. It's North Beach, it's South Beach, it's from the east to the west." At the Hyatt, Gore was telling the crowd, "The scientists have long since told us we have to change. But now Mother Nature is saying it with water in the streets in this city." In the southwest portions of Miami Beach where Bruce Mowry's pumps and street elevation had been implemented, all was dry, and Mayor Levine was ready to double down on that success. Regarding the state-maintained Indian Creek Road in the northeast part of the city, which was suffering the worst flooding, he said, "We need to put pressure on our state to raise this road, put in pumps and of course we're going to raise the sea wall as well."[48] Despite Levine's can-do attitude, it was starting to dawn on communities across Miami-Dade County that any such engineering solution would fail to keep them high and dry in the long term.

Confronted with a seemingly insoluble predicament, many locals have been looking in desperation to Dutch experts. After all, much of the Netherlands is even lower than Miami, and over centuries the country has developed an ingenious flood control infrastructure. The Dutch have been advising New Orleans and New York on dealing with their flood hazards. Most knowledgeable Miamians, however, seem skeptical. Philip Stoddard told us, "Yes, the Dutch are all over us. But they are still trying to figure it out. And they really want us to believe they can do it. Well, OK, they're good at making things float." Nicole Hammer said, "The Dutch are working hard to get some business over here. They came in, had a big workshop, and everybody was drawing diagrams and trying to figure it out, but they must have been annoyed, because we were like, 'Well, what about water coming up through the porous limestone we're sitting on top of? What about the Everglades? What about the heat? If we create more ponds, what about the mosquitoes?' So they're working on it. They're brilliant people, the folks who came down. So I'm sure they'll come up with some very interesting, very creative solutions." Pete Harlem viewed the Dutch invasion from a somewhat different angle, arguing that the best thing they could do for Miami is not to sell a particular technological fix but rather to explain how they had managed to get an entire country united behind an effort to deal with sea level rise—and then advise South Floridians on how to achieve their own consensus, even though the options are far more limited in Miami than in Rotterdam. At times even the Dutch seem to feel that in Miami they have met their match; Henk Ovink, one of the Netherlands' most prominent water-management experts, has taken to calling Miami "the new Atlantis."[49]

There is also much talk of embracing the water by becoming more like Venice, Italy, a city that has in recent decades been attempting to protect its canal-laced urban area with a huge flood barrier. Wanless isn't impressed. "Venice now has one-third the population it used to have. No Sandy, no Andrew, no Katrina, just rising seas, increasing frequency of flooding, rotting, disruption of infrastructure—that's made it an unpleasant place to live." Wanless concurs with the common characterization of Venice as a theme park rather than a city, adding, "And if it didn't have the history it has, it would be a closed-down theme park. People come into the city in the daytime for you to visit it, because it's a cool place. But it's not what it was admired for originally, a vibrant, vital part of civilization and commerce."

Few people we talked with in Miami believed that the city they all know and love would remain intact into the deep future. The question was not whether people will have to leave but when. When people ask Stoddard, "When should I think about selling my house?" he tells them, "It depends on whether or not you can afford to lose the capital in it. What happens to you? Are you ruined financially? It's a question of risk tolerance. If you can afford to lose the capital in your house, keep it. Enjoy yourself! But if you're counting on that house for retirement, or if you'll end up destitute if you lose it, I say *now* would be a good time to sell your place. Then you can either rent here or move somewhere else."

Wanless has a concrete recommendation: "Every community, from Phil Stoddard's South Miami to Miami Beach to Hialeah to Coral Gables, everybody needs to do a thorough mapping of the elevations of their businesses, of their communities, all aspects of their infrastructure, and then start to figure out what they need to do at each stage of sea level rise, to maintain connectivity of the communities." Furthermore, he says, in today's real estate market "there is no truth in selling, and there needs to be. If we plan, and if people buy property with the understanding that with another foot and a half of sea level rise the city will no longer be responsible, no longer maintaining infrastructure in your sector, then people will know what they are buying into. Cities, counties, states should start buying people out now."

Stoddard would like to see what's often called "managed retreat": devising a schedule to take land lying one foot, then two feet, then three, then four, above current sea level and pulling them out of the economy over time, turning them over to aquatic parks, protective wetlands, or other uses. For this, he prefers the term "rolling easements," because, he says, "Americans hate the word *retreat*." Today in Miami's political circles, he doesn't see as much resistance to the idea itself as he once did: "Now no one's even wincing now when you bring it up." He expects that the Miami that remains above the floodwaters decades from now, if it has managed to remain viable, will have abandoned its famed flashiness, excess, and bravado to adopt a culture more like that of the Florida Keys: "Down there, people just accept storms and floods as part of life. They just board their houses up and ride it out. If we go the way of the Keys, the place will be a lot funkier, no longer the playground of the rich. But it could save the tourism economy—except of course they'd have to keep resupplying the beaches with sand." By the time Stoddard's vision comes to pass, if it does, the old Keys will have gone under the waves and Dade County itself will be an archipelago, becoming the new Florida Keys in both geography and culture.

Harlem also recommended retreat but, like Stoddard, he steered clear of the word, using the term "strategic withdrawal," which he said he took from U.S. Marine Corps terminology: "It would be much better than a piecemeal retreat, which would be far more disruptive." He didn't necessarily think Stoddard's "new Florida Keys" scenario, the vision he himself called "Margaritaville," was the most desirable end point for Miami. But, he said, "that's probably the future."

If Miami does adopt an embrace-the-water culture, it will not appeal to the majority of residents, and there won't be enough land left for everyone anyway. Some are already thinking about the fate of the refugees fleeing Margaritaville. Henry Briceño knows Miami's predicament well, but rather than trying to spur action by creating awareness of the crisis, as Harold Wanless does ("I tell Harold, 'Hey, cool it, you'll scare the crap out of people!'"), Briceño emphasizes potential silver linings: "Everyone looks at sea level rise like Armageddon, but it's also an incredible opportunity! People will move out of the flooded zones, we'll have to found new cities with new engineering, architecture, materials. So what we need now is for people to sit down and think and then move ahead and try to solve things. We know, for example, that in a hundred years people will have to move out of this area. Where are they going to go? Where will we have the new cities? Where are we going to have these resources—freshwater, roads, telecommunications, schools, hospitals—where are we going to put those to be out of harm's way? They have got to be high, first of all! But it's a huge opportunity for the whole country, because you can have all the people and technologies and companies coming from all around, and we have to take advantage of it. The Netherlands has been the icon of this. They have the technology, they have the business. We can turn that around! The United States should be in the forefront of what countries should do to tackle sea level rise. We have the technology, we have the know-how, we have the entrepreneurial spirit, we have the greed of people who want to do business, and we have all different kinds of problems! It's really great to be here, at this point in history, it's great—whatever happens."

Despite Briceño's enthusiasm, a cloud of foreboding seems to be settling over Miami. At one point in our tour of Miami Beach's flood prevention efforts, as Bruce Mowry was steering his car slowly through the Flamingo Park neighborhood, his mood, which up to that point had been buoyant, became more reflective. He told us, "You know, I drive around a lot, looking at all these streets and trees and homes and thinking about what's coming. Even before we're underwater, within just twenty, maybe thirty years, the salt water's going to get those trees. This whole beautiful landscape's going to change." He couldn't stomach the thought that even if all of his engineering feats succeed in keeping the city's streets dry, it still won't be the same city. "We just can't make that sacrifice," he said. "We'll have to put our trees up in planter boxes."

Many share Mowry's desire to stay and make the most of a place they love, while the good times last. Could it be that what appears on the surface to be typical Miami recklessness may actually be a city adopting a Filipino-style culture of disaster—not in response to catastrophe but in anticipation?

Maybe. Stoddard says of his home, "Our floor is ten feet above sea level, so we have more time than some. But at some point this house will become unsellable. We could get a lot for it right now, but then I look at how pretty this place, South Miami, is, and I don't want to leave." Wanless agrees: "Compared with whatever is going on up in the rest of the country, this is just a wonderful place to live. We're all going to enjoy it as long as we can." Another Miamian echoed his city's *bahala na* outlook better than anyone: "People say Miami's a proto-Atlantis. I say maybe so, but if you'd had the opportunity to live in Atlantis, why wouldn't you?"

ENGINEER, DEFEND, INSURE, ABSORB, LEAVE

The world of policy seems to parallel the world of science with about a fifty-year lag.

—Deborah Stone, "Causal Stories and
the Formation of Policy Agendas," 1989[1]

The times between disasters can be as turbulent, risky, and confusing as the depths of any stormy night, but they are certainly better times in which to live. They are the times for solutions to appear—times in which to act, not just react. In countries and regions where there is easy access to finances, energy, and materials, there may be many options open for preventing or protecting against the next disaster. The best approach often seems obvious—and that solution often turns out later to have been inadequate or even self-defeating.

So in whatever time we have left before the next superstorm hits Manhattan or Miami, or a killer quake hits Seattle or even Oklahoma City, how can the risk of catastrophe be reduced and the inevitable losses absorbed? Here we will sort through some of the big prescriptions that the physical, biological, climatic, and actuarial sciences have offered for reducing risk of rich-world disasters. (Meanwhile, we'll reserve for Chapter 11 our discussion of proposals for the rest of the world.) Admittedly, we feature some illustrative failures. It should be clear enough that some big solutions do make people safer most of the time, but it may be more important to understand how big solutions can turn into even bigger problems.

The words of Deborah Stone above should also be remembered. Going by her estimate, most of the disasters written about in this book, while all relatively recent, occurred in a 1960s policy world—a world of concrete walls, earthen levees, dams, canals, pumps, bulldozers, helicopters, and water bombers. Large parts of the disaster research community began to drift away from a hard engineering approach decades ago, so that in reading disaster journals today it's easy to get the impression that concrete has gone out of fashion. It hasn't. Hard engineering solutions still hold a commonsense sway over societies that have accomplished so much with them.

DISASTER PREVENTION THE HARD WAY

In 2014, the Temple University physics professor Rongjia Tao proposed that tornado risk in the central United States could be reduced by interrupting seasonal atmospheric movement in the region. This, he wrote, would require three east-west barriers, each a thousand feet high and a hundred miles long, placed in North Dakota, on the Kansas-Oklahoma state line, and from south Texas into Louisiana. These "Great Walls of America," as the media were quick to christen them, would cost $16 billion all told and could, according to Tao, be "attractively" designed. His proposal apparently was offered sincerely; he presented it at a meeting of the American Physical Society in Denver on March 7, not April 1. Meteorologists and climatologists mocked the idea nevertheless, pointing out that while walls of that size could not prevent tornadoes (one critic said that would require barriers the height of the Alps), they could trigger weather chaos across the country's midsection. The Great Walls proposal became a rare case in which an engineered "cure" for a geoclimatic hazard was taken seriously by no one but its author.[2]

Other strategies for disaster prevention through engineering that, in retrospect, are almost as futile and risky as Tao's tornado walls are often enthusiastically pursued, even as the ultimate sources of recurring disasters are often left unaddressed. In his 1989 book *The Control of Nature*, John McPhee described three audacious attempts to physically divert geoclimatic hazards. All three projects—a 1973 attempt to steer the lava flow from an erupting Icelandic volcano away from a harbor, a long-term effort to block a natural course change in the Mississippi River and keep the river flowing toward the port of New Orleans, and the never-ending battle to prevent fire- and flood-triggered landslides from destroying houses in California's San Gabriel Mountains—proved partially successful, but with unanticipated costs and multiple dangerous unintended consequences.[3] Innumerable such situations have demonstrated that in the hard approach to disaster prevention, the verb "protect" usually translates as "deflect," as when armoring one stretch of coastline simply transfers the problem elsewhere along the shore.

On high-value coastlines, adoption of hard defenses against sea level rise and storm surge has become common. But once in place, the armoring of a shoreline disrupts the natural processes that keep it in its normal dynamic state. Seawalls promote erosion and the eventual loss of the beach in front of them. They cause erosion of beaches up and down the coasts beyond the barrier itself, cut off supplies of sediment that support mangrove forests and salt marshes, inhibit dune formation, and ruin wildlife habitats. Structures perpendicular to the shore, such as jetties, cause even more severe erosion downshore. Armoring of bays and estuaries can destroy wetlands and create hypoxic conditions, depriving marine life of oxygen.[4] All of these destructive processes can be expected to accelerate in the greenhouse era as oceans rise and fortifications are extended and strengthened to protect valuable real estate.

Protection through deflection raises thorny issues with almost every type of hazard. River levees usher floodwaters farther downstream to inundate unprotected areas. Physical defenses against landslides steer them onto other people's property. In McPhee's account, the heroic, ultimately successful effort to prevent a lava flow on the Icelandic island of Heimaey from spilling into and closing off its harbor ended up with a rogue river of lava streaming into the island's one town, destroying 20 percent of it. In Hawaii, protection against lava flows has traditionally been avoided or prohibited, because it would mean sending the unstoppable flow to another part of the island.[5] In the Philippines, lahars fed by ash from the 1991 eruption of Mount Pinatubo and propelled by heavy rains continued to plague parts of the island of Luzon for a decade after the eruption, eventually killing about six hundred people. On the southeast side of Pinatubo, a project called Megadike, completed in 1996, was designed to steer deadly lahars away from the provincial capital San Fernando and the town of Guagua; however, that meant sacrificing another town, Bacolor. With Megadike in place, the flows that struck the area almost every year were diverted such that they always made a direct hit on Bacolor. But because resettlement areas provided by the government were so undesirable, many residents of Bacolor—a long-established town that had served as a provincial and even national capital in colonial times—continued to live there, evacuating for the duration of each lahar season, then returning to the town to shovel it out when the rains ended.[6]

Even when individual projects seem to work, hard barriers in aggregate don't necessarily achieve the desired outcome. In his book *Acts of God: The Unnatural History of Natural Disaster in America*, Ted Steinberg showed that by the 1960s the U.S. Army Corps of Engineers had built two hundred reservoirs, 7,500 miles of channel modifications, and more than 9,000 miles of levees and other barriers. But even with $25 billion spent, per capita flood losses were almost two and a half times higher in the decades after 1950 than in the decades before.[7]

The heyday of dam-and-levee flood control may be over, but the hard approach is still the first resort in many prevention strategies, crowding out attempts to deal with underlying causes. Even in the decade following Hurricane Katrina—as obvious a case of system-wide failure as one could ask for—official attention focused on reinforcing the levees around New Orleans and the Army Corps of Engineers' construction of a 1.8-mile-long storm surge barrier.[8] But after 2005, with the nation's attention riveted on Louisiana, planners needed to show something fresher than a 1960s strategy. So they called in the Dutch.

THE DUTCH CURE

The engineers who flocked from the Netherlands to New Orleans were soon taking calls from other American cities—notably New York and Miami. After Sandy, it was Dutch engineers who drew up the short-lived proposal for a $6.5 billion barrier across New York Harbor, and Dutch designers and consultants worked on four of the six winning entries in the Rebuild by Design competition.[9] The competition itself was instigated by Henk Ovink, who became a senior adviser to the U.S. secretary of housing and urban development.[10]

More and more Dutch experts were visiting Ovink's "New Atlantis" Miami as well, and most came away feeling just as frustrated as he was by the plight of the city built on limestone. But there were plenty of other coasts around the globe needing protection. Countries as scattered as the Philippines, the United Arab Emirates, Mozambique, and Bangladesh have recently enlisted Dutch help. In one of the grandest projects, the Netherlands and its former colony Indonesia commissioned a troupe of top Dutch firms to build an immense seawall in the shape of the mythical eagle Garuda, designed to protect sinking Jakarta with its outstretched wings—while doubling as a luxurious new development district.[11]

For the Dutch, flood control is existential. A third of the Netherlands is already below sea level; more is at risk of flooding from the rivers that pour through the delta, which are becoming less predictable in their flows; and more than ever, given projected sea level rise, a cessation of flood control would cause most of the country to vanish. While the Netherlands traces its hydroengineering from the first dikes and windmills (which served to pump water off the land), the country entered the modern field after its own Katrina: a 1953 North Sea storm that wrought disastrous and deadly floods.[12] The government responded with a fifty-year program of unprecedented armoring, with higher dikes, sluices, seawalls, and moving storm surge barriers, under the collective name of the Delta Works. The last of the works was completed in 1997, just as climate fears bubbled to the surface elsewhere in the world. Combining its long hydroengineering history with Euro-cool design flair and a climate innovation pitch, the Netherlands was suddenly an undisputed brand—the Apple of adaptation.

The last great linchpin in the Delta Works was the titanic Maeslantkering storm surge barrier, a pair of automatic pivoting doors that form perhaps the largest man-made moving structure on the planet. Seen from a harbor boat heading toward Rotterdam, the Maeslantkering's open arms appear unreal, like the kind of lever Archimedes would use to move the Earth. But what's really incredible is that this monumental barrier was the cheap option. The Maeslantkering exists because the original Delta Works blueprint for Rotterdam harbor, which involved raising dikes up to thirty kilometers inland along this distributary of the Rhine, would have cost too much and buried too many historic town centers. And any solution that interfered with the channel in calm weather would have limited access to Europe's largest port and one of its busiest rivers.

The Maeslantkering is only the start of Rotterdam's water controls. "We want to become a resilient city from every brick to the barrier," says Arnoud Molenaar, manager of the city's Rotterdam Climate Proof program. Eighty percent of the city lies below the level of its own port (by more than twenty feet in places), and every drop of rain that falls and waterway that flows in behind the dikes has to be pumped out, including the entire Rotte River. Big rainstorms demand an immense holding capacity from the urban environment while the pumps do their work. Molenaar estimates that the city center must be able to hold 80 million gallons, the capacity of an oil supertanker. (To illustrate that quantity, there are useful visual aids plying into the Port of Rotterdam on a daily basis.) This is not to say that the whole city lies behind the dikes. "There is also a whole river zone developed on the 'wrong' side of the dikes," Molenaar explains, and there the solutions get more creative. "We see that zone as a chance to redevelop and get more dynamics in the center and to reuse and transform the older port areas." [13]

Russell Shorto's 2014 *New York Times Magazine* article "How to Think like the Dutch in a Post-Sandy World" described Rotterdam as a city "building floating houses and office buildings and digging craters in downtown plazas that will be basketball courts most of the year but will fill up with runoff during high-water periods." In 2014 at least, this was a bit of exaggeration. Rotterdam is a city poised between architectural rendering and reality. "We have been investing at the scale of objects," Molenaar says. He is referring to wonders such as the Floating Pavilion, a triple geodesic dome built on the harbor as a demonstration piece, and Water Square

Benthemplein, the sunken basketball court that Shorto oddly pluralized. Bigger district-level plans were stymied by the European debt crisis that struck in 2009. But the objects are working. The architects of the Water Square, De Urbanisten, parlayed the square's fame into a partnership in one of the successful Rebuild by Design entries in New Jersey.

De Urbanisten and fellow Rotterdam studio ZUS (Zones Urbaines Sensibles) joined with MIT's Center for Advanced Urbanism to enter the competition. In their presentations the team modestly referred to the task of defending New York and New Jersey as Delta Works 3.0.[14] Their winning design was an overhaul of the New Jersey Meadowlands, proposing a berm around the remnant wetlands built into a raised development band, a linear city of multistory mixed-use buildings where there are currently only warehouses. The logic of the so-called New Meadowlands is curious on the surface—protecting a vulnerable area by drawing in more assets—but it's classically Dutch. And it has translated well to growth-led cities in America, where the New York and Miami Beach waterfronts follow much the same logic. At the ZUS office, Kristian Koreman, leader of the New Meadowlands team, explained, "Yes, you can pour a lot of money into securing and making this area beautiful, but this whole plan would cost $1.5 billion. You could in a traditional way extract that from the federal government and pour it into this, but a much more resilient way would be to connect it to the local economic system, so it can be much more than just the investment." Koreman saw this coming up through pilot projects and incentive zoning inspired by Manhattan's privately owned public spaces (POPS, the best-known of which is Zuccotti Park).

For Koreman and his collaborator Florian Boer of De Urbanisten, the competition was a chance to design on a scale and with a freedom they can't at home. "The openness of the question and the process is unseen in the Netherlands," Boer said. Added Koreman, "Ultimately they asked us not 'Where's the problem?' but 'Where's the opportunity?' That's what we like about America."[15]

Rotterdam bills itself as the city that "connects water with opportunities," and the biggest opportunities are Rotterdam's.[16] As Molenaar explains, "We as a city have certain needs and we have challenges here at home, and we need innovative solutions to become more resilient. And if we do it in a way that is innovative and we implement it here, then we build our own showcase. We now have several examples where the companies who did things here, like De Urbanisten and DeltaSync, who developed the Floating Pavilion, can more easily get contracts in other cities."

As a showcase for the climate-ready future, however, Rotterdam carries every bit of that future's baggage. Two flows define the city: the 760 million tons of rainwater that enter and leave the city every year and the 240 million tons of cargo that enter and leave its port.[17] It's mostly because of the port that Rotterdam charts one of the highest per capita carbon emissions rates of any major city in the world, with 29.8 tons emitted annually per resident.[18] Beyond the port and its associated industries, Rotterdam is also the Netherlands' only highly car-dependent city, reconstructed after World War II on an auto-friendly American template.

In 2006, the city initiated the Rotterdam Climate Initiative (RCI), establishing an independent office within the administration to oversee a 50 percent reduction in these emissions by 2025. In 2008, it began a separate program, Rotterdam Climate Proof, which was focused on adapting the city itself to change. A year later, the two programs merged, just as the original mitigation goal was headed for trouble. The 50 percent target was based on a plan to capture carbon dioxide from the port and a new coal power plant and pump it into a depleted gas field out in the North Sea. It was a huge gamble: if the storage ever failed, the city's carbon contribution would have effectively been *increased* by 40 percent.[19] Seeing this risk, and a downturn in Europe's carbon pricing market, the companies involved decided to scrap the plan.[20]

In this environment it was the newly merged adaptation side of the RCI that came to the forefront. Mitigation efforts were scaled back to programs such as green roof subsidies, efficiency retrofits, and a network to pipe waste heat from the port to houses and offices. In 2014, Molenaar was "pretty sure" that the original 50 percent reduction target was still on the books, but, he said, "we communicate less with these figures of 50 percent, 100 percent, because of course the goal is to become more sustainable, to become the most sustainable port in the world. It's difficult to quantify these things." The more integrated, innovative approach seems to suit Rotterdam better, anyway. "In my opinion it's an organic or logical development, from mitigation, to adaptation, sustainability, and now a new thing is coming on the stage, and that's resiliency," he said.

In a grim 2015 headline, *Fast Company* declared Rotterdam one of "The 10 Most Sustainable Cities That Will Thrive as the World Crumbles."[21] It probably won't be quite so stark. But standing in the Floating Pavilion, seeing the bold new skyscrapers of the Wilhelminakade through the dome's transparent co-polymer skin, the city does look like the capital of an adaptive future—or at least its sales floor. Like all of Rotterdam, the view is a sampler of solid and liquid, old and new. The skyline was financed on the cusp of the European debt crisis and much of it is still vacant. Beneath the skyline, the pavilion is a cluster of soap bubbles on the Rijnhaven, a long-disused harbor of port days past. As part of the area's revitalization, the city is offering a thirty-year lease for the fifty-two acres of the harbor's surface to developers interested in further experiments with floating architecture. The bid book calls it a "playing field."[22]

Capping the end of the harborside row of towers are the Hotel New York and the New Orleans, the country's tallest residential tower. It's strange to see the namesakes of two American cities now seeking Dutch expertise in storm protection, but the connection belongs to an older Rotterdam. The names of the towers around this harbor come from the warehouses they replaced (or, in New York's case, the offices of the Holland America passenger line). A few of the old facades are still standing below the skyline. One bears names across it in brick lettering: SUMATRA—BORNEO—JAVA—CELEBES, major islands of the colonial archipelago that became Indonesia. The sight evokes the comments of Harold Wanless in Miami. "The Dutch are here to help everybody so they can make money to build and maintain their *own* structures so they can survive themselves," Wanless said. "It's sort of like the old Dutch East India Company."[23] This may be the most cynical description of the Dutch strategy, but the resonances are hard to miss in Rotterdam, ever a mainline conduit of the global market. Now the city and the country appear to have found a promising solution in marketing solutions. This sort of ingenuity will forever be a part of the Netherlands because of its existential commitment to engineered, development-led defense. Growth pays for protection and demands protection; defending capital with capital takes more of it all the time.

The Floating Pavilion was constructed in another harbor and brought here by tugboat, and after a few years it will be taken somewhere else, probably to other cities in other countries, a traveling floor model of Rotterdam green ingenuity. Inside, the attendant Duncan Vlag shows us its experimental solar climate control and water recycling. From our vantage point he also expresses a designer's admiration for the Wilhelminakade and the statuesque New Orleans tower, an Alvaro Siza–designed building with peach-colored stone cladding from a quarry in China. But Vlag acknowledges that these developments aren't really on the same page with the message of the Floating Pavilion. "It is kind of pointless to say we built this as an example for the city, and then right next to it you have this tower and four or five ships coming all the way from China full of stone. We could never balance that out."[24]

COME SAIL AWAY

The homeland of hydrocontrol has paid the price for its accelerated armoring of coast and riverbank, as more people and industries moved into better-defended land, raising higher the consequences of failure. The last years of the Delta Works construction were marred by economically disastrous floods in 1993 and 1995. In 2006, the Netherlands' water managers decided they couldn't keep raising their bets, so they folded. The next works, a $2.8 billion project called Room for the River, not only reversed the tactic of the previous sixty years but, in a sense, represented a retreat from centuries of Dutch protective strategy. Along the Rhine, Meuse, Waal, and Ijssel Rivers, dikes have been pushed inland, floodplains restored, built obstacles removed, and sacrificial overflows established. The residents of these zones—whole towns, in some cases—have moved. The shift from armoring to acceptance has dramatically increased the safe carrying capacity of the rivers (although that increase depends on future developments upstream in Germany, France, and even Switzerland) and restored environmental quality through the delta.

Such a quick about-face is impressive and the voluntary movement of entire communities even more so. The ecological turn shows a softer side of the Dutch. The country's regional Water Boards have existed since the thirteenth century, managing their respective patchworks of polders, dikes, and canals. They were the country's first democratic institutions, and they represent the deep pilings of Dutch civic spirit. Living within a giant engineering experiment, the medieval Dutch learned, meant that communal concerns always came first, backed by the weight of water. This didn't make Room for the River an easy task (the town hall arguments were long and vocal); it just made it possible.

From the moment it began, Room for the River was quickly absorbed into the Netherlands' self-image and sales pitch. It's a badge of environmental honor and engineering insight, but the ecological turn itself has not sold as well on the foreign market. Customers have reason to be skeptical of how ecological solutions—especially ones involving retreat—will fit with the free market. Is there a way to make room for the river, or the sea, without forcing a retreat?

Yes, there's a design for that, too. In 2014, the *Miami Herald* reported on the latest pitch for a city that had tried everything else:

> In the land of boom and bust where no real estate proposition seems too outlandish . . . a Dutch team wants to build Amillarah Private

Islands, 29 lavish floating homes and an "amenity island" on about 38 acres of lake in the old North Miami Beach quarry connected to the Intracoastal Waterway just north of Haulover Inlet.

The villa flotilla, its creators say, would be sustainable and completely off the grid, tricked out to survive hurricanes, storm surge and any other water hazard mother nature might throw its way. Chic 6,000-square-foot, concrete-and-glass villas would come with pools, boathouses or docks, desalinization systems, solar and hydrogen-powered generators and optional beaches on their own 10,000-square-foot concrete and Styrofoam islands.

Asking price? About $12.5 million each.[25]

As a new imaginative frontier in the engineering approach, floating architecture projects the perfect image. The Floating Pavilion in Rotterdam became an instant icon because it speaks to sacrifice without the sacrifice, a party that doesn't have to end but can continue on the water—even stormy water or rising water. Floating structures bob through sociologist Zygmunt Bauman's era of "liquid modernity," along with the hovering drones of war, drifting tides of labor, and buoyant liquidity of the neoliberal market.[26]

Which is not to say that floating buildings are a new idea. People have lived on the water for a very long time, not least in the Netherlands, where the barges and *woonboten* (living-boats) of city canals have long been a home for the outcasts, the eccentrics, and the just plain poor. In recent years this semi-outlaw mode of life has been rehabilitated into a real estate trend, apace with the country's renewed embrace of water. Studios such as Koen Olthius's Waterstudio.NL (one of the partners in the Miami venture) are designing hundreds of modern floating houses, floating apartment buildings, and much more. Anything that is able to float financially will be floated. The water hasn't entirely lost its role as a place of disposal and concealment: between 2004 and 2012 the Dutch government also used urban harbors in Rotterdam, Dordrecht, and Zaandam to imprison irregular migrants and asylum seekers on six hulking gray boats, each of which held hundreds of detainees. The government opted for floating prisons, including two designed by Olthius's studio, because they could be "made operational more quickly and with fewer administrative formalities than any land facility."[27] But the warehouse-like facilities were cramped, oppressively hot, and unsafe.[28] They became the targets of protests, blockades, and occupations, and one group of activists even took to the water themselves, playing a concert for detainees inside Olthius's floating prison from boats bobbing alongside.[29]

Today Olthius is sticking to prestige projects. Leaving aside the carceral archipelago for that of the tourist, his studio has designed floating wonders for clients around the world, including a master plan for the Maldives, the world's lowest-lying island nation—the facility includes a floating golf course and a 185-villa floating resort in the shape of a flower. He has also downscaled to the developing world with a suite of "floating city app" designs that could be installed and uninstalled, software-like, in wet slums around the world. The pilot "app" is a shipping container lined with Internet tablets and floated on a raft of plastic bottles, which will be sent to the slum of Namuwongo, along the main drainage canal of Kampala, Uganda. Olthius and his contemporaries regularly talk about floating cities in the future. It's an old science fiction idea that he believes is coming within decades of reality. For some, such as the California-based Seasteading Institute, it's a libertarian dream to seek autonomy in international waters outside any nation's jurisdiction.[30] For Olthius, only parts of cities need to float: plug-and-play neighborhoods and amenities, stadiums, emergency services, the occasional prison. "If we can convince them that it's also financially profitable and help governments change building regulations, we'll have a future where it's normal to see cities that are 95 percent built on land and five percent built on water—just enough to give them the flexibility they need for an uncertain future," he said in a 2014 interview.[31] The floating district in Rotterdam would boost this vision, taking the Netherlands from the hardest of hard solutions to the swelling crest of liquid modernity.

As the showcase for this future, Rotterdam's Rijnhaven harbor will combine the earnest with the absurd. An upcoming installation in the harbor is the Bobbing Forest, a collection of trees planted in repurposed buoys filled with soil to create a grove that floats on the harbor. It's a harmless folly with a lot of resonance. On our visit in 2014 there was just one tree—a prototype of an archetype—bobbing in the "Innovation Dock" where the Floating Pavilion was earlier built. It was just one scrawny Dutch elm sapling sticking out of the middle of a brightly striped buoy, rocking wildly when ships passed by. But the project is more conceptual than practical. It's based on an idea of artist Jorge Bakker, "evoking questions about the relationship between the city dweller and nature and how these two relate to each other and the world around them."[32] It does evoke questions, but it's hard to tell if Bakker is intentionally or unintentionally crossing wires in the narratives around engineering, ecology, hard, soft, trees, boats, nature, and resilience. If we can float cities to safety, can we float the forest to safety, too? If we can engineer ways to protect ourselves from the planet, can we engineer ways to protect the planet from ourselves? Answers to that question are being tested in many places, but perhaps most ambitiously in the world's eco-cities.

LIVING GREEN, ON THE EDGE

If there is any truth at all to the catastrophic predictions of climate models, then slowing global greenhouse gas emissions will make our planet less hazardous. This isn't just a solution; it's an imperative, the start of any sustainable solution. Several of the Miamians we spoke with, for example, said that the attention being paid to the city's need to avert future disaster was coming at the expense of efforts to reduce the very greenhouse emissions that are making disaster more likely. The same might be said of New York or even of Rotterdam. But what do you do if you're in charge of one city, but that's it—you have no control over the rest of the planet? Will greener be safer? At ground level, unfortunately, lowering emissions in a single location doesn't necessarily increase resilience in that same location.

Consider Tianjin, China. In 2007, the governments of Singapore and the People's Republic of China announced a joint venture to develop a "Sino-Singapore Tianjin Eco-City," which would, by 2020, be a "model for sustainable development" and home to 350,000 residents. Tianjin, on China's northeast coast about a hundred miles east of Beijing, is the country's fourth-largest city and includes a major world seaport (and, like Rotterdam and other global port cities, it ranks high in greenhouse emissions).[33] The eco-city, still under construction, is on the coast about twenty-five miles from downtown Tianjin. Its residential and commercial buildings have all the usual green features, plus some: green roofs, rainwater harvesting, something called a "vacuum-powered recycling system," good insulation, solar water heating (already ubiquitous in Chinese cities), natural lighting, "underground garage natural ventilation," "indigenous plants adapted to local climatic and soil conditions," water-conserving fixtures, and recyclable building material. More than 60 percent of solid waste will be recycled, 20 percent of the eco-city's energy is to come from renewable sources by 2020, and 90 percent of street lighting will be solar-powered. Ninety percent of transportation will be "green," that is, on foot, bicycle, or public transport. To avoid impact on agriculture or forests, the city is being built on what developers call "unproductive" land.[34]

There are more than 170 eco-cities of various sizes and descriptions being developed worldwide. At least one hundred of those are in China, and Tianjin is described as the flagship project. Unfortunately, it could one day be a sinking flagship. The eco-city is situated on the coast and along sea level canals. The "unproductive" land chosen as the construction site was largely salt marsh and sandy flats. The eco-city is even lower-lying than much of the city of Tianjin itself, about a million of whose residents are already dangerously exposed to coastal flooding. By 2070, it has been predicted, almost 4 million residents will be exposed, putting Tianjin at number twelve in flood exposure among the world's cities. Tianjin also ranks among the top twenty world cities in value of property exposed to flooding. By 2070, based on current sea level rise and storm surge projections, it will be the seventh most flood-vulnerable metropolitan area in the world.[35] We now must add on top of those projections the threat to the new eco-city, a monumental coastal construction project all of whose property and 350,000 residents will be exposed both to gradual inundation by sea level rise and to sudden destruction at the hands of a storm surge pumped up by that sea level rise.

In evaluating Tianjin Eco-City and another eco-city in Helsingborg, Sweden, an international team of experts concluded that Tianjin's planners had accounted neither for the risk to life and property that would result from the project nor for the fact that its construction would destroy wetlands that might help buffer the Tianjin metropolitan area against flooding. The report noted that "in the case of Tianjin Eco-City there is immediate threat from sea level rise for the population and urban area" and that "the issue of climate change and coastal resilience should have been considered carefully by the developers" but was not.[36] By 2014, with the eco-city's population having reached a mere six thousand, that risk to human life did not yet loom large. And, according to *The Guardian*, "while the developer claims that more than 1,000 companies have registered in the city, many storefronts on its main shopping plaza stand empty."[37]

Federico Caprotti of the King's College London Geography Department has studied eco-cities around the world and concluded that they typically end up being enclaves in which the more affluent can take refuge from pollution and other problems that plague the larger nongreen cities nearby. He writes, "Analysis of eco-city projects shows that they often form highly visible 'green' excrescences of 'industrial capitalism as usual,' emerald islands in highly oil-addicted wider regional contexts." The mechanisms of land acquisition for eco-cities—"land markets [and] reclamation, appropriation and dispossession"—are among their least green features. Furthermore, construction of China's eco-cities has been accomplished by large migrant populations of low-wage workers who live in ramshackle temporary quarters on the peripheries of the cities they are building.[38]

Often the eco-city's ostensible mission—to serve as a good example—is adulterated by its more urgent mission, one more attractive to developers: to be an economic stimulus. Tianjin's eco-city and port lie within a special economic zone, the Tianjin Binhai New Area, which, like most such zones, is devoted to hyperactive manufacturing and international trade. Binhai is now better known for a devastating manifestation of these forces: a series of titanic chemical explosions that ripped through the port on August 12, 2015, killing more than a hundred and setting off what the *New York Times* called an "apocalyptic fire" that churned out dense smoky clouds containing a "witch's brew" of toxic gases.[39] The eco-city, eight miles from the blast, sustained only superficial damage to buildings and one reported injury. And if anything, its intended role in Chinese society—to redeem the dirty image of Tianjin—became more immediate, if no more practicable.[40]

DESPERATE MEASURES

It is conceivable that the Earth, through degradation of its ecosystems, could reach the point at which affluence would no longer serve as a shield against catastrophe, and (outside the world of science fiction) there can be no managed retreat from an entire planet. Desperate to halt the degradation of the ecosphere before we reach the point of no return, a handful of highly credentialed scientists are suggesting a way out—a way that leads through the looking glass into the world of geoengineering.

Because the world's nations have so far proved incapable of restraining, let alone reducing, the quantities of greenhouse gases their economies generate, some prominent figures in the worlds of science and money have urged that we stop mourning our ability to influence the climate and instead embrace it, using our technological prowess to cool a planet that we've spent a century heating up. Geoengineering projects—defined loosely as "large-scale efforts to diminish climate change resulting from greenhouse gases that have already been released to the atmosphere"[41]—moved out of the realm of mad-scientist fantasy and achieved scientific respectability with a 2006 essay in the journal *Climatic Change* by Nobel laureate Paul Crutzen. He suggested that if efforts to control emissions fail, a last resort to head off climatic disaster could be the injection of aerosols of sulfuric acid into the stratosphere, where they would reflect some solar radiation back into space, leaving less heat to be trapped in our greenhouse atmosphere.[42] Crutzen was proposing one form of "solar radiation management"; others, including the launch of giant mirrors into orbit between the sun and Earth, have also been nominated as ultimate solutions. (The second most widely discussed class of geoengineering ideas involves proposals to remove carbon dioxide from the atmosphere by fertilizing the oceans with nutrients that would encourage growth of microscopic phytoplankton, which, in turn, would fix vast quantities of carbon through photosynthesis and take it with them to the seafloor when they die.)[43]

To those seeking an easy way to prevent disasters—a technological workaround that would reduce the extent to which greenhouse emissions must be cut—solar radiation management through aerosol injection is a very attractive idea. There is excellent evidence, provided by volcanic eruptions and industrial pollution, that the aerosols would have a strong cooling effect. The drop in global temperatures would happen very quickly, within weeks or months after injection.[44] And, compared with cutting carbon emissions, recapturing carbon once released, or putting gigantic mirrors in orbit to deflect sunlight, aerosol injection would be almost ridiculously cheap to accomplish; in their long-term calculations, economists are able to treat the injection process itself as costless.[45]

Both friends and foes have characterized geoengineering by drawing medical analogies. It has been compared, for example, to an experimental drug administered to a dying patient, or alternatively to a last resort such as the morning-after pill. Putting potential geoengineering technologies in such desperate contexts helps shield them from the critical scrutiny to which most major technological interventions are subjected.[46] A more apt symbolism might be that of an antipyretic drug that, while capable of bringing down the Earth's fever, does nothing about the disease that is causing the fever. Naomi Klein views geoengineering's increasing respectability as a typical case of the "shock doctrine" at work: "In the desperation of a true crisis all kinds of sensible opposition melts away and all manner of high-risk behaviors seem temporarily acceptable."[47]

For many who have examined the practicalities of geoengineering, the myriad geoclimatic disasters that it could potentially spin off are enough to disqualify it even under the most urgent scenarios. Aerosol injection in particular has far more critics in the scientific world than it has supporters. That's because, whether or not it is capable of bringing down the Earth's temperature, it would almost certainly have far-reaching unintended consequences. We now know, for example, that in the 1960s and 1970s, industrially generated aerosol pollution drifted from North America and Europe over the North Atlantic, shaded and cooled the sea surface, and thereby disrupted the complex climatic processes that produce rainfall over Africa's Sahel region.[48] The resulting drought led to widespread famine. Intentional aerosol injection could have similar catastrophic drying effects on both the African and South Asian monsoons. High-latitude countries such as Russia, whose agricultural productivity is largely limited by a shortage of warm weather, would also stand to lose if global warming is suddenly stopped or even reversed through blocking of sunlight.

Other predicted side effects of sulfate aerosol injection include increases in acid rainfall, increased eutrophication of lakes, drought in the Amazon basin, and a stirring of tropical cyclone activity in the Arabian Sea and South Atlantic. (Other approaches to solar radiation management, including space reflectors, cloud whitening, and ocean whitening, would lead to similar consequences.) Then there are those problems that solar radiation management would not correct even at its best. Because it would not reduce carbon dioxide concentrations, it would not curb ocean acidification. It also would not stop one of the most alarming phenomena caused by planetary warming: the disintegration of the West Antarctica ice sheet, which is melting from heat already stored in the ocean, not what will be coming into the atmosphere. Finally, there is the problem of addiction. Aerosol droplets launched into the stratosphere would not stay permanently aloft but would fall back to Earth, so injection would have to be repeated regularly, perhaps once a year. And that would be a perpetual commitment. If increases in atmospheric greenhouse gas concentrations are allowed to continue while their impact is being hidden behind an aerosol curtain, any cessation of the injection program would mean sudden, dramatic warming under an even more carbon-heavy sky. The climatic disasters resulting from this much more rapid warming would be even worse than those predicted under the more gradual warming we can expect if we carry on with business as usual.[49]

One of aerosol injection's biggest attractions, its relatively low cost, could also create one of the biggest potential nightmares. In the absence of global regulations, there would be nothing to stop the government of any medium to large country, or even a private corporation, from putting geoengineering into practice if the perpetrators saw an opportunity to benefit, regardless of the catastrophes it might set off in other parts of the world. The potential for international conflict, which could range from trade wars to a geoengineering "arms race" (that is, an intentional release of greenhouse gases by a rival country to boost temperatures) to actual armed conflict, is obvious.[50] Even if an international consensus on aerosol injection could be reached, and even if geoengineers found they really did have the ability to adjust the global thermostat to a desired setting, it would create a strong incentive for the world's governments to drop their much more contentious, costly, incremental efforts to reduce greenhouse emissions.

Solar radiation management's more judicious proponents are careful to stress that any such effort must be framed as an effort to buy time for the long transition period required to bring atmospheric carbon down to safe concentrations. But if the looming threat of an unprecedented rise in global temperatures has not been a strong enough incentive for the world's governments to enforce reductions in emissions, why would they possibly do it if geoengineering prevails, eliminating even that incentive? When hardheaded technocrats start weighing the profound economic disruptions that will inevitably result from emissions reduction efforts against the easy, cheap, business-friendly strategy of aerosol injection (with its bonus of instant gratification), the latter will become more and more appealing.[51]

RISK AND THE ART OF SPUD TOSSING

In the United States and other affluent countries where economic losses from geoclimatic hazards have escalated even as loss of life in disasters has fallen, property insurance has come to play a more and more important role. By pooling disaster risks across geographical areas or across different types of hazards, insurance makes it possible for individuals and communities to lower the probability that they will be wiped out economically by a single event, so that they can live with less uncertainty about the future. But it's well established that the security provided by insurance can also induce complacency. When shielded from some of the consequences of risky construction or behavior, homeowners, businesses, and local governments have less incentive to adopt preventive or risk-reducing measures.

Extreme weather events of the early 1990s, led by Hurricane Andrew's 1992 passage through South Florida, 1993's "Storm of the Century" in the eastern states, and the record 1993 floods throughout the nation's midsection, tipped the U.S. insurance industry into crisis.[52] Some companies went bankrupt, and those that survived set about dropping coverage in high-risk areas and raising rates elsewhere. But climate change awareness was beginning to blossom, so despite its woes the insurance industry was increasingly depended upon to play a prominent role in risk reduction. This could be done, it was believed, by firms providing information and analysis; making coverage conditional on the insured undertaking risk-reducing actions; giving customers or communities who reduce their risks a break on their premiums; and, most important, charging premium rates that accurately reflect risk.[53] But over the next two decades, the industry's record on risk reduction was spotty. For example, a worldwide survey of flood insurance programs found that two-thirds included no risk reduction provisions; in most of those that did attempt to cut risks, the only mechanism was through pricing of premiums.[54]

Free market advocates argue that insurance companies should charge every policyholder a premium that completely accounts for the estimated probability and cost of disaster damage faced by the insured household within the policy period. So-called fully risk-adjusted or actuarial premiums, it is argued, would encourage policyholders living in higher-risk situations to take measures to reduce their premium payments by reducing their physical risk. Precise estimates of catastrophic risks faced by individual people or businesses are difficult or impossible to achieve, however, and estimating individual risks would be very costly. Charging actuarial premiums would also be unfair to people who are stuck in exposed, vulnerable situations and cannot afford the actions required to reduce their risks.[55] For many of those living or doing business in low-lying coastal areas, in fire-prone landscapes, near geological faults, or in other places exposed to hazards, actuarial premiums would be impossibly high. And insurance is usually nonexistent in most low-income countries. Even when coverage is available in poorer regions or nations, people with exposure even to moderate risks cannot afford to pay risk-adjusted premiums.

To reduce the burden, governments can subsidize premiums, as has been done under the U.S. National Flood Insurance Program (NFIP). Subsidization helps address fairness, but it brings back the problem of complacency.[56] By its very nature, the insurance industry disproportionately serves those who own more personal property or physical capital, because they stand to lose much more. Wage earners who rent their residences face as much or more suffering in a disaster as wealthy homeowners and business owners, but most post-disaster recovery efforts, from Staten Island to L'Aquila to Miami, have provided renters little or no assistance.

Global surveys have found few examples of public or private insurance providers advancing mitigation measures in the face of climate change.[57] NFIP is a partial exception, having long attempted to promote prevention in several ways. For example, to qualify for the program, communities must adopt building standards appropriate to their degree of flood risk. NFIP, which is administered by FEMA, also has a voluntary Community Rating System that awards lower premium rates for prevention measures that go beyond minimum requirements. But good intentions can take a tumble when headline-making disasters crop up. The 2005 hurricane season sent NFIP deep into debt, and Congress struggled for years to make the program once again self-sustaining. The 2012 Biggert-Waters Act started that process by phasing out premium subsidies on policies for businesses, second homes, and properties that had had repeated flood losses. For those policyholders, rates would be raised by 25 percent per year until they reached the point of reflecting actual risk. Furthermore, rates would rise to meet actual risk immediately in the case of new policies, ownership changes, or payment lapses. Before those changes could even come into effect, the enormous covered damages from Superstorm Sandy prompted Congress to appropriate additional funds to keep NFIP afloat and to increase the agency's borrowing authority temporarily. Then, as subsidies started being cut over the following fifteen months, many policyholders were jolted by five-digit annual premium bills that more closely, and in some cases completely, reflected actual risk. The resulting outcry, much of it from affluent and influential constituents, became too much for Washington. In March 2014, Congress passed a bill co-sponsored by Staten Island representative Michael Grimm (see page 78) limiting annual premium increases for any policy to 18 percent and average increases to 15 percent.[58]

After the disaster outbreak of the early 1990s, insurers sought backup, and that meant turning more than ever to the reinsurance industry. The giant companies that insure insurance companies against extreme losses (led today by Munich Re, Swiss Re, Hannover Re, Lloyd's, and Berkshire Hathaway) were more diversified, both geographically and by types of risk, and therefore could count on surviving just about any calamity short of an asteroid impact. But with global climate disruption jumping out of the realm of computer models into reality, even the biggest reinsurers have begun wondering if they can afford to offer protection against rapid-fire, high-powered geoclimatic hazards while maintaining premium rates that insurers can afford. As the Earth warms, the hot potato of catastrophic risk that has been tossed from home and business owners to their insurance companies and then on to reinsurers can still burn. So where does a giant corporation go to quench even the scariest threats and maybe even turn them into opportunities? To the financial markets.

In the 1990s, hedge funds and other deep-pocketed investors began to see opportunity in the insurance industry's predicament. If they could purchase bonds that paid well and whose principal would be at risk only in the event of a major climatic or seismic disaster striking a given geographical area within a given time period, that would make an excellent buffer against other economic threats. Trading on the occurrence of rare geoclimatic hazards is a good hedge because they don't tend to coincide in time or space with purely financial crises that cause stocks or other investments to crash.[59] As one analyst put it, "Hurricanes don't care what the markets are doing."[60]

Because they are risking the loss of their entire principal if the worst happens, disaster investors are offered high returns on their money. Their potential losses, though huge, are extremely unlikely to materialize. And even if catastrophe does strike, the loss is not likely to coincide with large losses in the rest of the investor's portfolio. Meanwhile, reinsurers have the opportunity to dilute their risks in the vast ocean of world financial markets.

Catastrophe bonds have become the most well known among various types of so-called insurance-linked securities. As catastrophe bonds were originally conceived, if a specified type of disaster occurs in the specified region, the first layer of damages, up to a fixed dollar value, remains the responsibility of the "sponsor" on whose behalf the bond is issued (usually an insurance or reinsurance company). Losses beyond that so-called trigger level are paid out of the money that has been put up by the bond's investors and held in a kind of escrow by a "special purpose vehicle"—a company created for this purpose in an offshore location, often in Bermuda or the Cayman Islands. The sponsor's potential losses are confined below a fixed ceiling, and there are no worries that damages costing more than that won't be paid. If there are no losses above the trigger level, the investors continue to draw their premium payments and then, at the three-to-five-year maturity date, they get their principal back.[61]

Issuance of catastrophe bonds stood at between $1 billion and $2 billion annually in the decade between 1990 and 2000; by 2014, they had hit $10 billion. A bond covering Florida hurricanes, issued in May 2014 by Bermuda-based Everglades Re, was the largest one issued to that point. Worth $1.5 billion, it was sponsored by Citizens Property Insurance Corporation, a nonprofit entity created by the Florida state legislature in 2002 to provide insurance, mostly against hurricane wind damage, to property owners in the state who cannot obtain private insurance. The risk finance website Artemis, based in Bermuda, commented, "This deal saw Florida's Citizens Property Insurance Corporation massively increase its reliance on the capital markets as a source of reinsurance capital."[62] There have even been efforts to get the U.S. government into the business. In 2012, Frederica Wilson, who represents northern Miami in the U.S. House of Representatives, introduced H.R. 737, a bill to establish a National Catastrophe Risk Consortium, which would allow participating states to transfer disaster risk to capital markets. It would issue the equivalent of catastrophe bonds on behalf of states with the goal of protecting insured homeowners. But the bill languished, with no co-sponsors.[63]

With market uncertainties being compounded by a climate that appears to be behaving more chaotically all the time, one would reasonably expect that speculating against phenomena such as record wind speeds or extraordinary rainfall might eventually become too risky even for the big-time investor. Not so—the greater the calculated risk, the greater the return on investment, so a more hazardous future is a magnet for catastrophe bond investors. The chance of losing the principal still remains very small, and the market is open to playing for very high stakes. As one investor told Leigh Johnson, a geographer at the University of Zurich who has written extensively on insurance-linked securities, "The beauty of cat bonds is that you can put $500 million in!"[64]

Insurers and reinsurers use catastrophe bonds to protect them against the rare occurrence of losses so big that they can't be affordably covered through other means; the companies maintain control of the income from premiums and assume responsibility for managing the risks. But just as the pre-2008 mortgage industry pushed risky loans for the sole purpose of bundling and marketing them to derivatives buyers who had no relationship with the homeowners whose risk they now owned, some analysts are urging the insurance and reinsurance industries to begin creating catastrophe bonds to cover risks that they themselves would not be willing to cover, and then passing these "products," along with all of their income and risk, on to the financial markets. This, writes Johnson, would amount to "creating cat bond structures that would encourage insurers and reinsurers to underwrite new business expressly for the purpose of securitization—a further step toward true 'convergence' of the capital markets with (re)insurance." In the process, new kinds of previously uninsurable risks and regions would ostensibly become insurable, just as mortgage-backed securities had made it possible for people previously shut out of the housing market to obtain loans. Investments in a collection of different bonds could be bundled and marketed, putting greater financial distance between the geoclimatic hazard and the person holding the risk potato. When a catastrophe bond investment is packaged with other seemingly unrelated ones, the bundle as a whole is meant to appear safe. The market in mortgage-backed securities crashed because risks attached to apparently independent mortgages across the country were not actually unrelated at all; enough loans could go into default within a short time span to take down the whole bundle. Proponents of catastrophe bond derivatives claim that this time things will be different, because there can't be outbreaks of "natural" disasters of various kinds across geographically diverse regions.[65] But the history of the boom and crash of the 2000s still leaves a nagging worry that if this new attempt at risk spreading is adopted it could actually concentrate risk.[66]

Johnson doesn't believe that financial instability would be the only result of increased financialization of disaster insurance. It would also reduce incentives to mitigate climate change and geoclimatic hazards while making excessive risk more attractive:

> If cat bonds and insurance-linked securities in general do reach a "tipping point" of the kind some seek, in which underwriting to securitize climate risks becomes a principal strategy for insuring vulnerable landscapes, the prospects for climate-appropriate development and adaptation would seem dim. If high yield is a function of high risk, then it is credulous to imagine that [insurance-linked securities] will ever systematically encourage less risk taking.... [I]nvestors' ongoing interest in high-paying peak perils could ironically make more capital available for paying claims, rebuilding, and new underwriting in places like Miami, Florida, where the highest concentrations of value are being made even more vulnerable by climate change.[67]

THE NATION THAT RISKS TOGETHER INSURES TOGETHER

If protection from disaster is to be taken seriously as a right that we all share, then the time has come for the United States to consider establishing a publicly administered, universal mechanism to secure that right. With market-based pricing of insurance premiums having largely flopped as a means of encouraging disaster prevention, there may be good reason to examine the idea of social disaster insurance. Responding to disasters as we do now—through a cobbled-together nonsystem of private insurance, public insurance for "uninsurable" risks, reinsurance, presidential and gubernatorial disaster declarations, finance markets, and charity drives—makes recovery highly inconsistent, with lower-income areas and families often bearing the worst burdens. The trade-off between preventing catastrophic losses and achieving economic fairness has not yet been resolved. If the big goal is disaster prevention, and if disaster risks are going to be getting heavier in unpredictable ways in the decades ahead for some people while remaining the same or even lightening for others, private insurance providers will find themselves wholly incapable of sending market signals that will steer policy in a useful way. Government-sponsored plans, by contrast, don't carry the burden of turning out profits for investors; they can spread losses across broad swaths of time and space and are usually tax-exempt. Governments also have the unique ability to compel participation in insurance programs and risk reduction (including the prohibition of development in high-hazard areas), and that is absolutely necessary if such efforts are to succeed.[68] To work properly, though, they must be securely funded and not be riddled with loopholes.

The many problems that the National Flood Insurance Program has had should not be a reason to reject public insurance; instead, NFIP's experience can provide some ideas for improving public disaster protection. It already includes some egalitarian features: a $250,000 payment limit on structural damage (so that it doesn't subsidize full replacement of the most expensive properties) and $100,000 coverage for personal property that is available to renters. Over the past decade, Howard Kunreuther of the University of Pennsylvania's Wharton School, along with several co-authors, has been suggesting changes to NFIP that, he maintains, would make the program more fiscally sound while neither burdening policyholders with huge premium costs nor encouraging risky behavior through cheap premiums. Elements of Kunreuther and colleagues' proposal are (1) fully risk-adjusted premiums that inform everyone of the true degree of hazard faced by every property but, importantly, with a voucher program to assist policyholders of modest income who live with high flood risk to pay their premiums; (2) attaching policies to the property rather than the owner, so that when a house is sold the insurance stays with it; (3) multiyear rather than year-by-year contracts; and (4) incorporation of loss prevention loans to homeowners.[69] Those fixes would strengthen NFIP, but the program would still face two huge problems: it covers a single class of hazard—only floods—and its customer base is made up largely of property owners who are highly likely to suffer damages and file claims. Similar measures as part of a more comprehensive program covering diverse disasters would be better buffered and even more effective.

Were we to design a much more comprehensive public insurance program, it should cover all types of geoclimatic hazards, and it must be universal—so that, as with universal health insurance, risks and premium burdens could be shared broadly enough across regions, demographic groups, and types of hazards. That way, coverage could be made affordable for all. The view from a 2013 report published by Australia's National Climate Change Adaptation Research Facility is that universal coverage is possible, but that it would have to include "a high degree of compulsion" to ensure that measures are taken to reduce risks of future losses.[70]

Directing disaster insurance only toward properties at greatest risk is not viable in the long run, any more than providing health insurance only for those most prone to illness and injury. As with health insurance, sustainable disaster insurance must have universal participation. Bringing communities into the system that are at low risk of disaster would not be as odd as it might seem. The New Zealand government's Earthquake Commission, for example, provides nationwide coverage for damage from earthquakes, volcanic eruptions, tsunamis, storms, floods, and fire to all of the country's residential property insurance policyholders; the system is paid for through a surcharge on insurance premiums. In Spain, all holders of private property insurance policies are required by law to have coverage for "extraordinary risks," including geoclimatic hazards; the coverage is provided by the state-run Concorcio de Compensación de Seguros, which is funded by surcharges on property insurance premiums. France has a similar system.[71]

Would it be possible for a country such as the United States to develop a public, single-payer disaster insurance system that features universal coverage, low-income subsidies, payout caps, and enforced risk reduction—all of which would keep premiums down? If it did so, the government would put itself in a much stronger position to require tougher building standards and prohibit construction in high-risk areas while providing ample authority and funds for preparedness, property buyouts, and relocation. The system could be made progressive by setting strict payout ceilings for repair and replacement, as does NFIP already. There could also be separate coverage for personal property, and a three-tier premium structure for homeowners, landlords, and renters.

Such a proposal obviously would trigger a political volcano under today's conditions and would most probably meet the fate of single-payer health care proposals. But as the numbers of people potentially affected by disasters grows, we could see a majority emerge in favor of a comprehensive system. We outlined this idea to several experts in this area—Leigh Johnson; Lisi Krall, a professor of economics at SUNY Cortland; and Caroline Kousky of the nonprofit Resources for the Future—and asked them for feedback.[72] While they agreed that it would be necessary to deploy the stick of tougher regulation along with the carrot of universal insurance in order to avoid subsidizing risky behavior, some were not confident that regulation, along with our proposed premium subsidies and caps on damage payouts, would adequately solve the risk problem or fully ensure fairness to the least affluent communities and renters. But a bigger concern for all was that the system would not resolve a matter that, as we have seen, is at the heart of all insurance issues: the political dilemma of setting premiums. A uniform premium would be welcomed by those who live in hazard-prone conditions but, our experts felt, it would be unfair to those living in safer areas. On the other hand, charging fully risk-adjusted premiums for insurance that is mandatory would be harsh on people who live in potential disaster zones out of necessity rather than choice, and lower-income households would be hit especially hard. For example, a low-lying area prone to flooding may sometimes be the only place where a family can find housing that's close enough to a workplace and also affordable; others may not be able to bear the cost of moving out of a neighborhood near an oilfield that has started producing earthquakes. Furthermore, property-by-property or even community-by-community risk adjustment for all types of hazards is difficult or impossible (and very expensive) to achieve or maintain with any precision; it was pointed out that such estimates are easily distorted by political and special-interest pressures. (But in other ways, politics sometimes works. It took a buyout plan to make it financially possible for those Sandy-affected Staten Island residents to move out of their badly exposed neighborhoods, and that was certainly better than forcing them to stay and pay risk-adjusted premiums.)

We understand the difficulties involved in setting premiums, but we still believe the system we propose can be designed to work. In a universal disaster insurance system, premium rates could be smoothed out across the entire population—that is, they can be closer to uniform than to fully risk-adjusted. We suggest that premiums be adjusted according to income, with those subsidies backed up by increased revenues from the more progressive system of taxation that we need anyway. Universal coverage with universally affordable premium rates and a payout ceiling would be a forthright acknowledgment of the reality that disasters and, increasingly, many of the geoclimatic hazards associated with them are produced by societies as a whole. Most Americans are contributing more than their share to catastrophic events across the nation and world (with the more affluent contributing more), not only through climate disruption but also through the countless profitable alterations of the Earth's lands and waters that are setting the stage for fresh disasters.

There are additional reasons for universality. For example, as geoclimatic hazards become more frequent, more destructive, and/or less predictable, it will become harder and harder to say which communities are at high risk of catastrophe and which are at low risk. Furthermore, all of us, even those who live in relatively "safe" regions, have a stake in helping disaster-struck communities recover and rebuild; we do it already with our tax payments every time there's a federal disaster declaration. The government, with taxpayer funding, is already the nation's crucial backstop in times of catastrophe. We should have this system in place (to cite just one horrifying prospect) before there is a full rupture of the Cascadia subduction zone in the U.S. Northwest; the resulting quake and tsunami would cause total destruction west of Interstate 5, wreck the homes of a million or more people, and knock out public services and infrastructure for months or years. Given the likelihood of that event—a 10 percent chance sometime in the next fifty years[73]—and of other extreme catastrophes, it would be folly to continue depending on our current insurance industry and ad hoc federal assistance.

Insurance alone cannot patch over the increasingly diverse array of disasters that big capitalist economies are generating, much less halt the global ecological crisis. But if the United States were to follow the lead of New Zealand and other affluent countries to establish comprehensive public disaster insurance and prevention, it could represent a strong first step toward protecting all Americans, without encouraging or subsidizing construction in risky locations or extravagance anywhere. Meanwhile, other people and communities around the world, from the Philippines to Pakistan to Haiti, remain exposed and vulnerable to geoclimatic hazards, and solutions that work in the rich world may not provide either prevention or adequate response for the world's impoverished majority, no matter how resilient they are. People in Tacloban and Port-au-Prince need national and international disaster policies very different from those that might apply in Seattle or Christchurch.

10

THE ABSORBERS

Mumbai, India, and Kampala, Uganda

Yes, the flooding is caused by encroachment. It's the developers who are the real encroachers. And they are doing it only for the benefit of the elite few. Who is the SeaLink bridge for? Who is the Western Express Highway for? People with cars are a tiny minority.

—Sachin Kadam, Slum Rehabilitation Society, Mumbai, 2014[1]

It has been raining heavily in the past few weeks with some areas flooding to a point where houses are submerged. We see this on social networks, TV and close to home when our cars are submerged where waterlines are above the car tires. I see people driving through these floods like it's a competition to show who's got the more macho car. However, depending on how high the water level is, the smart thing is not to drive through flood water. Wait and later drive when the water has receded.

—Allan Rogers Kibaya, *Daily Monitor*, Kampala, November 19, 2015[2]

Affluent countries have available a thick catalog of disaster policies to debate and try out. Some policies are effective, some a waste of money, some potentially catastrophic. For others around the world, the options are far more limited. Are there real solutions for nations or communities who can't afford to build Brobdingnagian floodgates, floating golf courses, or shiny new eco-cities? What about families who can never imagine paying insurance premiums when they are hard pressed to buy their daily bread or rice? Or those who would like to retreat from a high-risk zone but have nowhere to go?

The global North has a long tradition of dealing with the South's geoclimatic disasters by providing funds aimed at rescuing disaster victims, alleviating their immediate misery, and maybe rebuilding some housing and schools. Now, with a complex international policy system in place, the focus is shifting to the resilience of those most affected. Any disaster policy will fail, however, unless the root causes of vulnerability are well understood and addressed.[3] In booming cities such as Mumbai, India, and Kampala, Uganda, those roots run fiendishly deep through the landscape. The most vulnerable residents are all too familiar with the origins of their disasters, but ripping those roots out from beneath the streets is more difficult and far more politically charged than most elected officials or city planners are willing to consider.

MAKING DISASTER INEVITABLE

For a geoclimatic disaster to occur, people and property obviously must be exposed to a hazard, and for a variety of reasons this happens far too often. When typhoons make landfall, they usually find people, making a living off a fertile landscape and crop-friendly climate; for example, without rains generated during cyclone season, the vast rice-producing areas of the Philippines and eastern India would be far less productive. But living on dangerous slopes, whether fertile or badly degraded, is not a voluntary choice for most of the highly vulnerable populations who do so. Across the Americas, Africa, and Asia, high-value crops claim the best valley landscapes, pushing subsistence farms onto slopes that should not be tilled in the first place.

Starting in the 1950s, farmers in Honduras were forced more and more from rich valley lands onto mountain slopes. Conversion of *ejido* (community) lands into private tracts for growing cash and export crops, steep rents, and summary eviction put agricultural lands in the hands of a wealthy few while creating large populations of impoverished, landless farmers. With lowlands increasingly devoted to bananas, cattle, melons, cotton, and sugarcane, farm families had no choice but to cultivate the highlands. Pressure to produce crops for export increased in the 1990s as Honduras struggled to deal with foreign debt and the demands of international lenders. Then Hurricane Mitch struck in 1998. Meandering over the Central American isthmus as if on a malevolent search for people and property, the storm killed an estimated ten thousand in Honduras alone, destroying or damaging half of all homes in the country and taking out 160 bridges.[4] In total, eighteen thousand lives were lost along the storm's path; with half of those deaths, victims' bodies were never found.

Mitch remains the deadliest Atlantic hurricane since 1780, when a storm killed more than 20,000 in Barbados and Martinique.[5] But the toll should not have been that high. The suffering was as much a result of economic relations in Central America as of wind and rain. In Honduras, according to anthropologists Bradley Ensor and Marisa Ensor, deforestation played a key role in the devastating effects of the intense rains borne by Hurricane Mitch. . . . The connection between unsustainable natural resource-use practices, environmental degradation, and heightened conditions of vulnerability supports the premise that "natural" disasters like Hurricane Mitch are fundamentally social in nature . . . [T]he increased concentration of lands in the hands of fewer wealthy landlords, the dwindling access to agricultural lands for the majority of peasants, and the lack of sufficient employment opportunities all led to high levels of unemployment and job insecurity. Many Hondurans were forced to seek livelihoods and residential locations on hillsides and mountainsides, resulting in deforestation that in turn resulted in unstable slopes, landslides, and mudslides in the presence of a natural catalyst like Mitch.[6]

Impoverished city dwellers fared no better: "Since land suitable for construction purposes is scarce and expensive, many of these shantytowns have been built on highly unstable hill slopes that are prone to landslides and mudslides during the rainy season. In fact, crowded squatter settlements built on the steep hillsides that surround the center of Tegucigalpa were among the neighborhoods most severely damaged by Hurricane Mitch."[7]

When they strike urban areas, geoclimatic hazards encounter an especially target-rich environment. In and around cities across the Earth's surface, three things are consistent: where landslides are the main hazard, slums cling to steep slopes; where the threat is flooding, slums are found on the lowest ground; and where earthquakes are likely, poorly constructed houses huddle on the least stable ground. In his book *Planet of Slums*, environmental historian Mike Davis takes just seven pages to prove that in poor urban communities disaster is inevitable—whether residents are being washed out by floods in Buenos Aires, Dhaka, Manila, Khartoum, or Vijayawada, India; losing everything to a landslide in São Paulo, Rio de Janeiro, Caracas, Algiers, or Ponce, Puerto Rico; or seeing their families crushed by an earthquake in Mexico City, Guatemala City, Lima, Istanbul, or Johannesburg.[8] Following Davis, we now turn our attention to two other major cities where the contours of inequality, vulnerability, and risk are razor sharp: Mumbai and Kampala.

DROWNING IN DEVELOPMENT

In July 2005, more than three feet of rain fell in a single day on Mumbai, India. More rain fell in eighteen hours than is received by most Indian cities in a year, and it was the heaviest one-day rainfall ever recorded in a country that contains some of the wettest places on Earth. Water depths reached twelve feet in several low-lying areas of the city and suburbs. An estimated 419 Mumbai residents died of drowning or other trauma. To wade or swim through the water to higher ground, people formed long human chains, if they could maintain their grip. Many could not. More than a million people were stranded overnight in cars, in offices, on tops of buses—wherever they could avoid the fate of those they saw swept away in the swift currents that ran through the streets.[9]

The city's vulnerability to flooding had long been a painful reality to residents of its slums, which occupy only 10 percent of Mumbai's land area but are home to more than half its population. Many of those areas are low-lying, poorly drained, and frequently inundated by even routine monsoon rains. But the sheer scale of this disaster and the fact that it brought an entire metropolis, India's biggest and richest, to its knees was enough to send storm drainage straight to the top of Mumbai's long list of urgent issues. Media attention in July and August 2005 focused on flooding in the historic city center, the financial district, the airport, and other cogs of the local, national, and global economy (much as with Sandy in New York City). But it was in the less glamorous working-class neighborhoods where the suffering was greatest.

Mumbai is squeezed onto a narrow peninsula reaching south from India's western coast into the Arabian Sea. The lower part of the peninsula, South Mumbai—the longtime seat of government and commercial power and chief destination for foreign tourists coming to see the Gateway of India and hang out at Café Leopold—was once seven small islands. Through dredging operations that began in the early 1800s and have never completely stopped, the islands were joined to one another and to the mainland. The former islands now constitute the city's slightly higher-elevation and relatively affluent neighborhoods, with Malabar Hill being the most well known. The reclaimed lands surrounding the high ground are vulnerable to monsoon flooding. Some northern parts of the city where the worst flooding occurred in 2005 are also reclaimed lands; they are home to some of the city's most valuable real estate as well as its largest slums. Those areas are drained, or are supposed to be drained, by the thirty-mile-long Mithi River, which emerges from the highlands of Sanjay Gandhi National Park north of the city and empties into the Arabian Sea through Mahim Bay.

In a report that is still the most comprehensive investigation into the causes of and response to the 2005 flood, a Concerned Citizens' Commission of nongovernmental experts emphasized the crucial roles that three of the city's premier development projects played in causing the disaster. Chhatrapati Shivaji International Airport was originally built on a wetland. Repeated expansion of the airport required extensive rerouting of the Mithi, to the point that the once nearly straight river was for a time forced to make four right-angle turns; it was later restraightened by running it through a tunnel under one of the runways. A little farther downstream, construction of an enormous business park began in the 1970s and continues today. The area, called the Bandra-Kurla Complex (BKC), is home to steel-and-glass office blocks containing 1.75 million square feet of space for government, commercial, and high-finance activities. It is also located on land reclaimed from a swamp, and it straddles the river. The National Stock Exchange is in BKC, sitting smack atop the original course of the Mithi. Then there is the spectacular SeaLink toll bridge, which spans the wide mouth of Mahim Bay and allows north-south commuters on the Western Express Highway to by-pass the traffic congestion of Greater Mumbai's midsection. (Critics say that SeaLink simply funnels more cars more efficiently into South Mumbai, in-tensifying the already horrific gridlock there.) The environmental clearance issued for SeaLink in 1999 allowed for 11.6 acres of new land to be dredged up for supporting the two ends of the bridge; instead, and in violation of the permit, 66.7 new acres have been created. The row of ninety pillars support-ing the 3.4-mile-long structure has reduced the remaining effective width of the river mouth by one-third. The land reclamation and pillars together have caused the Mithi River to back up and flood during heavy rains, block-ing inland flood waters from draining out.[10]

The commission concluded that future floods the size of those in 2005 could not be prevented without demolishing parts of BKC and the airport and stopping construction of the SeaLink bridge and further expansion of BKC into former mangrove areas. Those recommendations were ignored. Authorities have continued "training" the Mithi, making it easier for devel-opers to keep building while further reducing the river's capacity to handle runoff. The peninsula's natural drainage patterns have long been erased. At the time of the big flood, the Mithi River and intertidal zones no longer even appeared on the city government's land-use maps.

Because slum dwellings are constructed with more porous materials and often surrounded by bare soil rather than concrete roads and parking lots, the commission wrote, they are "generally more 'absorbent' and serve as a sacrificial buffer in heavy rain," helping to drain rich Mumbai and preserve the city in its fragile state. Now, according to the commission, "replacement of slums by water-repelling [multistory] middle and upper class develop-ments increases population and infrastructure requirements and further worsens the flooding problem." Slum settlements in Mumbai have tended to be built on low-lying land near water channels, which is often the only space available to families just arriving in the city to seek work. City officials and middle-class Mumbai residents often complain that the cause of the 2005 flood and more routine inundation is encroachment by slum settlements along the banks of rivers and creeks, where, it is argued, they constrict the water's flow and their trash clogs the waterways. Those who actually have to live in the "sacrificial buffers" don't see it that way at all. Slum dwellers' advocate Sachin Kadam told us, "Let people say whatever they want. Our concern is how to get our two meals a day. It's the developers, not we in the slums, who are doing most of the construction near waterways and on wetlands."[11]

After the big flood, disaster mitigation expert Aromar Revi wrote that for decades, "Mumbaikers, pragmatic, vital people that they are ... went about life as usual, all the way into the early evening of July 26, 2005. But by nightfall, it was clear to many million people in Mumbai that life may never be quite the same again. An exceptional rainstorm finally put to rest the long-prevailing myth of Mumbai's indestructible resilience to all kinds of shocks."[12] In the minds of many, an event that overwhelming should have served as a wake-up call. But while it unmasked Mumbai's increasing vulnerability, the flood appeared to stiffen the city's resistance to any policy changes that might undermine its famed moneymaking prowess. Heedless maldevelopment has continued. Meanwhile, sea level rise combined with increased rainfall will dramatically increase the extent and depth of flooding, doubling the likelihood that a flood on the scale of the 2005 catastrophe will recur.[13]

LIFE IN THE URBAN WETLAND

Dharavi is the largest slum in Mumbai, in India, perhaps in the world. Home to at least one hundred thousand families, the sprawling city within a city comprises hundreds of distinct neighborhoods; it gained international notoriety as the setting of the 2008 Academy Award–winning film *Slumdog Millionaire*. Much of Dharavi, which is situated near the Mithi River and adjacent to the Bandra-Kurla Complex, was deeply inundated by the huge 2005 rainstorm.

In early 2014, we talked with Shamshaad Sheik, Zahida Qureshi, Mumtaz Qureshi, Halema Sheik, Aabeda Qureshi, and several other women who live in Annanagar, a Muslim neighborhood of about five hundred families in Dharavi's Sector 3. They all lived through the big flood, when the water reached at least chin-high in this neighborhood; in some places, it rose above the women's heads.[14] In Dharavi, any heavy rainfall brings the contents of the open sewage gutters up into the streets and lanes. So in 2005, it was dangerously filthy water that filled the ground-floor homes. Disease spread through the neighborhood and city.[15] Community taps, the only water supply at that time, were rendered nonfunctional, but with the monsoon going at full throttle the women were able to collect plenty of rainwater.[16] Once the flood had receded to waist level, said Halema Sheik, she put a piece of plywood on top of her rain barrel and kept her kids sitting there for a couple of days while she stood half immersed in the flood. The sewage-laden waters didn't recede for three days.

The typical home in Dharavi consists of a single room in a long, fully attached, two-story row among innumerable such rows—all built without sanction from city authorities.[17] In many cases during the big flood, families fortunate enough to live in upper-story rooms gave shelter to lower-story families, but that was not always possible; after all, most homes were already packed to capacity even under normal conditions. Several women echoed the frustration of one who told us, "Nobody came to *my* rescue!" Another said, "The family living above me looked after some other people's kids but not mine." Complicating the situation, they all agreed, was that it was considered inadvisable ever to send young daughters to take refuge in others' homes even during a disaster, because, they feared, sexual abuse could occur.[18]

The impact of the flood persisted much longer than three days. Residents endured hassle after bureaucratic hassle. Those who had lost ration cards were unable to get replacements for at least two years. Without cards, they lost access not only to inexpensive foodstuffs but also to subsidized kerosene for their cookstoves; the fuel cost sixteen rupees per liter with a card and eighty without one. For two years, many had to cook over smoky, smelly coal, subjecting themselves and their families to potential respiratory damage because they could not afford nonsubsidized kerosene.

The 2005 flood was unique only because of its depth. The women were all too accustomed to dealing with routine monsoon misery, which one of them described as "four months of struggle every year." That struggle was becoming more like a siege. Greater Mumbai, which by official count had experienced 27 extreme weather events between 1960 and 1969, saw 131 such events between 2000 and 2009.[19] Dharavi has no storm drains, so even normal rainfall quickly fills the narrow streets and pathways. Some rooms lie below ground level, and in many areas water runs into every ground-floor room, through doorways and through the drain holes that are meant

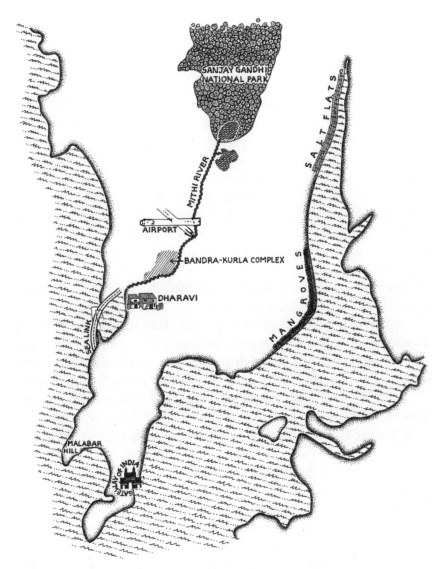

Figure 8. The city of Mumbai, India, highlighting the area along the Mithi River where the city's most severe monsoon flooding occurs. The airport, Bandra-Kurla Complex, and SeaLink have worsened the routine flooding and increased the probability that there will be a repeat of the catastrophic 2005 flood.

to let water out when washing the floor. During routine monsoon rains, everything between the buildings is hidden under brown water, so people often find themselves stepping or falling into the filthy gutters. The worst part, said the women we spoke with, is finding a place to go to the bathroom. The community toilet facility they would normally use—not too different from the type immortalized on-screen in *Slumdog Millionaire*—is flooded to overflowing during much of the rainy season.

Trading safety, health, and normalcy for an affordable home and a chance to make a living is a story endlessly retold in the Dharavis of the world. In Mumbai, this one slum has come to be embraced in recent years as a perverse source of pride: Dharavi has been the focus of films, celebrity visits, and some fairly ludicrous gushing by the business media over its unregulated, microscale entrepreneurial spirit. "By 2010 Dharavi was a well-established symbol, and what it symbolized was the capitalist dream: a wonderland of innovation in which resourceful economic actors deftly evade the interference of an overbearing government," wrote Daniel Brook in an analysis of this media trend.[20] Not just Dharavi but slums around the world are being rediscovered as veritable Galt's Gulches of scrappy ingenuity—particularly green ingenuity.

"Dharavi is greenest of all," Brook wrote. "How so? Because they're so desperately poor, Dharavi residents can't afford polluting private automobiles or much in the way of disposable consumer goods. Instead, like decomposers at the bottom of a food chain, they survive by recycling the things that richer people throw away. Dharavi is home to some thirty thousand ragpickers, scavengers who find and sort recyclable scraps from the city's garbage dumps." A typical reporter from Britain found in Dharavi not only "one of the most inspiring economic models in Asia" but also "the green lung stopping Mumbai choking to death on its own waste."[21] What he didn't see was evidence of the flood that had choked Dharavi residents to death less than two years before.

Dharavi may be the largest slum in the world, but its stories echo everywhere. A third of all city dwellers in the developing world live in slums, as defined by the United Nations.[22] Here one or another type of hazard is typically just a part of the neighborhood. Disasters are "the fine print in the devil's bargain of informal housing," as Mike Davis put it in *Planet of Slums*.[23] In his global survey, flooding is only one disaster risk that is combined in varying measure with many others, such as landslides and seismic risks. "Slums begin with bad geology," Davis declares.[24] Fire is a greater risk, too: "Slums, not Mediterranean brush or Australian eucalypti as claimed in some textbooks, are the world's premier fire ecology."[25] The convergence of these and many other dangers is pure economics—if not a devil's bargain, then certainly a bargain. "Squatters trade physical safety and public health for a few square meters of land and some security against eviction. They are the pioneer settlers of swamps, floodplains, volcano slopes, unstable hillsides, rubbish mountains, chemical dumps, railroad sidings, and desert fringes," Davis writes. "Such sites are poverty's niche in the ecology of the city."[26] And what they illustrate has implications far beyond the modern megalopolis.

If the city has an ecology, then these areas perform an ecosystem service. They absorb what others shed. The severity of a Mumbai slum's drainage problem is in direct proportion to the success achieved by the good drainage systems in the "better" neighborhoods upstream. Where efficiency meets inefficiency, water stays. Like the swamps they often replace, slums are human wetlands that soak up water, pollution, and risk. The people who live in them are being celebrated in policy circles as powerhouses of resilience for a reason: like the wetlands of southern Louisiana, the mangroves of Leyte Island, or the stands of *Phragmites* reeds on Staten Island, they have an ecological job to do, whether they like it or not.[27]

R.M.S. KAMPALA

Like many cities of Central and East Africa, hilly Kampala, Uganda, is not known for epic floods like the one in 2005 in Mumbai—just for a lot of flooding. It represents well the normalization of disastrous urban floods, the toll that such flooding can take in misery and hard work if not death and injury, and the opportunity cost of such everyday calamity: without the necessity of dealing with repeated floods, what could the people of lower Kampala be accomplishing instead?

The contours of Kampala can be seen from the top of Naguru Hill, over the roofs of the upscale neighborhood's villas and hotels, and the office compounds of USAID and the International Rescue Committee. If you visit Naguru Hill under a light rain, as we did in August 2013, and follow Old Kira Road along the ridge of the hill, you will see water flowing in the stone-lined drainage gutters that run alongside the road—not a strong flow, just a flat trickle in the deep drains that characterize established neighborhoods like this. At an intersection, the stream of rainwater branches off and flows abruptly down into Kamwokya.[28]

Kamwokya spills across the hillside from the ridgeline road into the valley. A classic Kampala slum, it's the childhood home of musician and "Ghetto President" Bobi Wine, who runs a recording studio there.[29] When the rain chases most people off the streets, young men take the excuse to duck into a wooden *kibanda*, an unlicensed video screening hall. The video halls of Kampala's fringe are special because of a local hybrid art form: the VJs who run the halls record their own Luganda dubs for English-language films, spiced up with topical humor. The claimed inventor of the form, VJ Jingo, recorded a perennially popular *kibanda* dub of James Cameron's *Titanic*, in which he rechristened the ship the *R.M.S. Bwaise*. Bwaise is a notoriously flood-prone slum located on swampland just downstream from Kamwokya.

Below the *kibanda* are numerous shops and restaurants. Their owners have to be cautious about flooding, so most of their establishments are found along paved streets on higher ground; lower-lying areas are the domain of food carts that can be moved out when the floods come. Here the paved gutter becomes a ditch with mud embankments on both sides; eventually it converges into a larger stream that flows through and out of the slum. The stream is crisscrossed with bridges, one of them made from an old truck frame. Where streams converge, large amounts of trash gather. In news accounts of Kampala's slum floods, the problem of trash clogging ditches is usually foregrounded as the primary cause. Celebrities periodically come down for community cleanup drives, reinforcing an essentially moral account of slum problems that portrays residents as drowning in their own trash. Whatever its origins, there certainly is a lot of trash down in lower Kamwokya, piling high on both sides of the stream. Crews of men with shovels hang out here, collecting donations from pedestrians crossing the truck-frame bridge in exchange for keeping the channel clear.

High-water marks far up the bare brick walls of lower Kamwokya's houses indicate the depth of past floods. Every yard, doorway, and alley is partitioned off with a concrete-lined threshold a foot or two high, so the area's regular nuisance flooding can be kept at bay. It's a complicated topography that traps as much water as it keeps out, so every flood is followed by a mass bailing-out. Even in light rainfall parts of the system can give way, and we encountered crews of men submerged waist deep, shoring up the channel. Ten days after our first visit, on September 3, 2013, the hardest rain of the year put the area fully underwater.

On another, sunnier day in the dry season of 2015, we met Madelena Iko sitting on her stoop next to one of Kamwokya's canals. The level of the black water in the canal was low, but by mid-September, she said, "this place will become a lake." Iko had lived here for five years, having come with her husband, now deceased, and children from the town of Narua near the border with war-ravaged South Sudan. In 2014, they had pitched in with a few neighbors to build a long zigzag diversion channel—two brick walls about three feet high and six feet apart—to guide incoming water around their houses. "If only we'd built it higher," Iko said; as it is, floodwaters easily rise high enough to overtop the wall and rush right in her front door. The house then fills to a depth of three or more feet. So during the rainy season everything must be placed on high shelves or hung from the rafters. When the flood comes, Iko, children, and grandchildren crowd onto the top level of their bunk bed set to sleep; when it recedes, they grab buckets and the great bailout begins.[30]

Farther along the canal, a man sat in his doorway. Asked about the flooding, he wouldn't be drawn in. "It's water," he said, along with a Luganda phrase equivalent to "What are you gonna do?" A woman sitting nearby overheard us and beckoned. "Come here, I'll tell you about this flooding." She introduced herself as Janet Kankwasa. ("But if you come back," she said, "ask for Mama Allan. That's how everyone knows me.") She pointed to a line on her tiny house's front wall below which the stucco texture had been washed away: "The floodwater comes up to here." As at Iko's place, it was about three feet above ground level. Mama Allan had had similar experiences, keeping belongings up high, bailing out the filthy water. But despite her best efforts, much had been lost. "Like when I was away at my village during one of these floods—all my kids' books were ruined," she recalled. When asked why she stays in Kamwokya and puts up with the misery, Mama Allan, a charcoal vendor, said the answer is simple: "The rent's fairly cheap here. It's fifty thousand shillings," a little more than $15 monthly. "The government has tried to evacuate this area, but without giving compensation. We wouldn't move. Some people have sunk their life savings into buying their houses, and we renters couldn't afford to live anywhere else in this city."[31]

The roots of Kampala's environmental problems run deep. The journey from the opulent villas on the hill to the homes of Madelena Iko and Mama Allan in the wetland bridges the interface between high and low efficiencies, revealing a chain of causation that is far less simple than pointing fingers at the absorbers in the valley. Geographer Shuaib Lwasa at Makerere University grapples with this in his modeling studies of Kampala. His team simulated flooding in the city under ten-year floods, the last of which was in 2007. (He said that "the increasing frequency of smaller-scale flooding says that we should start calling these six-year floods," however.) The city's Drainage Master Plan relies on one big drainage channel taking water out to the west, but Lwasa has shown that the change in elevation to the west is too gradual for this to do much, even if they keep widening and deepening it. In his team's models, they can cut down on some of the flooding just by replacing the system of stone-lined culverts higher up on the ridges with stepped, zigzagged, and unpaved channels. But those are the very types of channels that already exist down in the slums, much to the sorrow of their residents. Today, with the upgraded channels mostly in upscale and uphill areas, shedding water and trash rapidly into the irregular dirt channels below, it all accumulates in exactly the wrong place. "We are going to need to follow an integrated citywide approach," Lwasa said, but the problems are all political. "The plan needs to start from the hilltops, which are the most valuable land, of course." That could just mean encouraging more grass in backyards or catchments with controlled release, and he thought such steps could be incentivized (they would need to be, since the developers won't care otherwise; they aren't exposed to any of the risks they're creating downslope).[32] When we asked Mama Allan about the problem of runoff, she regarded the hilltops with all the hydrological expertise of a slum resident, and all the realism, too. "Those are rich people," she said. "If we had the money, the resources, to challenge them, don't you think we would?"[33]

At about the time we were speaking with her, the World Bank was publishing a seventy-page report on the causes of Kampala's flooding and potential solutions. The bank noted that the amount of impervious roof area in the city had grown 262 percent just since 2004, and that the increased rain runoff from roofs, streets, and parking areas was contributing to the dramatic escalation of flooding. The report also echoed the common charge that trash accumulation was a major culprit, even though the garbage collection rate had risen from 40 percent to 65 percent just between 2009 and 2013. While it was a stronger-than-usual analysis, the bank continued the tradition of pointing the finger downhill more than up. It resorted to conventional victim blaming with the statement, "Without intentional urban planning interventions, informal settlements will continue to be a primary source of environmental degradation."[34] Within a few weeks of the report, heavy rains began again. The World Bank's office on Nakasero, Kampala's most prestigious hill and central business district, stayed dry, while the slums were once more awash in the city's runoff.

HARD SLOGGING

Being flooded out in Kampala, Mumbai, Jakarta, or Asunción isn't usually about the threat of death or injury; it's about having to stay up all night bailing out your house and your neighbor's house, and then digging out gutters and rebuilding embankments when it all dries out. The reality of both vulnerability and resilience often translates into never-ending uncompensated labor. This extends from rescue and reconstruction to the communicative labor that produces and sustains information flows, and on to the extra work generated when labor-saving natural resources can't be utilized.

Such labor shadows the everyday economy, which not only draws on the precarious poor for its own workers but also feeds disruption back into the natural world—disruption that has to be absorbed. And demand for absorption is growing steadily. The IPCC forecast in 2014 that for the remainder of the century, climate change impacts would "slow down economic growth, make poverty reduction more difficult, further erode food security, and . . . create new poverty traps, the latter particularly in urban areas and emerging hotspots of hunger." Writing in the language of economics but describing an ecological mosaic of labor, the IPCC announced, "Climate-change impacts are expected to exacerbate poverty in most developing countries and create new poverty pockets in countries with increasing inequality, in both developed and developing countries."[35] While these pockets and traps may sound like a crushing inevitability of the climate models, they're not. They have a particular geography that is determined by the politics and economics of adaptation. The richest countries can (and do) fail to address their enclaves of risk, while the poorest countries can (and do) take action to smooth out the map of inequality. Yet we shall see next that the world of affordable resilience building—which, at times, looks poised to take over from the discipline once known as development—comes with traps of its own.

11

VULNERABILITY SEEPS IN EVERYWHERE

Disasters are never natural. They are the intersection of factors other than physical. They are the accumulation of the constant breach of economic, social, and environmental thresholds. Most of the time disaster is a result of inequity and the poorest people of the world are at greatest risk because of their vulnerability and decades of maldevelopment, which I must assert is connected to the kind of pursuit of economic growth that dominates the world; the same kind of pursuit of so-called economic growth and unsustainable consumption that has altered the climate system.

—Naderev "Yeb" Saño, Philippines delegate to the 2013 global climate negotiations in Warsaw, five days after Typhoon Yolanda's landfall [1]

Ecological culture cannot be reduced to a series of urgent and partial responses to the immediate problems of pollution, environmental decay and the depletion of natural resources. . . . To seek only a technical remedy to each environmental problem which comes up is to separate what is in reality interconnected and to mask the true and deepest problems of the global system.

—Pope Francis, *Laudato Si': On Care for Our Common Home,* encyclical letter, May 24, 2015 [2]

Urban flooding is one of many varieties of disaster—large and small, unprecedented and everyday, headline grabbing and totally invisible—that loom over the world's poor majority, threatening not only lives but also health, economic survival, and prospects for a better life. So far, national and international disaster policies have only nibbled around the edges of the problem. In the struggle to reduce death tolls in less affluent countries, we can find some outstanding examples. But preventing the destruction of homes and livelihoods, and ensuring recovery, will require radically new approaches.

SAVING LIVES ON THE CHEAP

When disaster strikes, low per capita GDP does not inevitably lead to a high body count. In disaster lore, Bangladesh is one of the most frequently discussed examples of success in saving lives on the cheap. The medical journal

The Lancet observes that while disasters can deepen poverty, thereby increasing the risk of future disasters, "Bangladesh has focused on breaking
this cycle by integration of disaster recovery and resilience in its overall
strategy of poverty reduction."[3] It wasn't always this way. In terms of flood
exposure, Bangladesh is the Netherlands without a budget, and major cyclones used to leave behind inconceivably high death tolls every time they
pushed massive storm surges up through the Bay of Bengal into the Ganges
delta. The Bhola cyclone of 1970 topped them all, killing more than half a
million people. In its aftermath, the Bangladesh Red Crescent Society set
up a Cyclone Preparedness Programme (CPP), the role of which was to
communicate cyclone warnings as soon as they are issued by the country's
Meteorological Department, to participate in relief operations, and to coordinate disaster management and development activities. Later improvements in early warning systems (including everything from more extensive
telecommunications to volunteers' efforts in going through their own villages with megaphones and hand-cranked sirens), along with the construction of fifteen hundred storm shelters and coastal embankments, improved
the situation greatly. In addition, a Comprehensive Disaster Management
Plan (CDMP) was formulated. By the time Cyclone Sidr, equivalent to a
category-5 storm, hit in 2007, satellite warning systems had improved, and
CPP had deployed 44,000 community volunteers to spread information
and urge people to evacuate to shelters before the storm and then assist in
search, rescue, and first aid after. The government and nonprofits together
distributed food and water to 11 million survivors, along with fuel, matches,
and alum-potash tablets and lime for water purification, while the army
provided medical services. In the end, half a million homes were destroyed
and there were 4,234 deaths—still far too many, but more than a hundredfold reduction from the toll of the 1970 cyclone.[4]

Since 2010, the number of CDMP volunteers has grown to 5 million,
and there has been deeper integration of disaster vulnerability reduction
with poverty alleviation programs. These efforts have been strengthened by
Bangladesh's large endowment of "community capital," strong networks of
mutual aid that tie family and community members. Meanwhile, the number of cyclone shelters in coastal areas has grown to 3,500, the goal being
that no one in hazard-prone regions should live more than a mile from a
shelter; the structures now double as schools, community centers, or other
type of public buildings, to ensure that they are well maintained and that
their location is already well known to residents when a cyclone approaches.
All are equipped with sanitation facilities and are provided with clean water,
food, and medical supplies to avert the very real threat of disease outbreaks

during and after floods. *The Lancet* observed, "Instead of focusing solely on individuals, Bangladesh has put more attention on households and communities in its prevention, response, and recovery strategies, which reduces the need for temporary shelters and camps (which can themselves create disease risks) and encourages rapid rejuvenation of markets and livelihoods. . . . One of the greatest drivers of Bangladesh's success is the residents themselves. Informal sharing of resources and collective activities of rebuilding are the rule rather than the exception."[5]

It is well recognized in Bangladesh that there remains much need for improvement in preparedness for and the response to cyclones and other hazards. Far too many people still do not receive evacuation orders or refuse to follow them. To help correct that, literacy rates, especially among women, need to be raised dramatically, while innovation is needed in emergency communication, since few households in coastal areas have radio or television. Distribution of megaphones to twenty thousand volunteers in remote areas has helped, but their audibility can be affected by high winds. Release of brightly colored hot-air balloons carrying warning messages has been suggested. But rapid expansion of access to mobile phones even in the most vulnerable areas may provide the best opportunity for effective early-warning systems. Protection of livestock also needs be provided; in the past, villagers have refused to evacuate to shelters for fear of losing their animals. And in low-lying parts of the country, building codes ideally would require floors to be at least three feet above ground level.[6]

Like Bangladesh, the nation of Cuba has learned from grim experience, and it has taken its hurricane preparedness strategy even further. In the Caribbean, Cuba is second only to Haiti in the frequency of disasters that it endures, but Cubans have earned international praise for achieving remarkably low death tolls in major hurricanes—storms that have killed far more people as they passed through other Caribbean nations and/or the United States than they killed in Cuba.

The shock that led to the creation of the nation's disaster preparedness system was delivered soon after the Cuban Revolution, in 1963, by Hurricane Flora, which killed 1,157 people. The results of the country's subsequent risk reduction efforts are apparent in the declining death tolls associated with recent major hurricanes: Georges killed five Cubans in 1998; neither Ivan nor Jeanne, both in 2004, killed a single person in Cuba; Ike, the country's costliest storm ever in monetary terms, killed seven in 2008; and Sandy killed eleven in 2012. On the other islands they passed over—Puerto Rico, Haiti, the Dominican Republic, Grenada, and the Bahamas—those same five storms caused a total of more than four thousand deaths.

Cuba's success in protecting life during hurricanes involves no magic formula. At the heart of its casualty-reducing strategy is the same high degree of social organization that has placed Cuba among the ranks of much more affluent countries in terms of health, education, and food security (while drawing criticism from some quarters for restrictions on civil liberties). The following elements of Cuba's hurricane preparedness and response system are well known: (1) disaster risk reduction is given a very high, very explicit priority at national and local levels, with detailed plans for evacuation and recovery; (2) the Cuban Meteorological Institute has state-of-the-art equipment and an excellent track record for storm tracking and prediction; (3) the institute issues successively more urgent warnings at seventy-two, forty-eight, and twenty-four hours before the predicted landfall of a storm, with announcements becoming continuous as the landfall hour draws close—and when evacuations are ordered, they are compulsory; (4) almost all Cuban homes have electricity, with television and radio providing almost universal access to preparedness and evacuation information; (5) Committees for the Defense of the Revolution, characterized as "all-purpose block associations," working with the Federation of Cuban Women, organize preparedness, cleanup, and rebuilding activities, even training children as young as nine years old to serve in emergency capacities, while flagging in advance all households with people who face problems with their own protection or evacuation; (6) a nationwide two-day hurricane preparedness drill is held every year; (7) roads and transportation systems are adequate to handle heavy evacuation traffic, and all modes of transport are mobilized for evacuation in an exhaustively rehearsed procedure; (8) soldiers patrol evacuated areas, so people don't fear burglary or looting and thus don't refuse evacuation in order to protect their property, as often happens in other countries; (9) the island has an extensive system of stormproof shelters, with detailed procedures for stocking and delivery of water, food, and medical supplies; and (10) in the aftermath of storms, communities see a great deal of mutual assistance with recovery and rebuilding.[7]

While disaster preparedness has saved many thousands of lives in recent decades, the effort has also helped Cuba, like Bangladesh, avoid falling into the kind of vicious cycle in which losses from major storms bring increased vulnerability to future disasters, creating a poverty trap. Citizens in hard-hit Cuban communities, while far from affluent, are not already living in debt or on the brink of bankruptcy, homelessness, and malnutrition at the time a storm makes landfall, so it doesn't push them over the threshold of destitution. Although half a century of a U.S.-imposed economic embargo and the withdrawal of Soviet assistance in the 1990s had devastating effects on its economy, Cuba has not had to deal with the kinds of fiscal hardships imposed on many other nations by the World Bank, International Monetary Fund, and other financial institutions that, as a condition of assistance, often force governments to undercut their own ability to provide services and reduce vulnerability.[8]

It is widely held, on good evidence, that the key to making Cuba's disaster reduction system work is a literacy rate of nearly 100 percent and a high overall education level, making it much more likely that everyone can understand risks and participate fully in disaster preparation. (The importance of education is not unique to Cuba. The role of women is crucial. A strong worldwide trend shows that countries with higher rates of female secondary education tend to have lower fatality rates from climatic disasters.)[9] According to Tania López-Marrero of Rutgers University and Ben Wisner of University College London, "Cuba's achievements have been attained because disaster and risk management are not seen as different entities; rather, they are viewed as an integral part of the development of the country and its people. Universal access to services such as health and education (both in urban and rural areas), policies to reduce social and economic disparities, investment in the country's infrastructure (including rural areas), and social organization have been among the priorities in Cuba's overall development over the years."[10] In countries where such policies are not pursued, the converse is true: disasters reveal the deficiencies in development policies.[11]

Of course, preventing death and injury during disasters does not necessarily prevent destruction of infrastructure and property. Over the decades, Cuba has repeatedly had to scramble back from economic crises caused by hurricanes. Hurricane Sandy, the second-strongest recorded storm ever to strike Cuba, made landfall at its peak strength on the eastern part of the island in the early morning hours of October 25, 2012. Sustained winds of 110 miles per hour blew there for six hours, leaving the nation's second city, Santiago, in ruins and its half million inhabitants in dire straits.[12] In passing over Cuba's landmass, Sandy lost some strength but remained a serious threat, so even before the storm had fully cleared Cuba, the rest of the world was already turning its attention to the U.S. East Coast. Residents of Santiago and the countryside were left with a mammoth cleanup and rebuilding job that would require years to complete. By the second anniversary of the disaster, the government announced that, out of more than 170,000 damaged homes, 49,000 still had not been repaired or replaced and 44 percent of Sandy's total damage to buildings and infrastructure had not yet been restored.[13]

MANGROVE DREAMS

Fending off flood hazards with hard technology in New York, Miami, Rotterdam, and Venice is very costly, and the chief rationalization for such expenditures is the high value of exposed property. Because most countries across the global South cannot afford such hard defenses on the massive scale that will be required as seas rise, many in the disaster risk reduction community have instead been encouraging the South to adopt the "ecosystem approach" or "social-ecological resilience."[14] These are conceptual steps very much in the right direction, modeled as they are on natural systems, which, per C.S. Holling (see pages 18–20), show great resilience in the face of natural disturbances. Ecosystems acquire that resilience not only by fending off hazards but also, when necessary, through compromise, sacrifice, limitations on growth, and retreat.[15] Those last four strategies, however, are poorly compatible with capitalist economics, so in practice the ecological approach to disaster prevention—or "working with nature," as it is often called—typically comes down to tweaking the details of this working relationship. This may involve putting monetary values on ecosystem services, conducting cost-benefit analyses, studying traditional and community knowledge, fostering local livelihoods, revegetating, and taking "soft" technological approaches such as beach nourishment.

We have seen how ecological approaches to disaster prevention have received attention in the context of fire, with efforts in Australia, the United States, Canada, Russia, and a few other countries trying to work with rather than stamp out natural fire regimes. On landslide-prone hillsides and in flood-prone watersheds the ecological connection is even better established, a leading narrative in the war on deforestation fought in regions such as the southern Philippines. But because they typically exclude compromise, sacrifice, caution, and retreat, none of the ecological approaches employed to date has fully achieved the resilience of natural ecosystems. Now there's also a vegetational solution being proposed for oceanic hazards: "bioshields" against storm surges and tsunamis. This is a welcome focus, yet most such efforts feature only a partial attempt to emulate the ways in which entire native ecosystems interact with hazards in integrated systems. The most prominent lesson learned by the international community from the 2004 Indian Ocean tsunami was that, in some places, coastlines still occupied by intact mangrove ecosystems fared better than those that had been disturbed by human habitation or activity.[16] Then Hurricane Katrina struck the U.S. Gulf Coast and reemphasized the benefits of natural bioshields, which in this case were coastal marshlands.[17] Growing recognition of the need for ecological resilience merged with observations made around the Indian Ocean and Gulf of Mexico to set off what might be called the Great Mangrove Rush.

Mangroves—wetland forests growing between the sea and shore—have long been recognized as having numerous ecological benefits. Since the 1970s, there has been much discussion of their role in protecting coastal lands, property, and people from destructive ocean waves. There is indeed good evidence that stands of native mangrove species can partially break the force of waves and, conversely, that destruction of mangrove forests to make way for shrimp farming or other economic activities has left many coastal areas, especially in South Asia and Southeast Asia, more exposed to damage from big waves.[18] After the 2004 Indian Ocean disaster, development agencies and aid groups more vigorously promoted mangroves in tropical and semitropical regions as having shielding power not only against storm surges but against tsunamis as well. The initial evidence that mangroves had protected parts of the coasts of India and Sri Lanka against the 2004 tsunami was based not on controlled experiments, however, but on anecdotal reports and surveys, along with some remote sensing and modeling. A series of papers published several years later with Rusty Feagin of Texas A&M University as lead author argued that there was no reason, in theory, to expect bioshields to reduce damage from the kinds of extreme surges produced by tsunamis or the strongest tropical cyclones.[19] A 2014 review of the previous decade's burgeoning research literature on mangroves' potential for coastal protection (including Feagin's work) concluded that (1) partial protection is possible against "more common low energy events" such as more moderate storm surges but not against tsunamis; (2) mangroves not situated directly between settlements and the sea will not provide protection and can even increase damage and mortality by increasing the amount of dangerous debris in the water; and (3) degradation of the mangrove stands through economic exploitation of forest products will reduce the protective potential.[20]

Despite such overall findings, in some situations mangroves do appear to have provided a degree of protection to even the most destructive surges, including those from the 2004 tsunami. On the Indonesian island of Sumatra, which was much closer to the earthquake trigger than South Asia and was hit by a series of waves with much greater force and reach, destruction appeared to be nearly total, and any protection provided by mangrove stands was much less likely. Some evidence of an effective mangrove buffer came nonetheless in a 2011 study based on a large data set collected in villages along the Sumatra coast. On average, settlements with native forests between themselves and the ocean suffered 8 percent fewer casualties; with agroforestry plantings instead of native vegetation, the reduction was only 3 percent. Both of those reductions were just at the borderline of statistical significance. Casualty rates depended far more strongly on people's distance from the shoreline when the tsunami hit than on the presence of mangroves, illustrated by a graph in the paper showing the dramatic decline in casualties as distance from shore increased. Coastal forests did add some degree of safety to that—about the same as the effect of a person moving another two hundred yards farther inland.[21] Thus, better tsunami warning systems could potentially save more lives than would tree planting.

In India more than anywhere else, the rush to plant coastal bioshields has been prompted by dual motives: to defend against storm surges and tsunamis and to generate economic products. Old-growth mangroves help maintain coastal fisheries, provide useful products in modest quantities, and perform many important ecological functions.[22] On paper, there is no better symbol than the mangrove for win-win-win development that meets the needs of resilience, the environment, and the economy all at once. Next to the post-tsunami Aceh peace process, the world's rediscovery of mangroves became perhaps the second most celebrated silver lining to come out of the disaster. Yet the mangrove as a symbol grew much faster than the mangrove as a tree. Because of the long establishment time required by native mangrove tree species, governments, nongovernmental development organizations, the World Bank, and other bioshield promoters have helped to establish plantations of fast-growing exotic trees in some coastal areas instead. But such plantations, it turns out, are incapable of significantly blunting the impact of sea surges, while their negative impacts have gone much further. Indigenous coastal residents have been displaced to make way for the plantations. Establishment of tree belts in some areas has meant the destruction of sand dune systems that had long proved resilient in the face of extreme events; one result has been loss of sea turtle habitat. The introduced trees have invaded and damaged existing native mangrove forests, and the plantings have failed completely to restore other ecological functions that native mangroves provide. In many projects, tree stands were established to the side or behind villages rather than between villages and the sea, so they could not even serve as bioshields and in fact, by contributing extra debris to the flood, could add to destruction when a storm surge or tsunami hits.[23] Finally, according to Feagin and colleagues,

The advocacy of bioshields also devalues the many other non–
"extreme event protection" functions and services that native vegetation provides, ignoring the more difficult work of defending these
ecosystems for their other benefits. For example, mangrove ecosystems are valuable for ecosystem services such as fisheries support,
water filtration, carbon sequestration, nutrient cycling, medicinal
and food sources, habitat and cover for a wide range of species,
land-building processes, tourism support, and aesthetics. Yet, there
is a risk of losing these ecosystems if we overvalue the protection
service at the expense of the many other ecosystem services. If direct
protection is recognized as the most important service that an ecosystem can provide, then society may eventually choose to replace
it by armoring of the coast, that is, seawalls, bulkheads, levees, etc.[24]

In view of the unintended effects caused by hard protection and the
difficulty of managing soft protection, is long-term defense of coastal areas
even possible? Addressing the long-term issue of coastal erosion, Andrew
Cooper and John McKenna of the University of Ulster have argued that the
only long-term answer is not to advance, defend, or even engage in strategic
retreat but to evacuate permanently: "We contend that only the conception
of allowing sufficient space for natural processes truly achieves the goal of
working with natural processes and permitting natural ecosystems to flourish." They conclude that "it is possible to either protect property or protect
the ecosystem but not both."[25]

BOOMERANG SOLUTIONS

Often the very enterprises that communities count on to lift them economically after disaster can make them more vulnerable to future disaster. Tourism, for example, turns up repeatedly in this book's stories, where it plays
any of three roles: a producer of disasters, an activity that disasters threaten
to choke off, and an industry that is counted upon for post-disaster economic stimulus. This last role is becoming increasingly prominent. Hardi
Prasetyo is pursuing his dream of having the Lusi volcano and Mud Island
become a world-class attraction for people around the globe. L'Aquila will
keep its damaged buildings propped up for decades until they, and tourism,
can be restored. The Philippines has turned to voluntourism. Miamians
are expecting tourism to remain even when most other economic activities
have fled north to escape the rising seas. Likewise in Thailand after the tsunami and New Orleans after Katrina, restoration of the tourist industry was

one of the first orders of business.[26] And in the concluding chapters, we will examine more situations in which communities have turned to tourism as a route to prosperity in disaster's aftermath.

It's well known, however, that dependence on tourism is a Faustian bargain. If an economy comes to be dominated by this single industry, it courts susceptibility to global economic slumps, seasonality, and, naturally, disasters. And international tourists aren't even a very good deal. The UN Environment Program estimates, "Of each $100 spent on a vacation tour by a tourist from a developed country, only around $5 actually stays in a developing-country destination's economy." Tourism, even "ecotourism," can be ecologically destructive; it can deprive local people of basic resources through depletion or inflation; local cultures morph into commodities; the influx of businesses that support tourism can be a homogenizing force, making every place look like every other place; as cultural goods become commodified, authenticity is lost or can come to be staged; and with more low-wage jobs, greater corporate involvement in the economy, and heavier dependence on affluent travelers, societies become less egalitarian.[27] Those side effects of tourism are enough to set up a small economy for big catastrophe when hazards hit.

Conventional wisdom says that economic development will prevent people, communities, and nations from suffering the worst effects of disasters. But the very process of building wealthier, faster-growing economies has increased the number and strength of hazards that threaten those same people, communities, and nations. Bigger, stronger economies handle geoclimatic hazards better than poor countries can; at the same time, breakneck economic growth not only helps fuel hazard creation but can also increase the fragility of a country or region in the face of some types of hazards.

In 2008, four months before the Wenchuan earthquake, unusual atmospheric circulation associated with melting of ice in the Arctic, combined with the formation of a particularly warm air mass over Tibet, brought China an entire month of the coldest weather in fifty years, along with unprecedented snowfall.[28] The biggest blow by far, however, was delivered by a record-shattering ice storm that struck China's south-central heartland, taking down power grids, paralyzing transportation, collapsing half a million buildings, wiping out 40 percent of the country's winter crops, damaging 77,000 square miles of forests, and disrupting the lives of 100 million people. With a price tag of $22 billion in direct damage alone, China's 2008 ice storm ranked globally as the world's eleventh-costliest geoclimatic disaster of the decade.

The ice storm delivered a terrible blow, but it was the fragility of a rapidly growing, technology-dependent economy that brought the entire midsection of the world's most populous country to a complete halt for weeks. According to a post-storm report published in the *Bulletin of the American Meteorological Society*, "The same storm would not have had the same impact 30 years ago. During the past 30 years, far-reaching economic reforms in China have dramatically changed the structures of socioeconomic and managed ecological systems in the affected region."[29]

The storm had some of its worst impacts on China's forests, and this was far from a natural phenomenon. According to the *Bulletin* report, development of the forests had not accounted for the fact that "biodiversity is a form of long-term insurance of sustainable forestry against extreme events." As a result, destruction was most severe in single-species plantations: "Plantation of pines that had been excessively tapped for oleoresins and exotic tree species suffered whole-stand destruction in much of the region. Oleoresin tapping through bark chipping, which was common in the region, created a weak spot in the stem." Stands of exotic trees, which were highly vulnerable to ice accumulation, had been planted widely because of their fast growth. Where forests were destroyed, soil erosion, landslides, insect infestations, and wildfires followed. Together they wiped out the results of a thirty-year reforestation effort in the region. Before the storm, those forests, through photosynthesis, had accounted for 65 percent of China's capacity to remove carbon from the atmosphere—a goal of growing importance in the country. But as a result of extensive ice damage, decaying vegetation on those forest lands would instead send more carbon into the atmosphere than the landscape would take out for years to come. Furthermore, noted the report, "the sudden loss of forest plantations created uncertainties for the livelihood of tens of thousands of households who depended on the plantations for cash. These families drew their incomes from both timber and no-timber products (e.g., oleoresin tapping). Because most plantations were destroyed altogether, these families will have to look for new sources of income for decades to come even if they can manage to replant the plantations now."[30] An analysis of 2008–9 satellite data showed that green cover recovered quickly across the ice-damaged native forests, thanks largely to the region's rich diversity of shrubs and other understory plants and the unusually warm, moist weather that followed the storm. Heavy salvage logging left extensive areas to fare much more poorly, however. Even with optimum management, the south-central region would not recover its impressive pre-2008 carbon balance for decades.[31]

FALSE SECURITY

Whereas an estimated 40 percent of geoclimatic-hazard-related losses in the global North are insured, only 3 percent are insured in the South.[32] As the economies of Latin America, Africa, and Asia grow, risking greater and greater property losses in the process (the eighth-costliest disaster in history, for example, was the Bangkok flood of 2011), insurance is becoming a high-priority issue in those regions as well.[33] Yet today in the global South, the majority of households or businesses could not afford to pay insurance premiums even if coverage were available, leaving local and national governments (along with foreign aid) the chief sources of help when catastrophe strikes. And those sources falter more often than not. Governments in countries such as the Philippines are hard-pressed to build up reserves for emergency disaster response, much less for rebuilding. In good economic times, governments tend to increase spending on food, housing, education, and employment assistance. But when disasters strike, especially large or repeated ones, they wreck the state's ability to spend at the very time when the need for spending is greatest. These kinds of predicaments can cause countries to fall into poverty traps.[34]

The Caribbean region was long considered mostly uninsurable because of its huge exposure to a range of climatic and seismic hazards, its socio-economic vulnerability, and the inability of individual states and populations in the region to pay the high premiums that private insurers would require in such a high-risk situation. In 2012, Lauren Brooks, then a doctoral candidate at the University of Arizona's College of Law, described the fiscal predicament in which the region's governments typically find themselves: trapped in a "liquidity gap" (a lack of access to cash) after a disaster and unable to provide basic services such as water and power, keep social services going, process tax payments, provide police and fire protection, or even fulfill executive, legislative, and judicial functions.[35] The purpose of the Caribbean Catastrophe Risk Insurance Facility (CCRIF), created in 2007 by a group of international organizations including the World Bank, is to help governments avoid post-disaster liquidity gaps by pooling resources, costs, and risks across the entire region.[36] Sixteen member governments pay premiums into the CCRIF, and they receive payouts if an event, usually a hurricane or earthquake of a specified intensity within a specified area, occurs. The payouts are narrowly targeted at the liquidity gap, however, not recovery and reconstruction. Therefore, the CCRIF interacts primarily with governments' fiscal agencies, rather than its disaster management agencies. Premium payments make up only a small portion of the CCRIF's income.

The additional funds required to cover risks arising from, for example, a hurricane season like the one in 2005 come from international lending agencies, reinsurance purchased from big "re" companies, and weather derivatives that the CCRIF sells primarily to the World Bank. Through the CCRIF, Caribbean governments are packaging and trading their extraordinary disaster risks on world financial markets.

Aberystwyth University's Kevin Grove argues that what is being traded in these markets is something very unlike commodities such as oil or wheat. This commodity is an imagined post-disaster future in which any of the region's small nations could become badly enough damaged by a specified disaster to become a "failed state." Investors who have rarely shown interest in efforts to protect ordinary people's property and livelihoods in the Caribbean region do become interested when political stability and security are at stake.[37] Not surprisingly, the system is more popular among economists, investors, and politicians than it is in the broader society. Grove notes that the premium the Jamaican government paid into the CCRIF in its first year amounted to about eight times the budget of the nation's national disaster management agency that year—a lot of money, none of which would go toward direct relief or rebuilding.[38] Grove refers to the CCRIF's approach as the "financialization of disaster management," which, he writes, "values infrastructure not in terms of human livelihoods and well-being, but in terms of the role infrastructure plays in maintaining a docile and ordered population."[39]

Meanwhile, international development agencies have not given up on individual disaster insurance policies. CCRIF recently participated in the launch of a "Livelihood Protection Policy" in St. Lucia, Grenada, and Jamaica that would insure smallholder farmers and laborers against excessive rainfall and wind damage through microinsurance and education.[40] In a 2008 report, Barry Barnett, Christopher Barrett, and Jerry Skees argue that one way to provide access to insurance and credit is through "index-based risk transfer products," individual insurance policies that pay out with the occurrence of extreme (usually weather-related) events. Premiums for such insurance policies could be kept low because they apply to rare extreme events, and the payout to get a Grenadian farm family on its feet would be much smaller than, say, to replace a vacation home on the Jersey Shore.[41]

There is much enthusiasm for such programs, but poverty alleviation strategies designed by experts in finance often appear to have much better prospects for expansion of financial markets than for reducing poverty. Most analysts argue that in the global South in the era of climate disruption, insurance and related financial mechanisms, if used at all, must be embedded in a much broader array of economic, social, and environmental policies designed to improve people's quality of life and reduce their vulnerability to disasters. A 2013 report to the World Bank summed up the predicament of the poor in the greenhouse era and recommended "climate-responsive social protection."[42] "Protection" measures such as cash transfers, social pensions, and public works programs, it said, are necessary and are to be distinguished from "promotion" measures, such as microfinance, that are intended to improve livelihoods. Because climate disruption increases the vulnerability of the majority in a country and community simultaneously, the report stresses, "a household-level focus will be insufficient to address that vulnerability."[43]

A problem resides in the trickle-down approaches taken by most proposals for preventing or blunting the impact of disasters. These strategies aim to stave off economic breakdown and promote overall growth while enmeshing low-income urban and rural working people more tightly into the national and global economy—all in the hope that everyone will share in the growth. But relying on corporations and investors to protect vulnerable communities from disaster, provide governments and communities the right incentives for collective action to reduce vulnerability, and at the same time generate what they see as an acceptable profit without going bust is an opaque, overly complicated, undemocratic, and high-risk approach to dealing with hazards.

TIME TO PAY THE BILLS

Provision of relief and programs to reduce vulnerability in economically distressed societies must be improved and adapted to the new reality. But more important, the rich countries must do even more by doing a lot less to cultivate fresh disasters. That will mean deep cuts in greenhouse emissions in accordance with the long-discussed goal of "contraction and convergence," whereby the North's emissions would be cut deeply over time until they converge with emissions from the South, all at a low level that can permit a decent life for all.[44] On top of that, the "climate debt" that the North owes the South for having produced the vast bulk of nineteenth- and twentieth-century greenhouse emissions must be repaid.[45]

Such demands on the North have both a legal basis and a philosophical one. The right to disaster mitigation and relief has a firm basis in international law, and a well-conceived principle of disaster justice has been developed as well.[46] These not only require disaster assistance and climate mitigation but also will mean ending exploitation of people and resources in the tropics while deeply cutting the North's consumption of products such as meat, biofuels, minerals, and manufactured goods that are obtained through deforestation, air and water pollution, triggering of seismic hazards, and degradation of soil and waterways in communities that simply can't afford the next disaster.

With all of that in mind, and in parallel with the social disaster insurance we advocate for the United States (see pages 219–23), we urge that the nations of the global North agree to create an international fund to break the cycle of disaster. The fund (1) would almost fully underwrite disaster prevention, vulnerability reduction, and socially just retreat from hazard zones in the South and (2) in the event of major disasters of the types discussed in this book, plus drought, would quickly and automatically provide substantial assistance for relief, recovery, and rebuilding. National governments in the South would pay nominal annual dues, analogous to an insurance premium, to be "members" of the fund, but the collection of dues would in no way be expected to be large enough to cover the cost of annual payouts. Governments in the North would cover the bulk of the funding, and there would be no token contributions; if the amounts deposited aren't large enough to cause noticeable pain to the donor governments (or, better, the dirty industries that could be taxed to support the fund), they probably won't be sufficient.

Development of a program for disaster vulnerability reduction and disaster compensation would have to run alongside, not within, the global Framework Convention on Climate Change. No such strategies were provided for in the framework's 2015 Paris Agreement, which encourages *but does not require* the rich nations to collectively provide $100 billion annually by 2020 for climate mitigation and adaptation in less developed countries. Prospects are poor that any such amount will be forthcoming. Meanwhile, the agreement quashed more explicitly the notion of compensation for past and future climatic disasters, stating that its language on loss and damage "does not involve or provide a basis for any liability or compensation." But in a last-minute edit, negotiators for small island states succeeded in getting that disclaimer moved out of the main text into the nonbinding preamble (the same fate as the $100 billion suggestion), leaving it slightly more susceptible to being changed in future deals. In time it may prove more significant that the language of "loss and damage" so fiercely opposed by the rich countries was included at all—and in the main text, no less. In Article 8, parties "recognize the importance of averting, minimizing and addressing loss and damage associated with the adverse effects of climate change, including extreme weather events and slow onset events, and the role of sustainable development in reducing the risk of loss and damage." Nevertheless, the only help offered to threatened countries was through existing insurance and financing channels—a route they would already be taking if they could afford to.[47]

Whatever the practical prospects for mobilizing adequate funds, the nearly two hundred nations that signed the Paris agreement are now on record that they expect climate mitigation and adaptation to require $100 billion in aid annually, and that the amount will increase over time. We expect that our proposal for reducing vulnerability to and compensating for geoclimatic disasters, running in parallel to climate mitigation and adaptation, will have a similar cost; however, it would require extensive calculations by experts in the field to determine more precisely how much would be required. Whereas wealthy nations, led by the United States, purged any compensation provisions from the Paris agreement in order to avoid the appearance of accepting blame or liability, they might actually be more willing to consider contributing to a comprehensive disaster fund, because it addresses not only climatic events but also disasters such as those that followed, say, the 2015 Nepal earthquake or the 2004 Indian Ocean tsunami— hazard triggers for which blame cannot be assigned.

As in our proposed U.S. disaster insurance plan, the provision of compensation would provide leverage for enforcing risk reduction. An international agency, with majorities of its geographically diverse members coming from the most threatened regions, would ensure that funds are used to reduce exposure and vulnerability, not to subsidize construction in risky situations or enrich elites through corruption; that community preparedness is emphasized; that people living in hazard zones out of economic necessity be supported in their efforts to find better livelihoods and homes in safer locations; and that payouts for individual properties be capped and targeted, so that policies are explicitly pro-poor. These funds would in no way be intended to substitute for the resourcefulness, solidarity, resilience, or *bayanihan* that can be seen in places all over the world whose people are struggling back from disaster. This program should support and augment their efforts.

Communities on the front lines of geoclimatic calamity didn't ask to have their risk wagered upon in commodity markets. They need the opportunity to escape the vulnerability trap. That means safe, comfortable housing and workplaces in less exposed locations, achieving good quality of life in rural areas to reverse the rural-to-urban migration, and security should disaster strike nonetheless. Just as social disaster insurance in the United States would eliminate the need for ad hoc presidential disaster declarations, the international fund could supersede the current post-disaster routine in the South. That routine included the kind of charade we saw in the Philippines after Pablo and Yolanda, when UN agencies made larger and larger appeals for aid money and only an inadequate portion was ever delivered (while part of what *was* delivered was lost to corruption), as well as what happened after Haiti's 2010 earthquake, when the major economic powers and the big aid groups and foundations gathered to pledge funding that at best would cover only a portion of relief needs, and then the actual money—only a fraction of the total amount pledged—trickled out only after a long delay, and to the wrong places.[48] Based on the North's failure to deliver on its promises after the Haiti quake, Jonathan Katz has expressed strong doubts that anything like the $100 billion per year for climate urged by the Paris agreement will be forthcoming.[49] And it is because of that rich-nation stinginess that our proposed disaster fund would have to include compulsory contributions.

At the time we were developing this disaster fund idea, it turned out that the international community was discussing the concept of an international disaster insurance program. At their June 7–8, 2015, meeting, the Group of Seven economic powers stated their aim to provide 400 million people in vulnerable communities around the world with "direct or indirect insurance coverage against the negative impact of climate change related hazards" by 2020. This was needed, they declared, because "mobilization of private sector capital" was crucial for "building resilience against the effects of climate change," so that "finance models with high mobilization effects are needed." The coverage would be subsidized by the wealthy nations, but it was not clear to what extent, and no explicit commitments were made. The United States and other countries in the group refused to allow the insurance plan to be included in the agreement that would come out of the international climate talks to be held in Paris later that year, presumably to avoid having the insurance program take on the appearance of compensation for the climatic hazards being created by the North's previous emissions. Instead, as a freestanding program, it could be seen simply as charitable aid provided to people suffering unpreventable "natural disasters," with no assignment of responsibility. Climate activist Harjeet Singh was not impressed, telling reporters, "They're trying to portray insurance as a magic bullet. But you need to clarify who's going to pay for it, especially when you talk about the poor."[50]

We aren't naive; we acknowledge that in the current global economy, our international disaster fund is no more likely to become a reality in the near future than our disaster coverage proposal for the United States. We could expect the world's economic and political titans to argue that capitalism would be crushed under the weight of any such initiatives aimed at reducing vulnerability to disasters for all people. But if that claim *is* true, the problem resides not in the universal need for disaster risk reduction but in capitalism itself. Affluent nations of the North, which refuse either to accept responsibility for having created the climate crisis or to compensate the South for having done so, are wholly averse to offering a stronger, more straightforward helping hand to victims of even the most devastating disasters. Instead, the fantasy documents coming out of Northern think tanks, universities, and philanthropies call upon communities who can't even afford bootstraps to pull themselves up and out of catastrophe through complex arrangements helpfully arranged by Northern financiers. But to expect vulnerable families to buy insurance or go into debt in order to fund their efforts to recover from disaster is to ignore the fact that their deep vulnerability was created by the wider economy and is not primarily a result of simple behavioral responses to risk. Small nations might want to think twice before sinking scarce public funds into catastrophe bond premiums or microinsurance programs rather than into resolving the factors that make their citizens vulnerable in the first place. Market-based programs would deal with that vulnerability by offering financial instruments that draw people more deeply into the very system that made them vulnerable, rather than by converting the system itself into one that gives highest priority to producing the essentials of life and providing the kinds of livelihoods that permit people and communities to shed their vulnerability.

In a 2010 book chapter titled "Vulnerability Does Not Fall from the Sky," University of Illinois geography professor Jesse Ribot argued that for economically stressed populations,

> their risk-minimizing strategies can diminish their incomes even before shocks arrive; and shocks can reinforce poverty by interrupting education, stunting children's physical development, destroying assets, forcing sale of productive capital, and deepening social differentiation from poor households' slower recovery. The poor also may experience threats and opportunities from development or climate action itself, such as efforts to reduce greenhouse-gas emissions in such sectors as household energy, land, and forest management. The inability to manage stresses does not fall from the sky. It is produced by on-the-ground social inequality; unequal access to resources; poverty; poor infrastructure; lack of representation; and inadequate systems of social security, early warning, and planning. These factors translate climate vagaries into suffering and loss.[51]

Inequality, poverty, and vulnerability don't need an insurance policy; they need fixing. The only meaningful measure of disaster risk reduction is a reduction in suffering and loss, whether suffering and loss come in days of clear sky or nights of dark violence. When searching for solutions to mounting disasters, the world needs to remember this truth. We all must hold decision makers accountable by it. Solutions that simply stitch back together the pre-disaster economic fabric can neither improve the lives of the majority of disaster victims nor stop the production of new disasters. Solutions that slow down or offset the carbon economy can hurt the poor, and we must be careful of this; however, solutions that are solely *adaptive* to climate change *guarantee* that the most vulnerable will see its most painful side. Vulnerability doesn't fall from the sky; rather, like the water table in Miami, it seeps into everything. A whole world of seawalls can't keep it out.

This truth—that when disaster policies don't reduce long-term suffering or improve quality of life, they are a failure—is all too familiar to far too many people who live within the borders of wealthy, powerful nations. The fate of the people of New Orleans's Lower Ninth Ward illustrates that truth for our generation, but it is evident in countless other, often surprising, times and places.

Now come the stories of two small communities, governed by two of the world's most powerful governments, that were nearly wiped off the map by geoclimatic hazards and have since struggled through harsh economic times to reinvent the building-back-better concept, each in its own way. The people of Montserrat, a British overseas territory in the Caribbean ravaged by a volcanic eruption, have fought through the fallout of colonialism, racism, and bad disaster policy for two decades, while those of Greensburg, Kansas, in the heart of the nation's economically stressed Tornado Alley, have labored creatively in disaster's aftermath but still are not sure how they can keep their community alive.

12

KEEPING THE LIGHTS ON

Montserrat, West Indies

Acceding [to the notion] that there is a silver lining to most clouds, it took deep thought before Mr. Kelsick could find that lining in the volcano cloud hovering over Montserrat. "What is the bright side?" he mused. "Let's try to look for one . . . let's say that this situation has tested our characters; yes, that is the bright side; we've pulled through."

—Sharon Williams, *West Indian Lawyers*, September 14, 2015[1]

Trust No Cloud

—Decal on taxi, Brades, Montserrat, 2015

Big trucks covered in gray dust rumble through a gray landscape, over the top of a lost city, to a gray pier. Each is hauling a gray cargo of gravel, sand, or ash, mined and sifted from vast gray volcanic deposits, to be loaded onto ships and barges and exported for use as construction material. Looming over the scene is the source of these exports, the Soufrière Hills volcano. On this day, it's producing only billowing clouds of water vapor, sulfur dioxide, and other gases; the road to the pier may be smelly, but it's safe for now. En route from gray mine to gray pier and back, the trucks roll past farms, homes, and hotels, now also ash-covered. They are overgrown with ferns and bushes and sometimes visited by their former occupants. Here and there, tourists poke around in the abandoned structures and crane their necks to get a good look at the gray ex-city Plymouth, which is off-limits to everyone but the truck drivers and, according to big warning signs, other "authorized personnel." On their way to Plymouth, trucks also pass a couple of geothermal exploration wells, heralds of a new electric utility that could generate plentiful green power—harnessing the same deep energy source that produced the export-quality sand and gravel and, in the process, almost destroyed the island of Montserrat.

For anyone who likes to root for the underdog, there could be no better home team than the people of Montserrat. Their papaya-shaped island 250 miles southeast of Puerto Rico in the Lesser Antilles has so far endured a

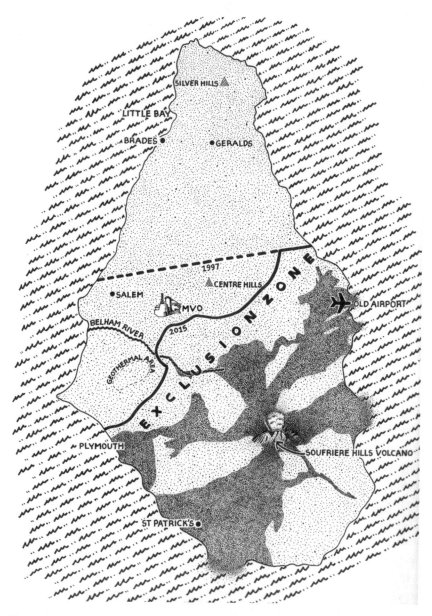

Figure 9. Montserrat, where stippled areas are those buried under pyroclastic flows, lahars, and other debris from eruptions of the Soufrière Hills volcano. Access to the exclusion zone, which in early 2015 constituted the area south of the solid line, is restricted to authorized personnel; the dashed line delineates the maximum extent of the exclusion zone, which existed in 1997.

quarter century of disaster. And the ordeal still hasn't ended. Climatic and geologic hazards have come and gone and come again, while political and economic disasters have lingered on and on. With Superstorm Sandy or the L'Aquila earthquake, one can easily speak of the time before and the time after the disaster; in contrast, Montserratians, like the mudflow victims in East Java and residents of the central Philippines, are living after, during, and before disaster all at once. The long-running headline act, the volcano, began erupting in 1995, and twenty years later the eruption still could not be declared over. As a result, a society long accustomed to living on an island with about two-thirds the land area of Staten Island now struggles to re-create itself on just one-third of that former territory.

In clawing their way back from catastrophe, Montserratians have found a few, but only a few, glimmers of silver in the volcanic clouds that still drift over the landscape where they once lived and worked. They are looking to their old nemesis, the island's volcanism, as a source of energy, exports, and attractions for tourists interested in nature, science, and catastrophe. Those efforts to turn threat into advantage were what drew us to Montserrat's story in the first place, but the island's bigger struggle lies in overcoming a history of exploitation, institutional racism, colonial politics, and political mismanagement. And the geographic predicament of this unlucky Caribbean island may hold lessons for us all if someday humanity has to retreat from large regions of the Earth should they be rendered uninhabitable by changes in climate, sea level rise, or ecological degradation.

Montserrat is an overseas territory of the United Kingdom. The elected government, headed by a chief minister, runs the island day to day while a governor, appointed by the queen, is responsible for national security and foreign relations. Members of a new political party, the People's Democratic Movement (PDM), won election in September 2014 and took over the reins of government on an island that was still a long way from being safe and prosperous. Their number one task was to convince the British government to fund the rebuilding of a self-sustaining Montserratian economy. London's position was that Her Majesty's Government had been more than generous over the years but that previous Montserrat governments had squandered the funds that were provided. The new chief minister (often called the premier locally), Donaldson Romeo, and his administration maintained that the admittedly poor performance of previous governments was something that the PDM had the desire and ability to correct, but that they could not succeed unless the British Parliament acknowledged and made amends for the neglect (amounting at times, they claim, to abandonment) of the small colony during its desperate ordeal of the 1990s.[2]

PROSPERITY POSTPONED AND RE-POSTPONED

Amerindians first came to live on Montserrat about three thousand years ago. The island received its name from Christopher Columbus, who sailed past on his second voyage to the Americas; its sawtooth mountain tops, remnants of ancient volcanic dome collapses, apparently reminded him of a mountain he knew in Spain. Columbus didn't drop anchor, however.

The first European settlers, Irish Catholic dissidents, arrived in 1632. (Although they didn't realize it, they had missed a major event that had occurred a few decades before their arrival: an eruption of the Soufrière Hills, something that would not be seen again for another 363 years.) By the mid-1600s, the island had become a British colony, sugar plantations had been established, and slaves were being shipped in from Africa. Settlement of the island was focused on the young volcanic soils surrounding the Soufrière Hills in the south. The Silver Hills, which occupy the north end of the island, had last erupted 1.2 million years ago, and the Centre Hills had done so a half million years ago; both areas had more highly weathered soils and steeper landscapes than elsewhere on the island and were less desirable for settlement.

By the early 1800s, Montserrat's slave population was about 6,500, well exceeding today's total population. Emancipation from slavery came in 1838, but white merchants and planters retained complete political control, and the former slaves found themselves living under an oppressive share-cropping system that persisted until 1959. Finally, in 1961, the island's majority wrested political power away from the white rulers.[3]

William Bramble served as the first chief minister through the 1960s. Setting aside visions of independence in favor of a quest for economic prosperity, his big move was to sell off six hundred acres of agricultural land to wealthy white foreigners, mostly North Americans, who began building vacation villas in the southern part of the island, where most of the population lived. The advent of "residential tourism" indeed brought prosperity for Montserratians. In her indispensable book on Montserrat's ordeal, *Fire from the Mountain*, Polly Pattullo writes of the changes that followed: "Instead of work with a cutlass in the hot sun, there were now jobs in the construction industry and part-time work in the service sector, as maids, gardeners, pool cleaners, and so on." (The pool cleaner seems to have remained the island's icon of economic stimulus, often mentioned in conversations on the topic. An American homeowner who told us that "the new Montserrat was created by residential tourism" explained, for example, that "when you buy a house down here, you need garbage pickup, you need lawn care, and you need a pool person.")

In 1979, former Beatles manager George Martin opened a branch of the recording company AIR Studios in Montserrat. Over the next decade, Dire Straits, the Police, Paul McCartney, Stevie Wonder, Lou Reed, Eric Clapton, and many others recorded there. An unintentionally prescient album produced at the studio was Jimmy Buffett's 1979 *Volcano*, the cover of which featured a fanciful painting of the Soufrière Hills. Though at that time the volcano had been dormant for three and a half centuries, the painting showed gases billowing from the summit.

By 1987, most Montserratians were still far from wealthy, but the island had the highest per capita income in the Organization of Eastern Caribbean States and superior health and education indicators to go along with it.[4] In 1989, Montserrat stood on the verge of economic self-sufficiency. Then on September 17 of that year, Hurricane Hugo hit the island head-on. More than 50 percent of its houses were severely damaged, with 20 percent completely destroyed, leaving almost a quarter of the population homeless. The colonial capital, Plymouth, was in shambles. AIR Studios was abandoned. London responded to the disaster with a large aid package, and much residential rebuilding was done with insurance settlements. Storm damage gave the owners of one of the island's tourism mainstays, the luxurious Montserrat Springs Hotel, an opportunity to do some renovations. An article titled "Lazy Days in Montserrat," which appeared in the *New York Times*'s travel section fifteen months after Hugo, painted a cheerful picture of an island making a comeback, ready again for the kind of tourists who wanted something different. But it contained one ominous passage:

> The road to a volcano known as Galways Soufrière is windswept and climbs into the clouds. At the spot where tropical sunlight and low misty clouds mingle is what remains of the 18th-century Galways slave plantation—the caved-in remnants of the sugar boiling house, the round turret of a wind-driven sugar mill and the circular wall of a cattle-driven one. It is a haunted place, dark and forbidding, with the bright Caribbean Sea behind it in the distance. . . . The road to the volcano comes to an end and the air is filled with sulfur. Here you leave your car and climb over yellow, red, purple and white rocks to Galways Soufrière, which is labeled a dead volcano. But if dead, why does it bubble and hiss and belch steam? Why can the molten earth be heard rumbling below the surface?[5]

By the summer of 1995, the island had fully recovered. Plymouth sported new governmental buildings, a public library, and a hospital. The expatriates were back, the new government chambers were almost ready to host meetings, the hospital was ready for its first patient, and it was almost time to begin filling the library with books when, on the afternoon of June 18, the Soufrière Hills started blowing up.

The eruption began slowly, but a month into it, huge repeated ejections of steam, water, rock, and ash known as phreatic explosions prompted the evacuation of six thousand residents from areas to the south and east of the mountain. Two weeks later, as the explosions eased and authorities debated whether to allow evacuees to return, it became clear that a fresh threat, Hurricane Luis, was headed straight for Montserrat. The evacuation was declared over, so that families being housed in tents could get back to the shelter of their own homes. For a time, the risk of death by hurricane had eclipsed the risk of death from an eruption.

In December 1995, with the mountain now producing much more dangerous pyroclastic flows—rivers of superheated gas and rock that flow downhill like water, but at speeds that can reach hundreds of miles an hour—a second evacuation of six thousand people was ordered, and it wound up lasting a month. The following April, in the face of much larger pyroclastic flows, authorities ordered the complete evacuation of Plymouth and surrounding communities in the south. Some people had family in other areas with whom they could stay, but a large portion of the population, an estimated 1,366 people, had to be housed in emergency shelters. This time, the evacuation of the capital would be permanent. Some Montserratians began leaving for Britain. Their exile was supposed to be temporary, but most have never returned to the island. The next year, 1997, brought further escalations of the eruption, and the island's population dropped to a mere three thousand; two and a half years before, almost eleven thousand people had lived in Montserrat. In the years that followed, repeated ashfalls and lahars buried Plymouth completely.[6]

Population numbers slowly crept up to nearly five thousand over the next decade and a half, thanks largely to immigration from other Caribbean countries. But now everyone was living on one-third of the island's landmass, many in its northern reaches, which before the eruption had been terra incognita to many Montserratians (despite the fact that no place is more than a few miles away from any other place on the island). And every time the mountain seemed to be dying down for good, fresh bursts of activity came. Massive eruptions cut loose in 2003, 2006, and 2010, burying much of the evacuated south even deeper in debris while raining grit and ash on inhabited areas and at times prompting temporary evacuations. Then things got quiet. Gases—but only gases—continued to stream from the top of the Soufrière Hills. As 2015 began, Montserratians prepared to commemorate the twentieth year of their volcanic troubles, as well as the passage of five years without a major eruption.[7]

"HOW MUCH LONGER DO YOU THINK IT WILL TAKE?"

On the last night of January 2015, we met with Montserrat premier Don Romeo in an apartment near the island's north end. That week he had been engaged in intense budget negotiations with representatives of the British government's Department for International Development (DFID), which had been supplying the island with funding for disaster recovery and development since the start of the crisis. People around the island had told us that with the PDM now in power—and with Romeo, a longtime activist and journalist, at the top—two decades of what they saw as mismanagement by local officials and neglect by the British government would end.

Romeo's first words to us that evening were "I never expected to become premier this soon." Now, he said, he faced his most difficult job: "I was happy to be in opposition. From there I could highlight everything that was going wrong around here and help put pressure on this government. I could go to London too and meet the Montserratians—there are six thousand there now—talk to them as an activist, try to motivate them to motivate the British government." He had even gone to visit the chairperson of the parliamentary committee that oversees DFID, telling him, "DFID should have been able to go in there for ten or fifteen years, get the job done, and come out—not indefinitely. Stay until the economy's viable, and then you can leave. But it's twenty years now. How much longer do you think it will take? Do you want us dependent on you forever? Your and my constituents both must understand the fact that you have to spend now to save later."

Now, as premier, Romeo had suddenly ended up with the full-time task of trying to revive an island in crisis by cooperating with the very institutions he had been challenging for so long. He and the other members of the new government were looking to London to provide the funds for more housing, to allow more Montserratians now in exile to return, and for infrastructure improvements that could finally get the economy back on its feet. They needed, for example, to expand and complete the geothermal project but, most important of all, they also needed the construction of a new port in the north of the island. Those and other projects would not only generate short-term jobs but also stimulate commerce and tourism. First, though, Romeo had to convince the British that the mishandling of funds and projects by the island's previous governments was all in the past and would not recur. Before the budget meetings, Romeo and DFID had taken a step toward that goal, agreeing to suspend all British-funded infrastructure projects then under way; the move was triggered by an investigation that had found evidence of "nepotism, collusion, bid rigging, and the payment of bribes and kickbacks" under the previous government.[8]

The Soufrière Hills may have been quiet for almost five years, but to those living in Montserrat in 2015 the disaster seemed far from over. Down deep, the volcano was still showing distinct signs of life, but the seemingly endless economic crisis and still-heavy dependence on British support drew most of the attention. Officials in the new government saw no path to a solution that did not involve confronting the early mishandling of the disaster in the 1990s and what Romeo calls "deliberate deception" by the British. There were two big issues from the past that had to be dealt with, they said, before progress could be made: a failure to make the northern third of the island livable for evacuees fleeing from eruptions in the south and the related (in some cases resulting) failure to ensure public safety.

Monitoring the eruption, delineating safe zones, and making evacuation decisions during the critical years 1995 through 1997 were extremely complex problems that have been picked over extensively in the academic literature.[9] Scientists were still striving to figure out the erratic volcano at the same time that officials of two different governments tried to formulate policies for saving lives, based on what the scientists told them—echoing the dilemma of L'Aquila. This meant creating controlled-access and exclusion zones covering the majority of the island's landmass and adjusting the zones' borders with each major change in the mountain's behavior. And the population they were trying to protect was really two populations: a majority living increasingly close to the edge of subsistence as the crisis deepened, along with a mostly affluent minority of part- and full-time expatriates. With hindsight, it is generally agreed that both the UK and Montserrat governments could have performed much better, but there is disagreement about what lies behind their poor performance.

JUNE 25, 1997

From the early days of the eruption, the Montserrat Volcano Observatory (MVO) held the responsibility for monitoring the Soufrière Hills and advising the two governments. Rod Stewart, a seismologist with the University of the West Indies, took over the directorship of the MVO in 2012. He stressed to us that while every volcano is different, Montserrat's is an extreme outlier: "Eruptions in the Caribbean normally last only a year or so. If in '95 MVO had said to people, 'Twenty years from now we'll still be worrying about eruptions,' people would have laughed." The longevity and erratic behavior of the volcano have made the task of communicating risks to the government and the public extremely difficult. Matters are further complicated by the contrast between individual and societal risks. As Stewart puts it, "When you think about risk to yourself, you're thinking about yourself dying. But the government wants *no one* to die. And this often leads to conflict. The government says, 'We want to stop people doing this,' and the individual says, 'It's my right to do it. I am responsible for myself.' And that is what has caused conflict here in the past." [10]

Don Romeo, on the other hand, argues that if the Montserrat and UK governments had wanted to ensure that no one died, then they have failed—that in fact their very response to the disaster, or often their lack of response, put lives at risk. Back in the 1990s, in the process of filming conditions in the evacuation centers, interviewing people in the shelters and the streets, quizzing public officials, and reading government documents, Romeo amassed what he believes is more than enough evidence to show that by providing for only the most meager resettlement and employment opportunities, the British government forced Montserratians into a harsh choice between living in intolerable shelters or leaving their home island, unintentionally creating conditions under which people would rather take a third path: risking death by trying to eke out a living in the exclusion zone. [11]

These issues were brought to a head on June 25, 1997. With the potential for volcanic activity on the north side of the mountain escalating rapidly, MVO scientists standing watch at W.H. Bramble Airport on the east coast, north of the volcano, had been warning that the facility could be hit at any moment by a pyroclastic flow. There would, they said, be very short warning, maybe as little as a minute and a half. On June 25, however, the airport was still open. A group had assembled, awaiting the return of the governor, Frank Savage, from meetings in Barbados. Local officials and MVO scientists wanted to confer with Savage before any evacuation was announced. But just before 1:00 p.m., very shortly after the flight carrying the governor had landed, the pyroclastic flow that scientists had feared burst from the mountain and hurtled toward the airport. Crew, passengers (including Savage), and airport staff were rounded up and piled into vehicles, while the pilot who had just landed prepared to escape by air. The evacuation was accomplished by 1:03, and at 1:07 the deadly flow arrived, coming within two hundred yards of the airport terminal. Everyone who had been in the airport area escaped without harm. A succession of eruptions that started three months later, however, wiped out the airport for good.[12]

What marked June 25, 1997, as the most significant moment of Montserrat's disaster was not the action-adventure scene at the airport; rather, it was the tragedy of nineteen people killed by the same pyroclastic flow as it raced through the countryside toward the airport. Some outside observers have been quick to ascribe the deaths to recklessness of the victims themselves. For example, an academic paper published in 2012 provides this too-concise statement: "Nineteen people were killed on Montserrat in 1997 (all of them in the exclusion zone illegally and despite several warnings.)"[13] The more complete account in Pattullo's book includes the heartbreaking personal stories of farmers, airport employees, and others who had been driven by economic necessity to venture into the exclusion zone.[14] There is no way of knowing whether any of the victims drew a false sense of security from the knowledge that the airport was being kept open at the time, but an inquiry into the tragedy by a British magistrate and jury deftly separated the obvious causes of death from the less obvious ones. In every case, naturally enough, the jury found the cause of death to be "natural catastrophe." But for victims Alister Joseph, Alicia Joseph, Joseph Tuitt, and Rueben Boatswain (Tuitt and Boatswain were off-duty airport employees who had gone to tend their animals in the area), the jury found that "a contributory cause of deaths was the continued operation of the airport despite the elevated volcanic activity in the days immediately preceding 25th June 1997." As for seven farmers who died, "the contributory cause of deaths was the failure of the authorities both local and British to provide alternative lands in the safe area for farmers displaced from the exclusion zone." In one of those cases, that of seventy-one-year-old Benjamin Joseph Browne, there was an additional factor: "the conditions in the public shelters were so deplorable that Mr. Browne refused to turn to them after his initial experience" and went back to his home.[15]

Romeo told us, "People were so desperate that they would go back onto the volcano to grow food and keep animals. They were trying to earn some money to pay off loans on their tractors, et cetera, loans that were not forgiven after the disaster. That's why some of them got caught up there. And the government, which needed food supplies, was buying produce from farmers, knowing that they were growing it in the exclusion zone!" He said that the withholding of information on the one hand and the failure to provide opportunity on the other converged in a way that not only took the lives of nineteen people but also ruined the lives of many others for years to come: "Through all this time, people have lost the opportunity to plan and to secure their livelihoods. Had the British government decided to buy land, build homes, resettle people, ensure work for people, arrange relief for farmers with loans . . ." He left the conclusion unspoken.[16]

The June 25 calamity, coming just two years into the eruption sequence, affected everything that followed. Pattullo wrote,

> A kind of mythic quality began to surround these deaths. In many ways, the dead came to represent all that was virtuous about Montserrat and its people. They became symbols of an old-fashioned, God-fearing society in which the values of an emancipated peasantry—individualism, independence, devotion to land and home—triumphed over the circumstances of death. They had refused to accept dependency, and by inference, a colonial status; they had rejected the inadequate conditions of the shelters in the north. Some were reaping their crops in the field (the biblical echoes are strong here) to feed the island; others were tending livestock, and others had stayed resolutely at home. In a sense, they became the heroic dead, the victims of a colonial war.[17]

"WE WERE BEING COUNTED OUT"

Volcanoes exert an especially strong economic magnetism. Today, almost half a billion of the world's people live within sixty miles of volcanoes that pose a risk of eruption.[18] Young, fertile volcanic soils, dramatic landscapes, and, less often, deposits of precious stones or geothermal resources are the chief attractions.[19] Centuries can pass in such places without incident, but then, as in Montserrat, catastrophe strikes. In establishing Plymouth, St. Patrick's, and other settlements in the south of the island, white settlers were naturally attracted to the broad, gentle slopes that had been created by pyroclastic flows of the past. They also had a good excuse for building in a hazard zone: they didn't know that's what it was. Then, after being devastated by Hurricane Hugo in 1989, Plymouth was rebuilt in place, even though scientists knew by that time that the Soufrière Hills had erupted before and could do so again someday. A 1987 report to the Pan-Caribbean Disaster Prevention and Preparedness Program had weighed the risks that Montserrat's population faced from the volcano. The authors included as a worst-case scenario the very kind of deadly eruptions that would actually commence eight years later. But somehow the report vanished. Two years after its publication, Hurricane Hugo struck, and, according to one story, that was when the only copy of the report was lost, blown away in the storm.[20] While Plymouth was being rebuilt in the few years between Hugo and the start of the eruptions in 1995, there apparently was no discussion of relocating the city to safer ground. It's possible that had the 1987 report been widely read and discussed, the governments would have opted for re-location. But it's probably more likely that they would have decided to take their chances and rebuild in place anyway; the report had estimated the risk of total devastation at just 1 percent in a given century. Nevertheless, in the view of Parliament's International Development Committee, Montserrat could have been much better prepared for the eruption when it came had the report's conclusions been taken into account.[21]

Montserrat government officials don't necessarily hold London to blame for taking a 1-percent-per-century risk in the post-Hugo rebuilding. But they do believe that, having lost their entire investment on that bet, the Brit-ish had no desire to jump right back into rebuilding mode. Michael Jarvis, the director of information and communications, put it this way: "DFID is still treating this as an emergency. In twenty years, we've never made it to the response phase."[22] Since the earliest days, there has persisted a strong suspi-cion around the island that the lack of progress has been something more than benign neglect, that the unspoken goal of British policy in the 1990s was total evacuation of the island. The statements of UK officials them-selves sometimes hinted at this, as when the secretary of state for interna-tional development at the time, Claire Short, wrote, "It would be wrong to trap people on the island." Short and other UK officials emphatically denied all rumors of a depopulation policy, but without convincing many in Mont-serrat. Two chief ministers during the early crisis, David Brandt and Robin Meade, believed that total evacuation was coming.[23] In August 1997, pyro-clastic flows triggered the evacuation of Salem, a community where many people, businesses, and public functions had found themselves relocated after Plymouth was abandoned. This was an unexpected and devastating development, and it led to the announcement of a voluntary evacuation plan under which the British government would pay travel expenses for those wishing to leave the island entirely. Many of those who hadn't already been certain that Britain wanted everyone out now were convinced. Several months later, they would learn that a contingency plan for evacuation of the island, called Operation Exodus, had been adopted as early as January 1996 but was kept secret for more than two years.[24]

David Lea, a former missionary from the United States who lived through the entire volcanic crisis and has documented it extensively through a se-ries of video programs, says of those years, "We were being counted out."[25] Other expatriates, as well as current government officials and people from all walks of life, use similar words in describing the situation. Romeo is now the third chief minister to believe that, as he puts it, "the idea was to get us off the island." But, he adds, "It didn't work."

Pattullo has rejected the idea that British wanted to get everyone out for good, though. In her book, she wrote, "Belief in a 'conspiracy' theory perhaps said more about the Montserratian leadership's lack of familiarity with the ways of the British establishment than about the reality of the situation." [26] But Montserratian journalist Warren Cassell says that at the time, hard numbers—actual thresholds for evacuation—were actually being discussed: "The British said that if the population dropped below 25 percent of its former size, they would evacuate the whole island. So chief minister Brandt went to a meeting of the Caribbean nations and told them, 'Send your people. We are relaxing our law on work permits.' That kept our population above that 25 percent mark." But, he says, the exodus of so many Montserratians and the influx of many people from around the Caribbean "has damaged the social tapestry of the island." [27]

Those who resolved in the 1990s to stay in Montserrat found themselves descending into the nightmare world of the refugee. Pattullo wrote, "Some went to the shelters. Some even lived in sheds, boats, abandoned cars, or half-built houses without doors or windows. Families moved not once but two, three, four times as their original place of safety became hazardous or their options diminished or their money ran out." After the closing of the airport in 1997, a jetty at Little Bay in the north was the only way on or off the island (except, noted Pattullo, for the "elite," who could come and go by helicopter). [28]

"WE GOT THEM INTO OFFICE AND WE CAN GET THEM OUT!"

The ranks of British officials and other whites swelled during the early days of the eruption, and they were eyed with suspicion by local residents. Lea saw good reason for the widespread resentment he witnessed during that period: "Sixty or so DFID employees came to the island. They rented expensive villas. They were driving up and down the main road in their big rented SUVs, and here were people covered in ash, living in shelters, watching them go by." Yvonne Weekes, a Montserratian who served as the island government's first director of culture, recalled the mood of those days in her 2005 memoir *Volcano*: "I notice that all kinds of officials of various shades arrive and there are secret and seemingly high-level discussions, about which people are very suspicious. I say secret because whenever I am visiting a particular hotel, having a drink with friends or something, I notice that the Governor and a set of White scientists are sitting huddled together. Even though the volcano is monitored by the Seismic Research Centre, which is based in Trinidad, no scientist with my complexion is ever there." And when both governments, having evacuated Plymouth, had to find make-shift quarters elsewhere, Weekes and her friends enjoyed the discomfiture of higher-ranking officials: "The entire Ministry of Education is in a single room. Two of my colleagues and I decide to put our desks on the patio of the building which used to be some kind of guesthouse for the rich and fa-mous. There is a lot of tension as people juggle for space. High and low now have to rub shoulders. I am secretly amused. It is interesting how people bristle when they no longer have their Great House, when the servant and overseers now have to share the same quarters. There is no longer a master bedroom and it is a humbling experience for some. Me, I love it." [29]

As the numbers of displaced people grew, they soon overwhelmed ex-isting space in the public buildings being used as emergency shelters; a tiny elementary school with a single toilet was crowded with more than a hun-dred people at one point. More shelter space had to be created in the north. The worst of the new evacuation centers was a tent city erected in a place in the north called Gerald's Park (a large level field—a rarity on the island—that today is home to Montserrat's small airfield). The tents reportedly were full of holes and tended to blow down in the plateau's notoriously strong winds and rainstorms. Each evacuee got a cot, a blanket, and access to an open-air stove and a shallow pit latrine. The latrines were meant as a stop-gap to be used for a couple of days, but they remained in use for two years. Today talk of that tent camp still causes Montserratians to shudder. Hard temporary structures, which began to be built nine months after the first eruption in the interim capital, Brades, turned out to be small, hot boxes reminiscent of shipping containers. [30]

Montserrat's shelters were viewed as unlivable by the evacuees who were stuck in them. Weekes provided graphic descriptions, and Don Romeo took many hours of video footage to document the conditions and shelter resi-dents' stories. In a report to the governor, he wrote of "poor ventilation, un-hygienic and inadequate cooking facilities, minimal toilet facilities, lack of storage space, washing in buckets, health risk," and one of the most frustrat-ing problems, especially for women: lack of privacy. The temporary hospital was being run out of an elementary school, without much of the necessary medical equipment—the crucial items that had been abandoned back in Plymouth's brand-new yet soon-to-be-buried hospital. [31]

The British were frustrated. They have always pointed to the great uncertainty that hung over the whole situation. They had spent large sums over five years to rebuild Plymouth after Hugo, and that investment was lost. Now they weren't eager to invest much more until they knew how extensive the destruction from the notoriously fickle volcano would be. In his testimony to the International Development Committee, Governor Frank Savage said, "I wrote as early as 15th September 1995 to set out what I thought were the minimum requirements for the north of the island, and that included permanent accommodation for 1,000 people, a new hospital, a jetty, a police station with facilities. . . . Indeed, had the British Government acted upon my recommendation and had the volcano not gone to that level of activity, I conceive there might be another inquiry here now asking what idiot spent nearly £10 million of British taxpayers' money on something which was not required. So it was a tough call. . . . In retrospect, had I got what I asked for (and I do feel strongly as Governor that not sufficient weight was given to my views) we would have had what we needed at the northern end of the island, but had the volcano not gone that far I would have looked very foolish."[32]

Claire Short assumed her job as the international development secretary in the spring of 1997, as the volcanic crisis was tightening its grip. A Labour Party member, Short had announced that her department would focus its efforts and funds on eliminating poverty around the world, and she was not going to let the desperate pleas of one small colony's residents divert her from that global mission. She said at one point, "It would be weak politics if I said, 'They are making a noise and a row. Oh dear, give them more money.'" Short pushed back against requests for decent housing, a hospital, and infrastructure improvements, quipping that a "wish list" for Montserrat would include "golden elephants next." With that comment, she invited and received a battering in the media. She apologized, but didn't open the floodgates to aid.[33]

The IDC report that came out the next year implicitly scolded Short with this statement: "Our responsibilities to Dependent Territories citizens are of a greater and different order to our more general humanitarian responsibilities to the developing world and involve different priorities. That should be recognised in the structure of administration and funding."[34] In a report on the 1997 deaths addressed to the IDC, Her Majesty's Coroner Rhys Burriss stated, "Montserrat has many needs consequent upon the volcanic crisis but none is more pressing than that for many acres of land to be acquired in the north for permanent housing and for houses to be built on that land," but "the British Government response has been unimaginative, grudging and tardy."[35]

This history, which still looms over efforts to revive Montserrat's economy and society, is what the island's residents are eager to change. Janeen Lester runs the ramshackle but welcoming Sup's Bar and Grill along the main road in Brades. She immigrated to Montserrat from Jamaica after Hugo, just in time to live through the volcanic crisis, and has had to move twice during evacuations. "For the money that has come in, much more should have been done," she told us. "People with good salaries are sitting in offices while the rest of us work hard. Mismanagement is stifling us." She had no use for the previous premier, Robin Meade, who, she said, "thought he was a dictator." The PDM's 2014 electoral victory, she believed, had opened up a chance to get something done about the island's predicament. Asked about the new chief minister, Romeo, she said, "He's a good man. We'll see what he can do. But all the rest of the government people have to get behind him and work hard. If they don't, well, we got them into office and we can get them out!" But Lester recognized that, whoever is in power, Montserrat won't be able to hoist itself into prosperity by its own bootstraps. For that, she said, "the British have to step up." [36] Lester's views were echoed in various forms by almost every member of the community with whom we spoke.

CAN GRAY BE MADE SILVER?

Will Montserrat's long seismic ordeal ever end? Certainly, says chief volcano watcher Rod Stewart: "We have clear criteria for declaring the end of the eruption. But currently, our monitoring shows there is still something happening underneath the volcano. We need those things to go back to their baseline before we can think about calling an end to this eruption." For one thing, sensors show that the north side of the volcano continues to swell, indicating the continued rise of magma toward the surface. That doesn't necessarily mean there will be another big eruption, but it does mean that one can't be ruled out. Then there are the ever-present volcanic clouds laden with sulfur dioxide—350 tons of it per day, which, Stewart told us, is "well above what we'd expect if the eruption were over. If you are downwind from it, *it stinks*." [37] After so long, the mountain's behavior has become a part of everyday life in Montserrat; the island's one radio station, ZJB, includes the level of sulfur dioxide emissions in the nightly weather report.

But when the eruption is finally declared at an end, could the exclusion zone be resettled? Stewart would be far from certain about making that call. He said, "All volcanoes in the Caribbean are similar in having long gaps between eruptions. So we are expecting that if this eruption is declared over, it will be a very long time until the next one. On the other hand, this eruption is so incredibly unusual that it makes me want to be very careful. You can say it's been four hundred years since the last eruption. That doesn't mean it will be four hundred years until the next one. We're looking at a snapshot in time, which makes planning very difficult."[38] Even if the volcano goes completely quiet, some areas would remain hazardous. The kinds of lahars that buried Plymouth remain a serious threat to the Belham River valley, where the debris mining is going on. The ash mud in the riverbed is more than twenty-five feet deep. There is no way to cross safely during rainy periods. Stewart explained, "Such areas cannot be reoccupied for quite a while, until vegetation can establish on the slopes above and stabilize the soil so it won't wash down."

Some in Montserrat credit the clouds of geologic and climatic disaster with having produced a rainbow of sorts: the fortitude to face an unpredictable future. Back in the early days of their ordeal and soon after Hugo, Robin Meade, then the minister of finance and planning, had tried to summon up that disaster's potential for renewal when, according to Anthony Carrigan of the University of Leeds, Meade "claimed that the island 'needed a disaster just once to . . . bring us back into focus with ourselves' and retain a sense of community in the face of increased materialism."[39] The crucial words there are "just once." Meade doubtless reckoned that Hugo, surely, would be enough to bring Montserratians together. No one anticipated an eruption. But the post-hurricane transformation, if one might have happened, had no chance to take root before the eruption hit. Now it's the volcano that is seen as having boosted community spirit. In its broadcasts, ZJB adopted the slogan "The Spirit of Montserrat" in 2009 with the aim of "telling the world we are still here and life is happening in a beautiful and real way."[40] In 2015, we asked Rolston Patterson, acting director of infrastructure in the Ministry of Communications and Works, if the volcanic disaster had brought people together. He said, "That was inevitable. Folks had to evacuate and come north. Different communities now are one community. The people of Montserrat are resilient. Those who decided, whatever happens, to stay on here—that is *making* them resilient."[41] Not everyone shares Patterson's view. A Montserrat government official, Roselyn Cassell-Sealy, told Polly Pattullo in 2005, "I detest the word resilience . . . we're not resilient, we're stressed and burdened. You have to have an incredible will to live here, because sometimes we just want to run away." Citing such statements, Carrigan warned that "sustainable reconstruction" must "account for the entwined social, ecological, political, and historical dimensions of disaster," and that "involves building on but not romanticizing the positive elements of cultural responses to catastrophic events, which often include high levels of anger and disaffection." [42]

Despite all the attention the volcano gets, Don Romeo has not forgotten that Montserrat, like most Caribbean islands, still faces a diverse array of hazards. If there is to be a building boom, he said, "we want to have as many of our homes as possible be earthquake and hurricane resilient—and most of all, insurable! But I am still more concerned that people pay attention to the volcano. With hurricanes and quakes, it's simply a matter of having construction that can withstand it. With the volcano, it's a matter of buildings having to be in a safe place. If private people want to build homes where it's just somewhat safe, fine. But schools must be the safest place from the volcano, and be able to withstand other hazards." As a new government center was being planned for the very safe Little Bay area, the island's seat of government remained where it had been for seventeen years: on a hilltop in Brades, also unreachable by eruptions and overlooking the sea far below. Offices handling many of the territory's functions—the treasury, the customs office, the Ministry of Education and Labor, even the Supreme Court—were still housed there after all those years in squat, weatherbeaten prefabricated buildings.

Many, including Romeo, see expanded mining of the volcanic deposits as playing an important role in the island's economy for years to come. But the most urgent need—and on this there appears to be unanimous agreement—is for construction of a new port. In sharp contrast to the neighboring islands of Antigua and St. Kitts, Montserrat has a convex shoreline with no good harbors. Creating an artificial harbor at Little Bay in the northeast, where a small pier now provides the only sea access to the island, will be an enormous undertaking. Romeo told us, "The British know that the port is the key project that would unlock all the other investments, but they are delaying committing to it. They are saying we must find a private investor who wants to go 50-50."

Another costly infrastructure project will be the development of a geothermal power plant. But on an island that's struggling just to keep the lights on, the benefits could be great; currently, Montserrat's electricity supply comes from diesel generators—not so green and very costly. Rolston Patterson, who works with DFID on the geothermal project, told us, "We have made significant progress in redevelopment, but we're far from being sustainable. Geothermal is one mechanism to help bring back sustainability and forward development. From a pure engineering and scientific viewpoint, the project can provide all capacity necessary." [43] In an October

2014 progress report, DFID, which had spent more than £10 million on the project to that point, laid out the numbers. Two geothermal wells had been drilled, and a third was in the planning stage. One well is needed for recycling water back into the earth, and DFID's tests showed that the two other wells together could generate three megawatts of power, providing a good cushion over the two megawatts that would be needed to meet base electricity demand.[44] That cushion would probably be needed. Consumption is currently very modest, largely because electricity is so expensive; if cheaper geothermal power were available, homes and businesses would use more. Some are dreaming of even greater expansion, which could allow the island to host "green-powered" server farms or become an outsourcing destination for business. There has even been talk of becoming an energy exporter, running undersea power lines to Antigua and other islands. But to be a viable exporter, the utility reportedly would need to generate forty megawatts.[45] That much energy could be tapped only by drilling much closer to the volcano, in areas that currently are far too risky to exploit.

DISASTOURISM

In the shorter run, Montserrat is pinning its economic hopes on that old Caribbean standby, tourism. It's the chief reason that a new port is considered so crucial. The artificial harbor wouldn't be able to handle the mammoth cruise ships that dock over on St. Kitts and Antigua, but Romeo wants to see small ships and more yachts coming in, bringing "people who'd want to go on hikes through our hills and tour Plymouth and enjoy the quiet and tranquility of Montserrat." In the past, he said, "those people, who are generally quite wealthy, would buy homes here. In the future, more of them can visit in other ways." He is clearly envisioning a return to the pre-eruption tourism culture, when an official guide described the island as "The Way the Caribbean Used to Be." But now the biggest attraction will be the volcanic disasterscape in the south of the island. It will indeed look like Montserrat the way it used to be, at least periodically, during those long-ago episodes in geologic history when the hills would explode.

Romeo has concerns over a British requirement that the new port be half privately funded: "Once you build a port, and it's a 50-50 deal, companies need to turn a profit and costs go up. But we are thinking of a plan under which the private sector would have the tourist side of the dock, and the commercial cargo side would be run by the government. That could help." There was a small cruise ship with two hundred aboard due to arrive at Little Bay two days after we spoke. It was a trial run, a test of how the island could handle a group of the size that used to dock in Plymouth in the pre-eruption days. No single spot on the island could handle that many visitors anymore, so they were split up into five groups and taken in buses to the MVO, the evacuation zone, and David Lea's Hilltop Café, a combination eatery and nonprofit community center that displays a large collection of memorabilia salvaged from Plymouth. After the visit, Lea told us, "I think it went well all the way around."[46] Montserrat has always taken pride in being an unspoiled haven, free of the throngs of mega-cruise passengers, the trinket markets, and the gated resorts that crowd the coastlines of other Caribbean islands. How much tourism Montserrat can take on and still remain itself is an open question. There seems to be little concern at this point about reaching that threshold anytime soon. Lea said of the cruise trial run, "This size ship or smaller is perfect for Montserrat and would allow us to retain our 'off-the-beaten-path' feel. In my view, we should keep the big cruise ships *out* of Montserrat and create a proper harbor for the small ships and the sailing community, with all the things they need to avoid the upwind beat to Antigua. Anything over three hundred folks would be detrimental to Montserrat. The big-ship folks don't spend money anyhow. They would only want a plastic volcano and a T-shirt that said they survived the volcanic disaster. The small-ship folks are a whole different bunch."

Meanwhile, residential tourism, which gave the island its big economic boost from the 1960s through the 1980s and hung on through the volcanic crisis, is on the increase again. Romeo believes there's plenty of room for more homeowners from the cold north even in the truncated territory: "I intend to do my best to provide incentives, to encourage them to come. You own a home here, you spend maybe three or four hundred thousand [U.S. dollars] on it, and you either sell it or you maintain it. Sell it and our government gets taxes; maintain it and people get work. And you always have to have somebody to clean the pool." Most Montserratians don't seem to worry about deepening their economic dependence on wealthy foreigners such as the yachters and seasonal residents, viewing them instead with an attitude of casual tolerance. Warren Cassell is typical: "I have no problem with them. They create employment. And they don't get in the way, since they're not trying to do business here."

There are also hopes that someday the south of the island might become a major "disastourism" destination. The Montserrat Springs Hotel still overlooks the sea just north of Plymouth, and now that entry to its part of the island's abandoned sector is once again allowed, the curious can stop in at the hotel and walk through the ruins of upscale Caribbean tourism "the way it used to be." The red tile roofs have withstood the fallout from repeated eruptions well. The vault ceilings below, made of polished wooden planks, look almost new. But in the lobby and restaurant, the ceiling-fan blades, made of laminated wood, droop low in the humidity, and the floors are covered in thick layers of volcanic mud; someone has thoughtfully shoveled paths through the muck in some places. Ferns have taken over wherever a little light filters in. The pool deck affords good views of the gray expanse that was once Plymouth. But with no one to clean it, the pool has become a thriving wetland; more than half full of volcanic mud, it's topped by several inches of water and hosts a lush growth of tall sedge.

13

"WE DO THINGS BIG HERE"

Greensburg, Kansas, and Joplin, Missouri

Some people worry about these new green buildings with so many windows. "What about the next tornado?" they ask. So I ask them, "Do you want to live in fear or do you want to live your life?"
—Ruth Ann Wedel, resident of Greensburg, 2014[1]

It's obvious as soon as you reach the only traffic light in Kiowa County, Kansas, and take a turn south off U.S. Highway 54: there is something different about Greensburg. Small towns in southwestern Kansas tend to have dirt side streets, boarded-up storefronts on Main Street, and more weeds than lawn grass. Houses tend to be decrepit and in need of paint. In most towns, only the old shade trees seem to be thriving. But in Greensburg, everything is new. The two-block downtown area along Main Street is mostly bright new red brick. The city and county government buildings, along with the school, the hospital, the visitors' center, the arts center, the theater, and even the farm-equipment dealership, are strikingly designed. The ranch-style houses along the broad paved streets look as if they have been transplanted from the outer suburbs of some larger city. (There are a couple of notable exceptions: a cylindrical house in earth-tone stucco and a split geodesic dome home.) Another difference: you can crane your neck in almost any part of town and see most of the rest of the town. There are trees everywhere, but all are young saplings no taller than a person. Unlike other Kansas towns, Greensburg is as sun-drenched as the wheat fields and rangeland that surround it.

Greensburg is much smaller in area than Montserrat and is home to only one-fifth as many people. Like Montserrat, it was completely devastated by a disaster and lost more than half its population to out-migration. But whereas the Soufrière Hills continue to plague the Caribbean island two decades later, the storm that completely destroyed this small Kansas town came and went in a few minutes. Because of that, and because Greensburg had much greater access to resources for its recovery, its survivors had a much better chance to "build back better" and seek a silver lining in its tragedy—even though the lining they found turned out to be more green than silver. And despite some unique opportunities to make the most of disaster, this tiny town in the middle of a vast, economically moribund rural region faces a clouded future.

"A CRISIS IS A TERRIBLE THING TO WASTE"

Greensburg would still look like all those other southwestern Kansas towns had it not been flattened by a tornado of the strongest category, EF-5, on the night of May 4, 2007. Had that storm with its winds of more than two hundred miles per hour passed just a couple of miles to the east or west, Greensburg would still be differentiated from surrounding communities only by its two longtime tourist attractions: the world's largest hand-dug well and the world's largest pallasite meteorite.

On a turbulent Friday night when eighteen other, smaller tornadoes were spotted across Kansas, this funnel touched down at 9:00 about twenty miles south of Greensburg, stayed on the ground, and began churning north. Radar showed that this 1.7-mile-wide monster was staying strictly, it seemed maliciously, on a path straight toward the town.[2] It was to be, in the words of Thomas Fox—author of a book about the disaster titled *Green Town U.S.A.*—a collision of "a monster-sized storm with a pocket-sized city."[3] The tornado arrived at 9:46 with winds of 205 miles per hour, exited town to the north, did a final pigtail turn, and lifted from the ground at 10:05. Eleven people were killed and almost a thousand buildings were destroyed. Other than the Kiowa County Courthouse, a nineteenth-century downtown building, and the local grain elevator, almost nothing stood intact. Greensburg had been erased from the map, replaced by eight hundred thousand cubic yards of debris.

The federal, state, local, and volunteer response to the disaster was prompt. But once the rescue phase was complete, the recovery process veered sharply from the typical course. On Saturday night, twenty-four hours after the storm had passed through, the seeds of an effort to build back not just better but in a dramatically greener way were planted in the basement of the courthouse, where (as the community leaders of Gainesville, Georgia, had done more than seventy years before) Mayor Lonnie McCollum, city administrator Steve Hewitt, and a couple of others held an impromptu meeting to makes plans for the town's future.[4] One week later, Governor Kathleen Sibelius addressed the citizens of Kansas in a speech that is widely regarded as the kickoff of the campaign for a new and different Greensburg. As she framed it, "We have an opportunity of having the greenest town in rural America!"[5] Meanwhile, the day after the storm, Daniel Wallach and Catherine Hart, residents of the town of Macksville, an hour and a quarter's

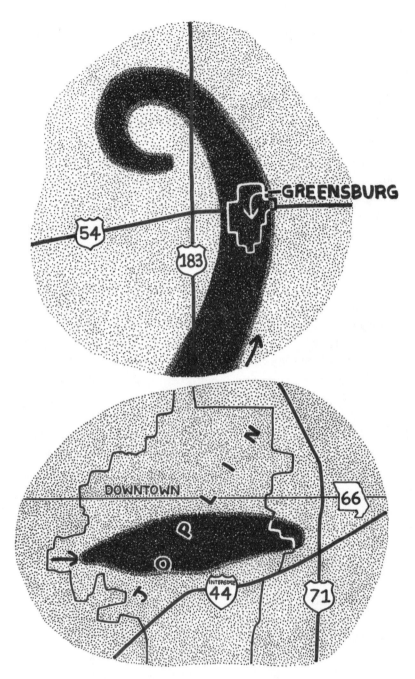

Figure 10. Paths of the tornadoes that destroyed Greensburg, Kansas, on May 4, 2007 (upper) and a swath of Joplin, Missouri, on May 22, 2011 (lower). The maps, which are drawn to the same scale, indicate that the tornado paths were approximately the same width.

drive to the northeast, had seen the devastation on television and also had begun discussing a green rebuilding effort. Within a few days, Wallach had written a statement that he later would present to a Greensburg town meeting. Wallach, Hart, and others eventually formed the nonprofit Greensburg GreenTown organization. Whatever the precise sequence, everyone in town seems to agree that the building-back-green plan was the inspiration of no single person or group.

Additional factors converged to give the environmental effort an extra push: a town name (a tribute to Donald "Cannonball" Green, a stagecoach-line operator) that could be repurposed;[6] the Big Well and meteorite, which had made Greensburg much better known around Kansas than most towns its size; the fact that almost the entire town had been destroyed, providing the opportunity to work with a clean slate; the fact that Greensburg had already been in steep decline at the time of the tornado, enhancing the appeal of a dramatic course change; and the prospect of ample funds for rebuilding, the bulk of which would come from the federal government. On this last point, the tornado's timing was a factor. The Bush administration had botched its response in Katrina-ravaged New Orleans twenty months previously—largely because, as Naomi Klein documented in her book *The Shock Doctrine*, economic and political powers in Baton Rouge and Washington saw Katrina as providing a blank slate for experiments in free market fundamentalism, not necessarily for building back better.[7] The administration, it was widely believed, was highly motivated to improve its disaster response image on what it saw as the politically and racially comfortable terrain of the Kansas plains.

The tornado brought a calamitous finish to a decline that Greensburg had been going through for decades. In the previous forty years, the town's population had dropped steadily.[8] At the time disaster struck, there were about fourteen hundred people living in Greensburg, a population three-quarters of what it had been in 1886. Economic activity had shrunk even more. Simply to rebuild the old Greensburg would only reset the downward spiral. Mayor McCollum later recalled, "We'd have to have something, or we'd be a newsflash for about three days, and then we'd be gone."[9] Architects Robert Berkebile and Stephen Hardy, whose Kansas City firm BNIM became deeply involved in the green building effort, wrote of the course change that "a crisis is a terrible thing to waste."[10]

The place that would become the new Greensburg had been buried under itself, and clearing the rubble had to be the first order of business. Anything that would burn—39,000 dump truck loads' worth—was hauled to a long trench and set alight. The fire burned for three months.[11] There was no housing left standing in the town's entire square mile, so some residents went to nearby towns or farms to stay with family members. Others moved away permanently. Many who wanted to stay and play a part in the rebuilding moved into a "FEMAville" comprising 298 trailers that were set out in a field beyond the town's southeast corner.

GREEN SLATE

Stacey Swearingen White, who chairs the Department of Urban Planning at the University of Kansas, has described how the green reconstruction effort began.

> FEMA developed the "Long-Term Community Recovery Plan for Greensburg and Kiowa County" in just three months following the tornado. FEMA worked together with a Kansas non-profit group, "Public Square Communities," to identify and prioritize the community's vision for its recovery. The expedited process included focused "citizen action team" meetings and larger community meetings where citizens discussed community assets and offered input on recovery ideas. Attendance at these four community meetings averaged 400 people, nearly 30 percent of the 2007 population and likely a large majority of those remaining in the area following the disaster. . . . One early step the Greensburg City Council took towards this desire to set a positive example was passing a resolution in December 2007 that requires all publicly-funded municipal buildings larger than 4,000 square feet to be built to "platinum" certification standards, the highest such certification level available under the U.S. Green Building Council's Leadership in Energy and Environmental Design (LEED) Program, which rates buildings according to a variety of environmentally-oriented criteria. Here Greensburg became an innovator on a global scale, as it is the first community worldwide to adopt LEED platinum standards for its city buildings.[12]

What came to be called the "LEED resolution," City Ordinance 2007-17, was not actually a strict mandate. It read, "It shall be the policy of the City that the design, construction, and operation of facilities and renovations with 4,000 square feet of occupied area, unless exempted by the City Council, . . . shall be designed to conform to the Platinum rating of the USGBC LEED Green Building Rating System."[13] In the end, no exemptions were needed, and all such public buildings were certified, most of them as Platinum,

LEED's highest rating. The first building to be certified Platinum was the nonprofit 5.4.7 Arts Center (named for the date of the tornado), which was designed by a group of graduate students in architecture at the University of Kansas.

The citywide "blueprint for a sustainable Greensburg" was the Sustainable Comprehensive Master Plan, which declared, "In Greensburg, protecting social equity and maintaining cultural heritage means establishing a framework for affordable, diverse housing, ensuring a mixed income range, and taking an inventory of the cultural qualities that made the town special. . . . By restoring the native environmental systems and utilizing natural capital, Greensburg can create a vibrant, sustainable rural economy." [14] Things got off to a promising start, as in these scenes described by Berkebile and Hardy:

> The first public recovery planning workshop set the tone for the recovery process. Hundreds of people gathered under a large tent erected on the east edge of town, eager to share their ideas for rebuilding. . . . The workshop was active; people moved around, looked at maps, and created their own drawings. . . . City staff, high school students, and other citizens presented the findings alongside the planning team. The planning team learned about the community values during this process, and just as importantly the community members reunited with their neighbors and formed stronger relationships. The workshop created a foundation for rebuilding; it was these relationships that constituted the solid bedrock on which Greensburg was rebuilt. . . . The tent remained a community gathering space throughout the recovery process, hosting several design workshops, community meetings, and even Sunday morning church service. [15]

Longtime Greensburg resident Ruth Ann Wedel was involved in those meetings. She told us, "We said, 'OK, we have a clean slate. If you could do anything you want, what would it be?' We decided we wanted a town with the kind of quality of life that would draw young people back and keep older people here." [16] Media coverage concentrated on the prospects for renewed economic activity and employment. White wrote, "Of particular concern to these residents in the weeks following the tornado were questions as to whether essential community services, such as a grocery store, would return, and if not, whether the city could even envision a viable future." Many people faced steep obstacles to rebuilding their homes, whether or not they followed the green path. Most residents had been insured only for the value of their existing houses, which had averaged about $46,500 in Greensburg in 2007. After the disaster, to build a new conventional house with a modest 1,300 square feet of floor space, three bedrooms, and two bathrooms would cost an estimated $120,000. The cost of any green features would come on top of that.[17] Higher property taxes would also raise hurdles; rates didn't increase, but higher home valuations boosted tax bills. Not surprisingly, Greensburg's post-tornado 50 percent population loss came primarily from the departure of elderly residents and lower-income residents. Many senior citizens decided they did not want to face as long as eighteen months in a FEMA trailer and then have to build a new, more expensive house from scratch in a town that bore no resemblance to the Greensburg they had known and loved. Renters, who had made up 30 percent of the city's population before the storm, had been able to work in the old Greensburg for as little as $10 an hour and still afford a place to live, with monthly rents as low as $300 to $350 for the town's aging housing stock. But they could not afford the $650 or more that it would cost to rent in the new Greensburg, with its new construction and tight housing supply. Many landlords who had been renting out old, often downscale houses declined to build modern replacements whose full cost would not be covered by insurance, and those who did rebuild were charging rents that covered their increased costs. Elderly people and renters could still find affordable housing in the larger town of Pratt, thirty miles to the east, or in other communities in the region, so many picked up and moved.[18] Few of the residents who remained in Greensburg were wealthy, and their new homes, while built to modern standards, were far from being large or deluxe.

No new building code was passed. There was no requirement that homeowners build back green, nor was there any monetary incentive to do so. But residents who chose to replace their destroyed houses, almost all of which had been old and drafty, with new, well-insulated houses saw an average 40 percent reduction in energy consumption.[19] Almost all residential construction was within the conventional range, however; it was nonresidential buildings that became explicitly green projects. The new public buildings, the arts center, and even the reconstructed John Deere agricultural machinery dealership were built to high environmental standards with architectural features that advertised their greenness. Structural materials were, as much as possible, highly energy-efficient, reused, or both. The exterior of the new Kiowa County School was composed partly of cypress boards salvaged from Hurricane Katrina. The 5.4.7 Arts Center used lumber from a decommissioned ammunition warehouse. The Kiowa County

Commons (which houses the public library, county extension office, and historical museum), city hall, and the SunChips Business Incubator (a facility funded by PepsiCo's Frito-Lay division, actor Leonardo DiCaprio, and the U.S. Department of Agriculture and designed to assist new businesses with their start-up) were all built with insulated concrete forms (ICF) in order to provide a high degree of energy conservation. The Kwik Shop on Route 54, a gas station/convenience store before the storm, was rebuilt as a full-fledged Dillons grocery store with green features. The local motel, the Best Western Night Watchman Inn, was rebuilt using energy-conserving structural insulated panels. The new Kiowa County Hospital was 40 percent more energy-efficient than a typical hospital of its size. Greensburg now claims to have the world's highest concentration of geothermal heat-pump systems. All three of its banks—a high concentration for a town its size—were rebuilt to green standards.[20] Finally, there are the ten wind turbines on the south edge of town with a generation capacity of about 12.5 megawatts. The entire town of Greensburg required a capacity of about 4.5 megawatts before the tornado, and that has dropped to about 3 since. (Coincidentally, the demand to be met by Montserrat's geothermal plant would be 3 megawatts, but for a population five times as large as Greensburg's.) The surplus power goes into the regional grid. The wind farm is privately owned, but the city gets renewable energy credits corresponding to 100 percent of its consumption. Dixson told us, "If we hadn't had the tornado, we probably could have gotten the financing and built a city-owned wind farm. But after the tornado, we just had way too much to deal with, so it's a private project."[21]

With its conventional-looking housing stock, the emerging Greensburg gave the appearance not so much of an eco-village as of a comfortably prosperous small town with an unusually large number of wind turbines. For the most part, the new look of the residential areas still does not shout "Green!" That's despite a vigorous attempt by the GreenTown organization to build a "chain of eco-homes" that would employ "both cutting-edge and 'retro' technologies." As of 2015, GreenTown occupied the town's one eco-home: the Silo House, whose cylindrical shape evokes that of the local grain elevator, one of the few structures left standing by the storm. Built with the support of sixty-five corporate and several individual donors, the Silo House has served as a model of green design and construction that could be toured by visitors; it even includes a bed-and-breakfast. GreenTown began work on a second eco-home, called the Meadowlark House, but as of 2014 the project had suffered a loss of funding, so windows had not been purchased and thus the exterior could not be finished. Viewed from the street, it looked like a dilapidated farm shed. Sitting in the Silo House looking across the street at the Meadowlark House, a GreenTown staff member told us, "I know, I know. Right now it looks like it's been through a tornado itself. A lot of the community regard it as an eyesore. We just need to get the funds to complete it. Some of the people here think sustainability is a one-time thing—build something and that's it. But it has to be an ongoing way of life."[22]

HUSHED CLIMATE

The town's green vision encompassed most aspects of sustainability, but some received more attention than others. The media outside Kansas pointed to the obvious connection between the impact of Greensburg's green campaign on its carbon footprint and climate science's projections of a stormier future. By reducing its emissions, Greensburg was potentially, in its own small way, helping reduce the annual number of tornadoes that residents of the Great Plains have to worry about.[23] But this build-back-greener effort was happening in western Kansas, where any discussion of climate change, let alone its relationship with fossil fuel emissions or increased storm activity, was contentious in the extreme. A survey of news stories on the rebuilding of Greensburg that appeared in the region's three major newspapers between 2007 and 2009 found no mention of climate change.[24] Instead the emphasis was on energy efficiency and renewable energy generation. Greensburg mayor Bob Dixson, like Miami's city officials, focuses his green efforts on issues other than climate; he told us this reflects the practical reality he's working in. "This is all about base values, strong, efficient buildings that maintain their value, conservation. You have to market the kind of terminology that will sell," he said.

And climate talk doesn't sell well in Kiowa County. There are oil and gas interests in the region, for one thing, but in the tornado year of 2007, other circumstances brought about the starkest polarization yet over climate change. Through that year there had raged a political battle royale over a proposal to build two new coal-fired power plants near Holcomb, a hundred miles west of Greensburg. Environmental groups across Kansas and the nation had mounted an extraordinarily effective campaign against what would in the past have been a routine application by Sunflower Electric Power Corporation to build the plants. The fight pitted climate activists in central and eastern Kansas, and from around the country, against western Kansas business interests and the local people who looked to them for jobs. Five months after the Greensburg tornado, Governor Sibelius and Secretary of Health and Environment Rod Bremby would make history by rejecting the Sunflower application on grounds that it would exacerbate global warming. The legal wrangling continued, however, and discussion of the global climate in any context remained particularly inflammatory in the region.[25]

Dixson, who was first elected a year after the tornado and has prominently led a green rebuilding effort in a part of the country where anything labeled "green" is eyed with suspicion, explained: "We stay out of these climate debates, because neither side is willing to compromise. We need a middle course, a commonsense approach. That's what we are trying to do here—to have an open, honest dialogue about things like efficiency." Dixson was serving on a presidentially appointed, twenty-six-member national group called the State, Local, and Tribal Leaders' Task Force on Climate Preparedness and Resilience.[26] This was no eco-vanguard, and Dixson's no climate warrior; he said the group's meetings focused not on controversial issues but rather "on how government can get better at customer service." The day before we talked in 2014, President Obama had announced a set of federal initiatives to cut carbon emissions and increase energy efficiency. Had the president relied on input from the task force in developing those initiatives? Dixson said, "Not at all. That was the White House on its own. That is something that makes me, I guess, uneasy."

The GreenTown nonprofit is giving climate and other environmental issues a much higher profile than the local government is. But even this group also holds back a little, not typically leading its arguments with the threat to the global climate. In the days following the disaster, Daniel Wallach wrote in the GreenTown mission statement, "Greensburg will have a unique identity that will appeal to new residents who are attracted to a progressive community dealing with issues of energy independence, innovative technology, and creative ways of living in more complementary ways with the natural environment."[27] Although it skirted climate issues, even this statement was a little too "green" for many Greensburg survivors, who were fine with energy conservation and maybe a few new technologies but who also preferred getting back to normal over living creatively with the natural environment.

The city government and the GreenTown initiative are following parallel but separate paths in the rebuilding effort. Mayor Dixson told us that he has been badly misunderstood as being a critic of GreenTown: "That is not the case. The green initiative is alive and well, and Greensburg is benefiting tremendously from their efforts. They are very important to this community. What I have referred to is how some perceive them here locally. It's a perception problem." The other problem, he said, is that it's very hard to form partnerships between municipalities and nonprofits. The most obvious difference is that the city has tax income while nonprofits have to raise contributions or get grants. People have complained that the city hasn't subsidized GreenTown's efforts. "But there are so many worthy nonprofits," Dixson says, "we couldn't support them all without having a big tax hike." Keeping a low climate profile may have been an effective tactic in marshaling local support; however, it remains to be seen whether in doing so Greensburg has severely limited the reach of its green message.

TWO YEARS OF ADRENALINE, MAYBE

Stacy Barnes, Greensburg's convention and tourism director as well as director of the 5.4.7 Arts Center (and Dixson's daughter), stresses a point one often hears: "The fact that every single person was affected by the tornado made for a stronger sense of community."[28] Stacey Swearingen White's study identified another strong theme to go along with that spirit of communitas. She found that "what media coverage of the months following the disaster made exceptionally clear is that the resilience of the community would be a pivotal force in its future. The notion of *resilience* is perhaps the most dominant theme among all the newspaper articles examined."[29] But even within a small local population suffering uniform, total destruction of property, those who had been better off economically before the storm could afford a better, more comfortable recovery—more resilience—than those who had been worse off. With the help of outside aid, the public and corporate infrastructure was built back better than ever, benefiting everyone in town. When it came to individual household economies, however, this storm was no more benign or egalitarian than any other.

Some residents don't like the fact that a few large tree stumps and a few pieces of foundation and wall have been left standing as memorials of a sort in the middle of town. GreenTown's Ruth Ann Wedel says that's an understandable reaction to what the town went through. She told us that in the three years it took to get Greensburg fully cleaned up, "we all had 'tornado brain.' There were far too many things to focus on. It became hard to get up in the morning. We were just so tired. So those reminders of that time really get some people down." There were also people in the community, according to Stacy Barnes, who initially did not want tornado exhibits included along with the well and meteorite in the new Big Well Museum. It would be too painful and disturbing, they said. "But, like it or not, it's part of our history now," added Barnes, and the tornado exhibit was included. "I realize that everyday people get tired of all the new stuff. They still have the feeling, 'I want the old Greensburg back.' I understand their fond memories.

But we *can't* have it back, can we? It was not a choice." On the other hand, Barnes noted, there are now some who have even become nostalgic about the days of the tent meetings that were held in the weeks after the tornado. In that period, there was no alternative; everyone had to pull together. There was a lot of adrenaline flowing, and the survivors who stayed formed bonds among themselves in a way that may have made the place seem even more insular than the typical small Kansas town. After the tornado, Wedel said, "the adrenaline lasted, I'd say, about two years. These days you hear some outsiders refer to Greensburg residents as 'unfriendly.' That's not fair. We're still dealing with a lot of emotion here. People have a feeling that if you weren't here at that time, you don't understand." She told us, "We all want things fast—we have no patience. If we're willing to stick it out, it's going to take new blood and a plan to reenergize people. We had a choice: to be a town or not be a town anymore. Daniel Wallach did a survey showing that small communities that thrive are ones that have something unique about them." That, to Wedel, was the storm's silver lining: "If Greensburg had one LEED building, we wouldn't be unique. But we have eleven! That makes us unique. So many positive things have come out of what happened here."[30]

One of those positive things has been the opportunity to exert influence far beyond the city limits. Since 2007, Greensburg has played a prominent role in an unofficial solidarity network of communities around the world that have suffered major disasters. City officials, GreenTown participants, and other residents have traveled to disaster sites and conferences across the United States as well as recovering disaster scenes around the world. In 2014, three months after Yolanda made landfall at Tacloban in the Philippines, two visitors representing the U.S. Agency for International Development made a stop at the GreenTown office the day before they were to depart for Tacloban. They were looking for information on the community involvement and planning processes that Greensburg had gone through; the impact of green planning on access to funding sources; the role of self-help, local leadership, and governance; and maintenance of morale and community spirit. They were especially interested in how Greensburg had organized its "public square meetings" after the tornado.[31]

AGAINST THE TIDE

In 2014, while talking with a GreenTown volunteer about the still-slow growth of Greensburg's population and economy, a reporter asked him, "So did you build back *too* green?" This is a common notion around Kansas— that the effort to build the "greenest town in America" has impeded the town's recovery. But Bob Dixson says there are other reasons for Greensburg still lingering in the economic doldrums. Once the development plan was in place, he told us, "some of the local people expected companies to be lining up at the city limits to come rushing in. We suffered from very high expectations that would have been hard to fulfill under any circumstances—and then 2008 happened." With the financial crash and Great Recession starting the year after the tornado, employers became even harder to rope in. But by 2014, he told us, "there has been some activity. We've had oil and gas leasing people here, and we continue to explore opportunities in that area. When the utility put through the big transmission line, it brought a lot of people to town, and they were spending here. Some are still here."

Bimal Paul, a professor of geography at Kansas State University, has studied the impacts of tornadoes not only in the United States but also in Bangladesh (which has suffered even more tornado deaths over the past century). He has concluded that today's Greensburg faces especially limited opportunities for growth. He says the town's biggest economic asset may be, of all things, U.S. 54. The highway, which runs a diagonal route through the nation's midsection from Illinois to the Mexican border at El Paso, brings almost continuous traffic right through Greensburg. Paul points out, "After all, this is the only green city within a thousand-mile radius. There are many people who would be interested. Ecotourism can draw many more visitors, providing development opportunities."[32] The Big Well Museum, the 5.4.7 Arts Center, and GreenTown's projects would be obvious focal points for such development. A local resident once said of the town's potential, "We do things big here. We have the biggest well, the biggest meteorite, and now we've had the biggest tornado. People will want to see it."[33]

If Greensburg does prosper, it will be bucking the region's economic tide, which is rapidly ebbing. Nature is not cooperating, either. The western Kansas climate continues to deteriorate, with severe drought and stronger straight-line winds creating Dust Bowl–like conditions at times.[34] Climatic models predict a worsening of drought in the region through the century. The rainfall that western Kansas does receive, some models predict, will come increasingly from powerful storms—the kind of storms that spin off tornadoes.[35] The odds of Greensburg taking another direct hit from an EF-5 twister in the foreseeable future will remain very low, but if the Great Plains climate becomes increasingly harsh, repopulation of the region will become even less likely. The region has never had enough rainfall for intensive agriculture; its lifeline has been the Ogallala Aquifer, a huge subterranean body of water underlying parts of six states. The water has been trapped there for eleven thousand years, since the end of the last ice age, and it is not replenished by rainfall. Parts of the aquifer have been badly depleted by the giant center-pivot irrigation systems that make it possible to grow corn and other feed and fuel crops in the region; the way things are going, the entire Ogallala will eventually be tapped to the point that it can no longer support agriculture. One by one, the center pivots will sputter to a final halt. If agricultural extraction were curbed, enough water remains to keep western Kansas's small towns going for centuries, however. The hitch is that without agriculture, the towns will have lost their economic foundation.

In its efforts to be sustainable, Greensburg is clearly dealing with many complexities, some particular to its time and place, others common to most small towns or even larger urban areas. In *Green Town U.S.A.*, Fox writes that the people of Greensburg have a much bigger mission ahead of them: to set an example for the rest of rural America, demonstrating how a sustainable rural community can work.[36] Yet Bimal Paul believes that for the typical small rural town seeking ecological sustainability, "Greensburg's model is not replicable." He, like others, cites the "clean slate" opportunity and the influx of public and private funds—each the by-product of catastrophe—as essential to the green building effort.[37] If other communities who have had the good fortune *not* to be 100 percent destroyed by a tornado fail to follow Greensburg's remarkable example, it's not their fault, and it's not Greensburg's fault. All communities on the Great Plains, including Greensburg, are enmeshed in national and world economies that tightly limit their options. Almost all of the avenues to employment and prosperity available to the new Greensburg are dependent on some of the very economic forces that are eroding the region's quality of life and disrupting the world's climate. The LEED-certified John Deere dealership brings to Kiowa County the very kinds of large-scale technologies that have industrialized and depopulated America's farm country and depleted the Ogallala. The green Kwik Shop feeds gasoline to the traffic coming down Highway 54, in a region where going anywhere means a long trip, and the personal vehicle is the only means of travel. Even as its wind farm has allowed Greensburg to reject coal as a source of electrical power, it is stuck with oil- and gas-related industries as its best hope for bringing in new jobs. Meanwhile, the foundation of the western Kansas economy, without which towns such as Greensburg would have no chance at all of recovery, remains industrial high-input agriculture, specifically the kind of industrial agriculture that exists primarily to supply corn, soybeans, sorghum, and other grains to supersized feedlots and the fuel ethanol and biodiesel industries—and which poses the biggest threat of all to the Great Plains' soil and water resources.

In a 1987 paper titled "The Great Plains: From Dust to Dust," published in the journal *Planning*, Deborah Epstein Popper and Frank Popper of Rutgers University sparked furious debate across the nation's midsection by suggesting that the days of agriculture in the western plains were numbered.[38] The depletion of the Ogallala was already a widely acknowledged reality, so the Poppers urged the establishment of a "Buffalo Commons" across the region. In their vision, most of the private land would be acquired by the federal government and become a commons owned by no one and everyone. They wrote, "The small cities of the plains will amount to urban islands in a shortgrass sea." If the Ogallala shrinks to a size that can support only sparse animal grazing, small industry, and a limited human population— or if depletion of that eleven-thousand-year-old water is restricted before that threshold is reached—the Buffalo Commons could well become a reality. The land could eventually restore itself, and the people who remain in towns such as Greensburg could achieve a pleasant though not affluent existence. In a second article, published at the turn of the millennium as bison populations were growing and human populations shrinking across the Great Plains, the Poppers noted that the Buffalo Commons was "materializing more quickly than we had anticipated."[39] But nothing had been done about what they (and Donald Worster in his 1979 book *Dust Bowl*) had identified as the chief source of the region's problems: the drive to wring excessive private profit out of an often parched landscape.[40] That problem persists, and until it is resolved, prospects for a smooth transition to a drier, quieter, less crisis-prone future remain under a dark cloud.

JOPLIN: A CITY CUT IN HALF

The Joplin, Missouri, tornado of May 22, 2011, offered what some saw as a natural experiment that might determine whether the "Greensburg model" could be replicated in a larger (but still not very large) city devastated by a tornado of similar power. Since 1950, on average, one EF-5 tornado occurs per year in the United States. But these strongest of windstorms come in streaks, so that average has little meaning year to year. Greensburg was the first EF-5 of the 2000s, coming seven years into the century. Parkersburg, Iowa, was hit by an EF-5 the following year.[41] Then came the extraordinary tornado season of 2011, when four EF-5 tornadoes hit northern Mississippi and Alabama on April 27, the most ever recorded in a single day. Then within a three-day period in May, EF-5 storms struck El Reno, Oklahoma, and Joplin. The storm that hit Joplin, three hundred miles east of Greensburg, remains the deadliest single tornado since 1947, and it left behind the United States' seventh highest recorded death toll ever from a tornado: 161.

The slow-moving storm took almost forty minutes to cut a west-to-east path almost a mile wide through the heart of Joplin and into the eastern suburb of Duquesne. It touched down near the Twin Hills Country Club and passed south of downtown, through a less affluent part of the city. Destruction along the main path was, as in Greensburg, almost total. The *Kansas City Star* summarized the results:

> The May 22 tornado took out a huge swath of the city's cheapest housing. With near surgical precision, it bore down on those who didn't have much to begin with—and Joplin already had more people living in poverty than the national average. The tornado killed 161 people and injured more than 900. It damaged or destroyed 7,500 houses and apartments. Many of those who lost their homes also lost their jobs, according to data assembled by the state Department of Economic Development. More than half the households affected by the tornado were low- to moderate-income families. Many of them, including seniors and the disabled, earned less than $10,000 a year.[42]

Although the tornado's path ran through Joplin's low-income midsection, it wasn't poor quality of housing that caused the high death toll.[43] In a storm that powerful, no one aboveground was safe, and the former lead-mining town's underlying geology precluded basements for most homes.[44] Many Joplin residents who were fortunate enough to survive the tornado had their lives turned upside down in its aftermath. A severe shortage of affordable housing, which persisted well after the storm, created some of the worst headaches for both residents and city officials. In the last half of 2011, the *Kansas City Star* reported, "Evictions spiked and rents soared. Scam artists are victimizing homeowners, and some landlords are taking advantage of renters. Residents have bickered over where to put low-income housing."[45]

Replacement of middle- to higher-value homes was remarkably rapid. Viewed three years after the disaster, the neighborhoods consisting of single-family homes had been largely rebuilt. At some high vantage points, it was possible to look around 360 degrees and see nothing but new structures, almost all finished, and few vacant lots. Street names that had been spray-painted on curbs after the storm—because it had obliterated not only street signs but all familiar landmarks as well—were still visible. The destroyed St. John's Regional Medical Center was being replaced by construction of Mercy Hospital in a new location, a new public elementary school was already in operation, and a new high school was being completed. All new schools would have EF-5-rated tornado shelters, with a total capacity of fifteen thousand, in order to shelter many residents without basements in addition to students. The schools are built with low-footprint materials, they can collect rainwater for irrigation, and their design makes use of natural daylight. The new public library was to feature plug-in parking spaces for electric vehicles.[46] A moving and powerful memorial to the tornado's victims had been completed in Cunningham Park, on a hilltop where the storm's winds had reached their peak velocity. Farther east, where the twister had crossed Range Line Road and taken out a long stretch of its commercial corridor, the entire lineup of fast-food establishments and chain retail stores had been replaced, with everything back where it had been before; residents said that just to look up and down the street, you'd think the storm hadn't even happened.

Despite the rapid rebuilding throughout the storm's path, however, construction of new apartment complexes had lagged. Work was still in progress in 2014, with a long way left to go. Many of Joplin's lost housing units had been apartments, and a large share of the deaths had occurred in those complexes. Elsewhere there were still some vacant lots, which had been sites of single-family rental houses that landlords had elected not to rebuild. An acute shortage of rental housing arose. Some landlords with undamaged properties outside the storm path took advantage of the tight housing situation to increase rents. Foreclosure and eviction rates shot up, as lienholders and landlords sought to find new occupants and take advantage of soaring real estate values and rental rates. A new moral low point may have been reached when some landlords refused to return deposits to tenants who had been forced to move out of damaged properties, claiming that the residents, some of whom had been left with little but the clothes they were wearing, had themselves caused some of the damage that was actually done by the tornado.[47]

In an effort to ease the situation, the State of Missouri approved $88 million worth of tax credits for private developers who would commit to building rent-controlled affordable housing. Minimum-income requirements for renters meant that many of the lower-income displaced families were not eligible for those tax-credit-supported apartments, however; they needed federally subsidized housing, and that was not easy to find. Meanwhile, proposals to locate some of the tax-credit housing units outside the debris zone, blending them into middle-income neighborhoods, were met with stiff resistance from homeowners worried about property values. In the end, six of the eight projects were built in what had been the poorer sections of town and was now the debris zone. Philip Berke, an expert in urban reconstruction at the University of North Carolina, told the *Star* that Joplin's reluctance to deviate from the pre-disaster status quo was not unusual, observing, "There's not a lot of motivation for change unless someone says, 'No, these people count and they are part of our community.' There's a window after a disaster when people have this kumbaya feeling and then indifference sets in."[48]

Instances of bad behavior notwithstanding, most in Joplin acted with compassion and goodwill. Soon after the disaster, a Citizens Advisory Recovery Team was formed. Led by resident Jane Cage—who, along with Joplin's citizens, was recognized for her efforts with the U.S. Department of Homeland Security's National Award for Resilience in 2012—the team sought to solicit and compile observations and recommendations from all of the city's citizens. The process bore no small resemblance to that undertaken in Greensburg's tent meetings. The group issued its detailed report of recommendations, *Listening to Joplin*, six months after the tornado. A long list of post-disaster goals in the report included encouragement of sustainable, energy-efficient construction; rebuilding of affordable housing in mixed-use neighborhoods; "adequate and properly placed" apartment buildings; school improvement; "emphasis upon bike and pedestrian use" in transportation planning; creation of parks and natural areas; "major, mixed use, anchor projects" and new commercial corridors to foster balanced growth; and retention of local businesses.[49] Progress has been made toward achieving those and other goals.

DIFFERENT TOWNS, DIFFERENT RECOVERIES

In Greensburg, we heard much about the spirit of cooperation and mutual aid that developed when residents of that town found themselves all in the same boat, with no neighborhood left standing. Mayor Dixson told us that the aftermaths of the tornadoes in Joplin and Moore (an Oklahoma City suburb hit by an EF-5 in 2013) were more typical. In those communities, he said, those living outside the tornado path struggled to understand what it's like to survive such a calamity and just wanted to get on with "normal" life. He said, "Too many people in Joplin were saying, 'Let's just clean this up and forget about it.'"[50] While acknowledging that some of that attitude was present, people in Joplin point out that the opposite reaction was also common. Andrew Whitehead, an accountant whose home lay a mere six blocks north of the zone of destruction and suffered no damage, says that he, like many others, was overcome with a case of "survivor's guilt" that led to a need to pitch in and help. Looking back from three years later, he said a cooperative, can-do attitude had indeed taken root throughout the city in the wake of the storm. President Obama also recognized this when he delivered a commencement speech to Joplin High School's first class to graduate after the tornado. Foreshadowing the "Spirit of Sandy" speech that New Jersey governor Chris Christie would be making the following year, Obama said, "And so, my deepest hope for all of you is that as you begin this new chapter in your life, you'll bring that spirit of Joplin to every place you travel, to everything you do. . . . That's the spirit that has allowed all of you to rebuild this city, and that's the same spirit we need right now to help rebuild America."[51]

Whitehead was particularly impressed by the work of the big-denomination local churches, which he characterized as having been rather inactive on socioeconomic issues before the tornado. In the aftermath, he said, "they became less elitist. They moved away from pure evangelizing and worked on providing the survivors the help they needed." Did that new approach persist as the years passed, though? "No," he said, "it's pretty much back to business as usual now."

Even with the national building-back-greener icon of Greensburg just half a day's drive away, Joplin has so far followed a more traditional rebuilding trajectory, restoring the zone of destruction as much as possible to its pre-tornado status, albeit with updated construction. A few residences were built with green features that included ICF blocks, structural insulated panels, recycled materials, and solar arrays. While the countless other new houses lining the streets of Joplin's south side appear sturdy and attractive and are mostly of modest size, they, like most of Greensburg's new homes, do not have explicitly green features. Nevertheless, they, too, are more energy-efficient than the older, often run-down housing stock they replaced.

Joplin's reasons for having kept its green side in the background seem fairly clear. Recall that the guidelines in Greensburg applied to public buildings, which were easiest to designate for green reconstruction. Although Joplin lost its hospital and twelve public school buildings (including the entire high school), its governmental office buildings lay outside the zone of destruction. For the many commercial and residential properties that were destroyed, highest priority was placed on rapid replacement with improved construction and safety standards, not high-profile attempts to reduce ecological footprints. This reflected more than just a desire to avoid political wrangling over climate change in a conservative community. In an economically stressed area like the part of town that bore the brunt of the tornado, people of limited resources, and their desperate need to get themselves a home and a place to work as soon as possible, took precedence over other considerations. For people of modest means to be able to build, buy, or rent super-insulated housing with solar electricity and hot water, geothermal climate control, natural water recycling, and other such features would have required beefy subsidies from the state or federal government—something that was not forthcoming in either Joplin or Greensburg.

Finally, Joplin, although not far away in highway miles, is a world away from the economic disaster sweeping the plains of western Kansas: it's an active economic hub of a region that encompasses the corners of four states. Furthermore, despite the terrible destruction and loss of life, Joplin was not wiped completely away in the way Greensburg was. Those who were displaced were seen as still being embedded in a large community with infrastructure and resources available. Although Joplin received relief assistance from many private and public sources, its fate was not the extraordinary blank slate framed by wheat fields that Greensburg's was; it was not the kind of situation that could draw the attention and funds of environmentally minded people from across the nation and world.

As Greensburg and Joplin illustrate, no two disasters are the same, even when the same types of hazards lie behind them. But there are some themes that arise repeatedly in times of disaster. These run throughout the preceding stories, from the Blue Mountains fires through the Lusi mudflow to the foot of the Soufrière Hills. There is the sometimes successful, sometimes wrongheaded search for silver linings. There is the consistent failure to find the brightest and most elusive silver lining of all, the wake-up call that inspires us to stop and reverse the global ecological crisis. There is the unexpected backfire from solutions such as engineered defenses, recovery through tourism, and the pursuit of wealth. There is the quest to make the most of ecological and social resilience, even if it has to be defined into existence and imposed on communities that may have different ideas. And there is the human production of geoclimatic hazards and vulnerabilities of all kinds. All of these and more converged in the Indian Himalaya in 2013, in a cataclysm that swept away almost everything that's supposed to be true about the nature of disaster.

14

WHEN MOUNTAINS FALL
Uttarakhand State, India

The leadership of the State has succumbed to the conventional model of development with its familiar and single-minded goal of creating monetary wealth. With utter disregard for the State's mountain character and its delicate ecosystems, successive governments have blindly pushed roads, dams, tunnels, bridges and unsafe buildings even in the most fragile regions.

—Ravi Chopra, *The Hindu*, June 25, 2013[1]

Before the flood, some of the old women who've been living in this valley for eighty years and have seen all these dangerous changes were warning us: "It is just a matter of time before all this leads to a big disaster!" Just three days before the flood, a local sadhu was warning, "Everyone had better get out of here!" They were right. When you disturb the Earth, you bring out a bad reaction from her, like she is regurgitating the destruction.

—Nandini, teacher, School for Natural Creativity, Guptkashi, India, 2014[2]

The Ganges River begins as four chief tributaries that spring from four glaciers atop the Himalaya. The sites of origin—from west to east, Yamunotri, Gangotri, Kedarnath, and Badrinath—are referred to collectively as the Char Dham, or "four abodes" of the gods, and their temples are among the holiest destinations for Hindu pilgrims. Each of the temples is at an elevation between ten thousand and twelve thousand feet, and they are approachable from the foothills only by long, bone-rattling drives through deep river gorges. The Char Dham pilgrimage has long been undertaken by the devout from all over India. But in the 2000s, in response to vigorous tourism promotion campaigns, pilgrim traffic into the state of Uttarakhand, home to the Char Dham, exploded. By 2012, the state was seeing up to 28 million pilgrims and tourists annually, twice its own population. Then the following year, the boom was brought to an abrupt halt by the deadliest disaster in the history of the Indian Himalaya.

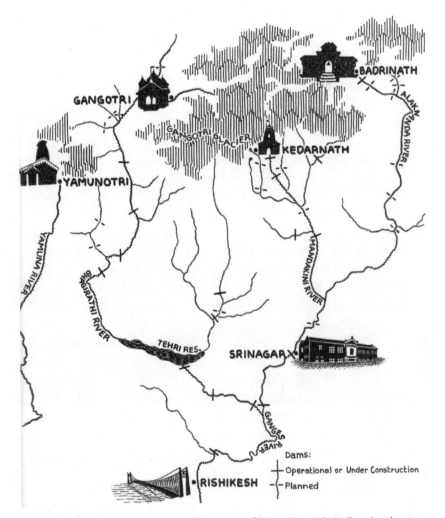

Figure 11. The Char Dham region of the state of Uttarakhand in India, showing temples situated at the four sources of the Ganges River and locations of some of the major hydroelectric dams that are either working, under construction, or planned.

TEMPLE OF DOOM

Between June 15 and 17, 2013, parts of the Char Dham region were pounded by nine to twelve inches of rainfall. This was vastly more precipitation than had fallen on the region in any other three-day period on record. The timing was also unusual; until recent years, the monsoon season did not normally begin until July. The results were dramatic. Walls of water roared through hundreds of miles of river valleys, and huge swaths of mountainsides loosened and tumbled. Writing a month after the event, geologists S.P. Sati of Garhwal University and Vineet Gahalaut of India's National Geophysical Research Institute took stock of the damage.

> Unofficial claims put the death toll at more than 10,000, damage to about 8000 km motor roads, and severe damage to 200 bridges. At least 30 hydropower projects under construction or completed have been either completely destroyed or severely damaged. About 30 governmental establishments including paramilitary camps, guest houses, food outlets, government offices and schools have also been destroyed either completely or partially. An initial assessment made by us suggests that more than 5000 hill villages and thirty urban clusters of the hills have been affected severely.[3]

Estimates that took into account "invisible" people, such as seasonal Nepali workers in the tourist industry, put the death toll at thirty thousand or more.[4] The worst-hit site was Kedarnath, the highest of the shrines in both elevation and sacredness. The temple, surrounded by a seasonal village, is situated about two miles downslope from the foot of the massive Chorabari glacier. The glacier is the source of the Mandakini River, which flows right by the temple and down through mountain gorges to join the other tributaries of the Ganges. At about 5:30 on the evening of June 16, a flash flood laden with sediment and rock roared into Kedarnath, burying parts of the settlement. The wall of water raced down the Mandakini as the downpour continued; together, they obliterated every trace of the village of Rambara, two miles farther downstream. But with the torrential rain continuing, a much bigger hazard was taking shape at the foot of the glacier. There Chorabari Lake, which lay behind a natural dam of debris and ice formed by the glacier, was reaching a crisis point. Days of rain were swelling the lake and melting and waterlogging the rock-and-ice wall that held the lake in place—a dam that was already badly weakened by unusually warm temperatures. A little before 7:00 a.m. on the seventeenth the glacial dam broke, and the lake—by then a quarter-mile long, seven hundred feet wide, and up to sixty feet deep—emptied completely within five to ten minutes. The surge was powerful enough to carry enormous boulders and a huge load of mud (and possibly the ashes of Mahatma Gandhi, some of which had been scattered in the lake in 1948) down through Kedarnath, Rambada, and far beyond, drowning, burying, or crushing to death any person or structure in its path. There had been an estimated 35,000 visitors in and around Kedarnath at the time, along with uncounted thousands of hotel and restaurant workers, porters, shopkeepers, and itinerant holy men. Some people, such as Ramala Khumriyal and his children (see pages 1–3), could make it up the valley slopes in time to survive the initial blow, but they were left soaked and shivering, with the roads and trail to the outside world cut off and all food, water, and shelter swept away. Many more died in the following days.[5]

Meanwhile, the torrent of water, mud, and rock was surging down the Mandakini, wrecking village after village. Far downstream, the flood merged with other rivers swollen beyond recognition: the Alaknanda, coming down from Badrinath, and the Bhagirathi, from Gangotri. Taking with it everything along the riverbanks, including goodly portions of the banks themselves, the river—now officially the Ganges—spilled out into the flatlands. At the holy city of Rishikesh, it ascended to twenty feet above flood stage.

"THIS ISN'T OVER"

We arrived in the Mandakini valley seven months after the flood, aiming to go up the gorge to Guptkashi, a mountain town that's the jumping-off point for the journey to Kedarnath. Much of the main road up the valley had been wiped out by the June landslides. With only temporary repairs having been completed, and with continued efforts at daredevil road reconstruction causing long delays, the 110-mile trip from Rishikesh to Guptkashi took nine hours. The roads were deeply undercut in some places, jutting out over nothing but thin mountkain air. After staying overnight in Guptkashi, where a few year-round hotels were still operating, we continued north on the road toward Kedarnath. In the dramatic landscape, the only horizontal surfaces to be seen were a few large concrete aprons cut into the steep slopes: helicopter pads. In recent years, a rash of new enterprises had set up shop in the region, offering to deliver well-heeled tourists and worshipers directly to the Kedarnath temple, so that they would not have to endure the hair-raising jeep journey to the crowded, chaotic village of Gaurikund at road's end or the traditional pilgrim's ten-mile trek on foot from there to Kedarnath. By 2013, the high-decibel racket from hundreds of chopper flights a day during

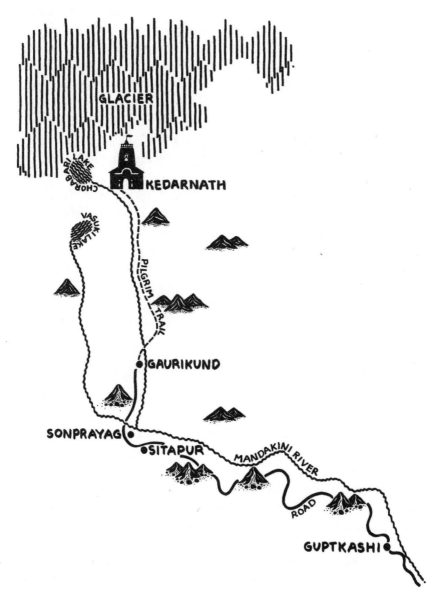

Figure 12. The Kedarnath shrine and the uppermost stretch of the Mandakini River, the area devastated by a glacial lake outburst flood on June 17, 2013.

pilgrimage season had become a serious nuisance for local residents, but it was tolerated because the "heli-resorts" around Gaurikund were providing well-paying jobs for several months a year. Now those jobs had been swept away, at least for a while, by the flood.

Just above Guptkashi on the road toward Kedarnath, we came across a crew working to repair one of the helipads used for airlifting pilgrims. Of about four hundred such structures built in recent years throughout the Char Dham region, only three or four had been left undamaged by the 2013 catastrophe. This pad had been hit hard by water and debris that cascaded from the top of the excavation scar. The structure, with a reinforced concrete landing surface supported by a stone-and-masonry wall, had been sandwiched between the road and the steep, bare face of the road cut in a fold of the mountainside, so it could not have been more badly exposed. The torrent coming down the ravine had undercut the wall on one side, bringing with it rocky debris that tumbled far down the slope, taking most of the road with it. Despite that, the pad was being rebuilt right back in the same spot. During the helipad building boom, the practice had been for companies to site the pads wherever they could get land the cheapest, not where the risks were lowest.[6] That was clearly the case here, and similar efforts were under way across the region; with roads so badly damaged and vulnerable, resumption of the tourist airlift, whatever the risks, was viewed as the best opportunity to get the mountain economy going again.

A few miles farther upriver is the town of Sitapur. There a dam under construction had reached a height of about thirty feet by July 2013, and the reservoir was still filling. A few minutes after bursting out of the glacial lake above Kedarnath, the wall of water reached Sitapur, instantly wiping out the unfinished blockade. The waters that had been contained behind the dam joined the flood. Now, strung along what would have been the shore of the empty reservoir there remained a phalanx of tall tourist hotels overlooking a vast gravel lot. Here and there objects, including the rusting remains of a bus, protruded from the stones.

Upstream from Sitapur lay the remains of Sonprayag, where every building was either gone or partially buried. More than half a year after the disaster, it was the last settlement along the pilgrim road that was accessible to even the most daring driver. Like a town in a film Western, Sonprayag's main street had a front but no back—between it and the river lay nothing but an expanse of rocky debris. Towering just upstream, still alive and dangerous, was the wound of a great landslide that had supplied much of that debris. A closer look at a row of buildings left standing revealed that they were actually one story taller than they appeared; the ground floors were completely filled and buried by sediment and rock. A few shops and a tea stall had reopened on the old second floors, now at street level.[7] Cars abandoned at the time of the flood—their owners either dead, missing, or alive with no intention of ever returning—lined the street, some simply parked and others sprouting at weird angles from the debris. Idle men stood in the street while working men were shoring up walls and scooping sand from the former ground floor of a building, excavating what was now a basement. Hauling the sand out in bags, they loaded it onto waiting donkeys. Hammer blows chirped across the valley as women worked out on the debris field, breaking up mighty rocks into smaller pieces that could be used along with the bagged sand in rebuilding Sonprayag.

Above Sonprayag, just beyond the giant landslide scar, we found a road crew belonging to the General Reserve Engineer Force under India's Ministry of Defence. The crew leader, Chaddar Daar Singh, sat with two coworkers atop a huge bulldozer. Dressed in brown coveralls and a stocking cap, he was taking a break, having a cup of tea without milk. The road's end was just ahead; beyond, where it had once continued up the gorge, there was now empty space. A little upstream, two rock faces constricted the river's flow into an extremely narrow gap. According to Singh, the water that roared down from Kedarnath on June 17 had hit the rocks and vaulted over, rising a hundred, maybe a hundred and fifty feet in what locals were calling a "tsunami." It had ripped away the river's banks and the road, leaving a smooth rock face. The crew had no choice but to cut into the rock face from both ends, employing a pair of excavators fitted with big pneumatic jackhammer-like attachments. The "rock breaker" on Singh's side had come up by road; the other had been brought up via the riverbed and then hauled up the bank on the upstream side of the rock face. Each was making a fresh road cut, but on an upward incline, so that they would meet at the center of the rock face at the top of a hump, then later grade it downward. That was the routine approach in the river gorges wherever the bank had been washed away—Singh said that having the grading equipment work over a hump reduced the potential for collapse of the slope. The upstream jackhammer rig was idle that day. It had been damaged in a landslide from above, and a new one was making its way up the riverbed to replace it. The crew, Singh said, had lost six or maybe seven of the big machines in the flood and its aftermath. One had recently tumbled into the riverbed, and the crew had spent four days hauling it back up. No one mentioned whether there were human casualties in these accidents. We asked a member of the crew if he ever felt frightened while doing this sort of work. Before he could speak, Singh answered for him: "The government pays us a salary to do this work, so we don't worry about being afraid. Anyway, we spend our whole lives in the hill areas. Our life is very tough as it is."

It was time to move on, but getting to Gaurikund from where we'd been chatting with Singh was not a simple matter. Following his men's advice, we picked our way down to the riverbed, clambered upstream about a hundred yards, and then ascended back to the road on the upstream side of the broken-down jackhammer rig. A short distance up the road, we encountered what locals call the "parking lot," where two tourist buses and a few SUVs had been stranded back in June; clothing and photographs from the tourists' luggage remained strewn on the ground. A few more yards ahead, the road again disappeared, so we had to climb straight up the slope to a pilgrimage trail near the top of the ridge. The old footpath had seen very little use in the decades that the road had been open. But after the flood, the Engineer Force had repaired and paved most of the path in rock to provide their only direct access to Gaurikund while they were rebuilding the road.

Gaurikund is where the final trek up to Kedarnath traditionally begins; pilgrims and tourists cover the ten miles either on foot or on the back of a donkey or human porter. The town grips one side of a narrow gorge, and its lowest row of buildings teeters on the edge of the riverbed. In the flood, the lower part of this town had met a fate like that of Sonprayag and Sitapur; at least seventy-five cafés, lodges, and shops had been scoured away. As in most of the towns along the route, year-round residents lived higher up the slope in what was known as "the village," while the now-obliterated riverbank area had been all either tourist infrastructure or the lodgings of Nepalese immigrant workers. Among the very few people we encountered in the town was a force of twelve police who had been posted there for a four-month stint after the flood. Their mission, they said, was to prevent looting, but up to that point there had been no looting or other crime of any kind. They were passing the time by playing carrom.

Beyond the police station, the pilgrims' footpath was lined with shops on both sides, all but two of them shuttered. A few men were hanging around, also playing carrom. No one had anything good to say about the disaster response. It had taken three months after the flood for even a single outside official to show up along the town's footpath, and now, four more months after that, government aid still had not arrived. Shopkeepers and hoteliers still had not received reimbursement for food they had provided to survivors who had taken refuge in Gaurikund in the week after the flood. Were aid organizations to offer assistance at this point, the locals said, the biggest help would be to restore some of the area's tourist attractions, such as a nearby temple and a hot springs, which had been buried and filled with mud. They needed somehow to be able to make a living. But so far not a single aid group had contacted them. As for the future, they said, farming was not an option; the slopes around Gaurikund were too steep even for terracing. The coming year would be bad for tourism, but the men were confident that in the next few years the pilgrimage season would surely bounce back—if the road up the gorge was fully restored.

As we were talking with the Gaurikund residents, two travelers approached from the south. Keeping their eyes on the footpath ahead, they silently brushed past us and continued north out of town, vanishing around a curve. The man striding along in front wore a light wool jacket and scarf, while the other carried an expensive-looking camera and led two donkeys. Ten minutes later, the travelers reappeared. The well-dressed man was now riding one of the donkeys. The cameraman paused briefly to explain to one of the locals that his companion was a representative sent by the government in New Delhi to report on the situation in and around Kedarnath. But it appeared that instead of actually making the difficult ten-mile ascent to the shrine on the still ravaged and now snow-choked footpath, he'd gone just around the bend to the trailhead to have his photo taken, sitting on the donkey with the picturesque mountains looming in the background. This he could show as evidence that he'd made the trip.

The strange encounter prompted one of the locals to cut loose with a string of obscenities about the government's failure to provide his town any relief of any kind. India would be holding parliamentary elections in a few weeks, and the man—who said he was seventy-seven years old but wouldn't give his name—told us, "We hate *all* of the political parties now, and if any campaigners come up here, they will get a hot welcome. If they bring a ballot machine up for the election, nobody will touch it." Speaking of the flood disaster, he declared, "This isn't over. More is coming!"

ROADS TO RUIN

In the months after the Uttarakhand floods, the role played by climate change in the disaster was almost universally acknowledged—and with it, the truth of the old man's warning. In one of the first articles to be published in an academic journal after the event, M. Balasubramanian and P.J. Dilip Kumar of the Institute for Social and Economic Change in Bangalore wrote,

> Some observers have commented that it was an expression of the wrath of nature at the disruption caused to a natural ecosystem by accelerated infrastructure development. There is a grain of truth in

this. We need to understand this while the horror of the situation is still upon us, lest we forget and go back to our usual short-sighted ways once the immediate crisis has passed.

First, the floods and devastation in Uttarakhand are probably linked to the effects of global warming and climate change, which until now was not seen as something imminent, but as part of a distant future scenario. We need, therefore, to treat this as a wake-up call and reformulate our thinking on climate change. No longer is it something that future generations are going to have to deal with, the initial stage is already upon us. It has attained the status of the immanent, or something that is already here, and we have to factor in climate change and its effects just as we were so far doing with climate itself.[8]

It had long been known that Kedarnath and the other shrines were badly exposed to both climatic and seismic hazards. They are located close to the geological juncture known as the Main Central Thrust, between the Lesser and Greater Himalaya. The Greater Himalaya—a towering wall that forms a backstop for the southwest monsoon—makes the Char Dham region especially susceptible to intense rainstorms like the one that hit in 2013. Furthermore, the glaciers that loom above the shrines have created moraines capable of damming up deceptively beautiful lakes such as Chorabari. There are at least two hundred such glacial lakes in the Himalaya that are very likely to swell during the summer under a warming climate; at the same time, their natural dams weaken, threatening to discharge the lakes' entire contents in so-called glacial lake outburst floods, like the one that struck Kedarnath. Finally, there is the constant threat of earthquakes. The uplift of the Himalaya that began 40 million years ago when the Indian subcontinent slammed into Asia is still under way. The greatest geological strains, along with the greatest risks of earthquakes and slope failure, occur along the Main Central Thrust. A magnitude-6.8 quake struck Uttarakhand in 1991, and eight smaller ones have hit since.[9] Then, on April 25, 2015, the worst-case scenario struck the heart of nearby Nepal. The devastating magnitude-7.8 quake left behind a death toll approaching nine thousand and a toll of misery that can never be quantified. Shaking was minor in Uttarakhand, but it frightened tourists away from what was supposed to be the Char Dham's first season of recovery.[10]

The potential for disasters has grown throughout the 1990s and 2000s. A commentary published by the Geological Society of India immediately after the 2013 tragedy stressed that the human hand of causation could be seen not only at the site of the outburst and not only in climate disruption: "The construction of dams, tunnels, roads and civil structures, drilling, mining and large-scale deforestation flouting all ecological laws of the terrain in the State's development spree, have all weakened the slope and loosened the soil," triggering landslides throughout Uttarakhand.[11] Across the Himalaya, two-thirds of landslides are triggered by such economic activities.[12]

We saw an unmistakable trend along the Mandakini River after the flood: while terraced wheat-growing slopes had held up very well to the record rainfall, landslides had occurred in countless spots where roads or helipads were cut into the slopes. The only landsliding we saw in terraced fields—and only in a small minority of them—was at the bottom along the river, where the deluge had wholly inundated and ripped away the entire riverbank, terraced or not. But at many road cuts and helipad sites, slides extended from near the top of the excavated surface all the way down to the riverbed. According to Vineet Gahalaut, a native of Uttarakhand, a senior geologist at India's National Geophysical Research Institute, and an expert on the 2013 disaster, our observations reflected a general reality in the region. Road construction, not agriculture, was to blame for the bulk of the landslides that occurred during the floods.[13]

It is a common misperception that the frequent occurrence of slope failures and related hazards in the Himalaya is a result of population growth, the accompanying scarcity of usable land, and the infiltration of people deeper and higher into the mountains, where they chop down trees to clear land for crops and to fuel cooking fires. Richard Marston, a professor of geography at Kansas State University, has demonstrated the falsity of this idea, one that he calls "a series of interlinked propositions based on the casual observations and conventional thinking of many." Countless studies in the Himalaya, including in the mountains of Uttarakhand, have, he says, shown that occurrence of landslides is not associated with deforestation or terracing but *is* closely linked to road building. He says, "Terracing improves the stability of slopes, even though they have been deforested. Heavy rains can cause breaks in some terraces, but farmers quickly repair the damage." On the other hand, when a middle section of a slope is carved away to expose a large, bare, and even steeper face (and this on a mountain range still being pushed upward by tectonic forces), collapse becomes much more likely.[14]

Most of India's border with its bitter rival Pakistan and uneasy neighbor China runs through steep mountains, so one reason for risky mountainside road building, according to the government, is to strengthen national defense.[15] But a stronger motivation, it would seem, was the surging tide of pilgrims and tourists eager to visit the sources of the Ganges. At the time of the flood, the tourist industry accounted for an astounding 27 percent of Uttarakhand state's gross domestic product, a dependency on tourism that's on par with the Caribbean island of Aruba. According to *Down to Earth*, India's leading environmental publication, almost 80 percent of visitors arriving in the state were undertaking what is called the Char Dham Yatra, the grand tour of all four abodes of the gods. That trip requires the driver of a bus or SUV to haul a group of pilgrims up through the river gorge leading toward one of the shrines for a brief visit, then descend all the way to the foothills, drive to the next river valley for another ascent, and repeat the process until all four shrines have been checked off the list. *Down to Earth* reported that in the years before 2013, "pilgrims were charged a hefty sum of money for a whirlwind tour of the four pilgrimage sites."[16] The economic impact has been dramatic. Not long before the flood, facilities at Kedarnath were augmented so that they could handle up to fifteen thousand visitors, and all the way down the river construction was extending willy-nilly into floodplains where only crops had stood before.[17] The state's transport department estimated that vehicle traffic spawned by the pilgrimage industry grew by an incredible 1,000 percent just between 2005 and 2013. That has required the building and widening of roads throughout the state, which has left bare faces of soil and rock all along the river gorges.[18] The pilgrimage industry, built on the barefoot wanderings of holy men, very quickly became a juggernaut that changed the face of Uttarakhand.

Alan Ziegler of the National University of Singapore and eight colleagues wrote of the Uttarakhand catastrophe in 2014 that

> throughout the area affected by the heavy rainfall, about 2,400 landslides occurred on steep slopes above the river channel, delivering an enormous and as yet unquantified load of coarse sediment into the stream, including huge boulders. The high energy of the "sediment-laden" floodwaters had a cascading effect. As riverbanks were eroded and landslides occurred on foot slopes undercut by the floodwaters, even more hillslope material washed into the river, possibly bulking it up and raising the flood height. In some locations, the deposition of coarse sediment during the flood elevated the riverbed 30 to 50 meters. The river channel now has a greatly reduced capacity to transport high flows and, subsequently, is more prone to flooding vulnerable buildings than before the event.[19]

They added that "while the disaster was unprecedented, the underlying hazards leading to it are not and will happen again."

HIGH ELECTRIC BILL

Seventy-five miles downstream from Kedarnath on the Alaknanda River, the city of Srinagar lies well south of the major 2013 landslide zones.[20] But the supercharged deluge did some of its most remarkable damage here, where it buried a large number of homes, schools, and other buildings in fine, densely packed silt almost a full story deep. Some initially believed the silt had been carried in from landslides farther up in the mountains, but that was not the case. According to S.P. Sati of Garhwal University's Department of Geology in Srinagar, most sediments from the hundreds of landslides that struck the upper reaches of the Alaknanda and Mandakini Rivers were redeposited well before they reached Srinagar. Any remaining sediments that came down the river would have settled out before reaching the newly constructed hydroelectric power project upstream from the city, which during the flood had backed water up for four miles behind the dam, slowing the current. Rather, says Sati, the silt that buried Srinagar was very much of local origin. In an e-mail, he told us, "The disaster was in fact caused by material that had been generated by the dam construction project and dumped along the river bank without any protective wall. About five hundred thousand cubic meters of this material washed away in a flash and was deposited just downstream in Srinagar. The chemical and mineral composition of the material has striking similarity to the dumped muck that remains near the dam."[21] Thus a catastrophic flood triggered at Kedarnath by fossil-fueled climate disruption buried Srinagar, seventy-five miles away, thanks to an effort to build hydroelectric capacity as a substitute for fossil energy. In the genesis of the Uttarakhand tragedy, the drive to satisfy demand for electricity was almost as important as the drive to satisfy demand for religious experience.

In 2013, there were ninety-eight hydroelectric projects with a capacity of 3,600 megawatts packed into little Uttarakhand. But an energy-hungry India saw room for more—a lot more. In the same year, there were 238 projects planned or under construction, with a capacity of 23.5 gigawatts.[22] If all were built, the rivers that form the Ganges would be interrupted by an average of one dam every eleven miles.[23] The steep rise in electricity demand is coming primarily from large businesses and affluent urban households. Much of the electricity generated by the planned hydropower development would be transmitted out of Uttarakhand to be consumed in New Delhi and other large cities of India's Ganges basin. Academics and environmentalists stress that dam construction, by requiring massive excavation, road building, and transport of materials, made the region much more vulnerable to flooding, landsliding, and other devastation, including the burial of Srinagar.[24] And there are plenty of other reasons, including protection of globally unique biodiversity, to exercise caution in dam building.[25] Vineet Gahalaut told us, "There is an argument whether dam construction increased or decreased the devastation. To me, dams saved some of the cities in the downstream side. But citing this as an argument in favor of dams is not a good idea." For one thing, he said, the region is overdue for that killer earthquake. "Imagine the bursting of a dam in the event of a great earthquake in the Himalaya, something we talk about all the time. It will cause havoc!" Gahalaut says that the last time a big quake released severe seismic strain in this region was way back in 1803, and he stresses that today "there is unequivocal evidence of strain buildup—*this region is ready to go.*" The accumulation of strain in the vicinity of the Char Dham makes that area especially fragile, and, he adds, "the slopes which are already steep and are then cut further by humans become more susceptible for failure. We are not taking account of the impact on ecology and environment. We are also not taking care of seismological inputs while constructing these big dams." (Gahalaut later noted that the April 2015 Nepal earthquake was much too far from India's Char Dham region to have relieved any of the strain he was concerned about.)[26]

TOURIST TRAP

Climatic or seismic hazards set off major disasters almost every year somewhere on the South Asian subcontinent. For more than a decade before and after the 2013 Uttarakhand floods, every year brought calamitous geoclimatic events: there was the Indian Ocean tsunami of 2004,[27] the Kashmir earthquake and Mumbai floods of 2005, the Gujarat floods of 2006, Cyclone Sidr in Bangladesh in 2007, the Pakistan earthquake of 2008, Cyclone Alia in Bangladesh in 2009, the historic floods in Pakistan in 2010, the 2011 Sikkim earthquake, the floods and landslides that plagued Bangladesh and Uttarakhand in 2012, Kashmir's epic 2014 floods, and, in 2015, the Nepal earthquake with its record death toll, followed seven months later by a catastrophic flood in Chennai, India's fifth-largest city, that rivaled the one in Mumbai a decade earlier. Meanwhile, killer heat waves and droughts strike somewhere on the subcontinent every year. Yet the ghastly drama that unfolded at India's holiest sites in 2013 prompted some especially intense soul-searching. An opinion piece in the newspaper *The Hindu* observed,

> The national media's focus on the plight of tourists has grossly distorted the true nature of the tragedy even in the Char Dham area.

It has not reported on the fate of the thousands—almost all male—who come from the villages in these valleys (and elsewhere) to earn a major part of their families' annual income on the yatra [pilgrimage] routes during the tourist season. They help run the dhabas [restaurants] that line the entire 14 km trek route from Gaurikund to Kedarnath; they sell raincoats, umbrellas, canes, walking sticks, soft drinks, water bottles, home-made snacks and other supplies. On their backs, they carry children, the old, the infirm and tourists who are simply unfit and out of shape to walk the entire route. They run along the path with their ponies or horses carrying yatris. . . . Last week's disaster not only spelt doom for thousands of household economies but also dealt a grievous blow to Uttarakhand's lucrative religious tourism industry.[28]

Men spend months at a time working for the pilgrimage industry along the Guptkashi–Kedarnath route, feeding visitors, selling to them, driving them, or serving as their guides and porters. Before the floods, those who saved or borrowed enough to buy a mule could earn more than $10 per day during the season, hauling goods or people up and down the mountain. (After the flood, countless of those mules lay dead—an economic disaster for their owners.) Everyone made good money during the annual tourist invasion, but there was little paid work the rest of the year. Men liked to say, "We're used to working for six months and then sleeping for six months." On the other hand, women work twelve months a year in the noncash economy, collecting and hauling firewood, fodder, and water; cooking; and taking care of all other household and sustenance needs—and that unpaid labor didn't end with the floods.[29]

It was not only tourism workers who suffered in the wake of the flood. The dam construction companies abandoned all of their projects in the affected areas and pulled out without paying local workers any of the wages they were owed. The companies expressed no intention of paying until they themselves were compensated by the government and could restart the construction work. Typical of the abandoned construction sites was a branch of the Phata Byung Hydroelectric Power Project just north of Guptkashi in a deep hollow along the main road. A gaping ten-foot-wide hole in the slope was the mouth of a tunnel running about half a mile into the mountain, where it intersected the five-mile-long main water tunnel. The longer tunnel, if finished, would carry fast-moving water from streams farther up the mountain, and the water would run through turbines to generate electricity. The shorter tunnel would be used to remove debris from the water flow. In early 2014, no construction workers were present at the site, and all of the heavy equipment sat idle. Two guards stationed at the tunnel mouth said that they worked for the Netherlands-based corporate security company G4S. They had had little to do for months, since Lanco, the construction company, had suspended operations after the flood. And they were working for only half salary; the other half, by contract, was to be paid by Lanco, which was refusing to restart construction until the company first received disaster relief from the government. The guards saw little choice but to stay at their posts and be cheated out of their pay. They were at least getting some money from G4S, and if Lanco did return, any guards who had given up and left would get nothing.

Despite the disappearance of livelihoods, most people in the area around Guptkashi did not migrate out. And despite the threat of floods, many residents were rebuilding their homes in place. People with education could remain in the area and get work with aid groups, and Nepalese immigrants, who often stayed on during the off-season to work as rock breakers, would find more of that work than ever if the building of houses, roads, and dams resumed. But most people were staying put because they saw no alternative. Adarsh Tribal, who represented the development organization iVolunteer in Guptkashi at the time and helped guide us through the region, said, "People are fearful of leaving for the plains or certainly the big cities. They do not feel they have the skills or knowledge required by the jobs there, and they worry that they will end up penniless, far from home, with no way to get back." So they stayed, but with the falloff in visitor traffic, there would be few options in the tourist economy.[30]

"REGURGITATING THE DESTRUCTION"

The state of Uttarakhand is relatively new, having been carved off the top of the vastly larger and more populous state of Uttar Pradesh in 2000. It was created in response to demands by the people of the Himalaya, especially women, that they be allowed to determine their own social and ecological fate. The week after the flood, Ravi Chopra, director of the People's Science Institute in Uttarakhand's capital city, Dehradun, gave readers of *The Hindu* a bittersweet reminder of the extraordinary activism, idealism, and optimism that had characterized the statehood movement:

> In the 1990s, when the demand for a separate State gained momentum, at conferences, meetings, workshops and seminars, Uttarakhandi people repeatedly described the special character of the

region. Consciousness created by the pioneering Chipko Andolan [literally, "treehuggers' movement"] raised the hopes of village women that their new State would pursue a green development path, where denuded slopes would be reforested, where fuel wood and fodder would be plentiful in their own village forests, where community ownership of these forests would provide their men with forest products–based employment near their villages instead of forcing them to migrate to the plains, where afforestation and watershed development would revive their dry springs and dying rain-fed rivers, and where the scourge of drunken, violent men would be overcome. Year after year—in cities, towns and villages—they led demonstrations demanding a mountain state of their own. Theirs was a vision of development that would first enhance the human, social and natural capital of the State. Recalling the tremendous worldwide impact of the Chipko movement, Uttarakhandi women dreamed of setting yet another example for the world of what people-centric development could look like. But in the 13 years after statehood, the leadership of the State has succumbed to the conventional model of development with its familiar and single-minded goal of creating monetary wealth. . . . Yes, wealth has been generated but the beneficiaries are very few—mainly in the towns and cities of the southern terai plains and valleys where production investments have concentrated. In the mountain villages, agricultural production has shrivelled, women still trudge the mountain slopes in search of fodder, fuel wood and water, and entire families wait longingly for an opportunity to escape to the plains.[31]

Despite the shock triggered by the horrors that struck the state in 2013, there appeared to have been little impact on policy. The state and central governments' post-disaster strategy seemed to consist of nothing more than a desperate push to revive the tourist economy. Just two days after the flash flood, the central Ministry of Tourism issued a press release announcing a new campaign, 777 Days of the Indian Himalayas, to promote the region as a world-class destination. Five days after that, the ministry pledged a measly $170,000 to assist the flood victims of Uttarakhand, before then announcing a "special financial package" of about $17 million—one hundred times as much as for victims' aid—to rebuild the tourism infrastructure in Uttarakhand.[32]

Nandini is a co-founder of the School for Natural Creativity, an ecologically based alternative education center for children and adolescents in

Guptkashi. She told us, "It's fine to talk about relief and rehabilitation, but few are asking what is the root of the problem. We had thirty thousand people at a time trying to crowd into Kedarnath. We were building right along the river, even out over the river. Why do people do this? It's our arrogance that tells us we can control nature." Asked if the disaster had at least served as a wake-up call that would lead to policy changes, Nandini was doubtful. "It is very difficult to get people, especially the youth, to think about that. They still just want to get into business, to get a piece of the action." [33]

As expected, the patched-up road system through the Mandakini-Alaknanda valley did not fare well when rains returned in June 2014. There were washouts here and there throughout the monsoon season, and roads were completely cut off for four to five days at one stretch. More than twenty-five people were killed by landslides in areas downstream toward Rishikesh. The Kedarnath temple, which had been attracting as many as six hundred thousand pilgrims and tourists annually in 2010–12, saw fewer than forty thousand in 2014. Reaching Kedarnath was now more arduous than ever, and that meant the numbers of visitors at less badly damaged Badrinath, Gangotri, and Yamunotri plunged as well; the idea of undertaking a pilgrimage to just three of the four abodes of the gods wasn't selling well.

Sixteen months after the flood, the state government's cabinet trekked to Kedarnath to hold a meeting, in what *The Hindu* called "a novel way to tell the world that the holy town's redevelopment after last year's devastating landslides was the government's top priority." A redevelopment "road map" approved at the meeting "would make Kedarnath an even better pilgrimage destination than it was before the deluge," promised Chief Minister Harish Rawat. The paper explained the context of the Kedarnath meeting this way: "The Cabinet's decision to meet at an altitude of over 11,000 feet has a parallel in the world's first underwater Cabinet meeting organized by the Maldives government in 2009 to highlight the threat of global warming." It was a highly imperfect parallel. The Maldives stunt had been intended to shock humanity into respecting the Earth's ecological limits, while the Uttarakhand cabinet meeting aimed to push those limits right back to the breaking point. [34]

The 2015 season began in the shadow of Nepal's historic earthquake and stalled in June and July when the monsoon-induced landslides once again left more than ten thousand people stranded along the pilgrimage routes. [35] The road from Sonprayag was still out, and pilgrims were further discouraged by the knowledge that, two and a half years after the flood, some bodies remained buried under Kedarnath's rubble. The government had attempted to spruce up the shrine site and make it look like what a local priest called a "tourist hot spot," but according to the *Indian Express*, the place remained largely a "ghost town." At least for the faithful, there was one new attraction: priests and pilgrims could now offer prayers to a massive boulder that on the night of the great flood had hurtled down the mountain and stopped just short of the temple, protecting it from the devastation.[36]

A boom economy, green electricity, natural refuge, the communitas of pilgrimage—these are many things to demand of any one landscape, let alone the precarious, chiseled gorges of the Himalaya. Visions of Uttara-khand as a hydroelectric generator, a land of myth, or an experiment in people-centric development have pulled the young state in multiple directions; they seem, at most times, irreconcilable. There's only one point where all the visions came together: the point of failure. The disasters of June 2013 damaged every hope for Uttarakhand. Now hope returns like the summer rains, sometimes nurturing, sometimes powerful, sometimes too much for the mountains to bear.

EPILOGUE

Rainbow of Chaos

We are the planet, fully as much as its water, earth, fire and air are the planet, and if the planet survives, it will only be through heroism. Not occasional heroism, a remarkable instance of it here and there, but constant heroism, systematic heroism, heroism as a governing principle.

—Russell Banks, *Continental Drift*, 1985[1]

In March 2015, representatives of 186 national governments met in Sendai, a city freshly reconstructed after the Great East Japan earthquake and tsunami four years before. The Third UN World Conference on Disaster Risk Reduction convened to decide on the final text of the so-called Sendai Framework, which would guide national efforts against disaster by setting goals for the year 2030. All of the countries discussed at length in this book—Australia, Russia, the Philippines, the United States, Indonesia, Italy, the Netherlands, China, Bangladesh, Cuba, India, Uganda—were in attendance. (Montserrat joined the drafting of the previous framework in Hyogo, Japan, in 2005, but in 2015 the island was represented only by its territorial master, the United Kingdom.)

The Sendai conference started off a year of major UN meetings that would lead up to the 2015 Paris Climate Change Conference, and from within the climate wars Sendai appeared as an oasis of optimism and easy consensus; after all, everyone agrees that disaster risk must be curbed, and everyone believes that it's possible to do something about it. The organizers hoped that disasters, as always, would bring people together. And like the climate conferences of recent years, the meeting coincided with a calamity: Cyclone Pam, a storm nearly as strong as Typhoon Yolanda, had bulldozed the Pacific island nation of Vanuatu the day before. The country's president came to Sendai to plead before the delegates, urging them to take real action.

Action proved not so easy to plan. Harjeet Singh, climate policy manager at the organization ActionAid, reported from within the conference:

The draft document . . . had taken a major leap by proposing specific targets for reducing disaster risks.

All [seven] targets had different options of specific numbers but six out of them have now ended up without numbers and just with the subjective language suggesting "substantial" progress in comparison to the last ten years.

The only target that is currently under intense negotiations is around international cooperation that includes providing financial and other support to developing countries.

Facing the heat at climate negotiations year-on-year, developed countries are in no mood to concede to the demands of "additional and predictable" funding for developing countries.[2]

The negotiations dragged late into the final night. The members of a children's choir, standing by to sing at the closing ceremony, were sent home to bed. Jokesters from one aid organization delivered cases of emergency food and water rations to the assembly hall where delegates sat trapped.[3]

Finally, fear of commitment won out. The final version adopted all of the watered-down goals and dropped the mention of "additional and predictable" in favor of "adequate and sustainable" funding, an objective rich countries could interpret at will. The United States blocked a section on sharing technology with poorer countries, fearing for intellectual property. Only the host country, Japan, put actual money on the table, pledging $4 billion over four years to support disaster efforts in developing countries. At the end of that year, rich countries would follow a similar playbook to soften the Paris climate agreement, deflecting further attempts to hold them accountable by, among others, a Philippines-led bloc called the Vulnerable Twenty. For the foreseeable future, it seems, the global South will see neither reparations for loss and damage nor an international disaster fund such as the one that we have proposed in Chapter 11.

What did shine through at Sendai was the doctrine of resilience. In the 166 pages of the conference proceedings, the words "resilience" and "resilient" are used 364 times, while "vulnerable" and "vulnerability" appeared only fifty times and even "hazard" only ninety-six.[4] Instead of action, resources, or technology, the people of nations such as Vanuatu were offered advice on how to become resilient on their own.

A MANAGED RETREAT

To live resiliently, we're told, communities have to abandon as a mirage the old promises of security and development and embrace a dizzying world of unpredictable hazards. They have to open themselves up to catastrophe; like the painter Cézanne, they must become one with the rainbow of chaos in which they practice their arts. The reward for this is not stability and security but nimbleness, adaptability, survival. For Brad Evans and Julian Reid, the most prominent theorists who have examined resilience with a critical eye, it also means a managed retreat from our hopes for the future. They argue that "accepting the imperative to become resilient means sacrificing any political vision of a world in which we might be able to live better lives freer from dangers, looking instead at the future as an endemic terrain of catastrophe that is dangerous and insecure by design."[5]

After publishing their 2014 book *Resilient Life*, in fact, Evans and Reid decided the whole discourse had gone down a cul-de-sac from which it could never emerge. When the journal *Resilience* dedicated an entire issue to analyses of their book, they responded with a renunciation:

> And to repeat, we are also exhausted by resilience. Its nihilism is devastating. Its political language enslaving. Its modes of subjectivity lamenting. And its political imagination notably absent. That is why we have decided after this volume to never write, publicly lecture or debate the problematic again. We will not engage with those who would have us brought into some dialectical orbit in order to validate its reverence by making it some master signifier in order to prove its majoritarian position. Yes, the doctrine of resilience at the level of policy and power is ubiquitous. And yet in terms of emancipating the political, it is already dead.[6]

The suspicion that resilience is simply one more scheme for protecting the status quo has dogged the concept ever since it entered the social realm. There have, however, been attempts to retrofit resilience with a capacity for radical transformation. In C.S. Holling's later social-ecological work, he has expanded on his adaptive cycle (the famous figure-eight diagram) with pathways called "revolt" and "remembrance," the former having the capacity to carry change up to larger scales of the system and enact abrupt or transformational learning. In this version, nested loops of resilience can be infected by change from below or contain it from above.[7]

Geographer Mark Pelling has promoted the related concept of

"transformative adaptation," a nonlinear change in a system caused by dis-satisfaction with the status quo. In a study with David Manuel-Navarrete of two coastal towns in Mexico, Pelling used the resilience lens to show the internal contradictions that maintain a rigid governance regime and devel-opment pathway pointing away from climate adaptation. The researchers observed that a direct hit from Hurricane Dean in 2007 instigated transfor-mation away from this institutionalized state, as residents organized them-selves newly during reconstruction—but, like Occupy Sandy years later, these were temporary transformations, far from the centers of power, and their achievements were quickly reabsorbed into the rigid whole. In this and other work Pelling identified different forms of resilience that can either hold a society to a set path or break it free. His notion of transformative adaptation was eventually included in the Fifth Assessment Report of the IPCC, of which he was an author.[8]

Meanwhile, Kevin Grove has described more subversive local interpre-tations of resilience in Jamaica, running quietly alongside the development doctrine. Subversive resilience reaches for an empowerment that exceeds the neoliberal definition of helping individuals make proper adaptation decisions. On the one hand, resilience-building programming facilitated communities' access to resources from state or international agencies by teaching them to translate their needs into the terminology of the donors, in much the same way that Don Romeo is urging Montserratians to block all British attempts to run out the clock. This is not a capture of local strug-gles by outsiders; instead—drawing on familiar themes from anticolonial resistance—it secures a negotiated self-determination while keeping alive the potential for alternative social and ecological futures. Another neolib-eral message, communities' responsibility for their own survival, resonated in Jamaica with Christian and Rastafarian belief in being "your brother's keeper," with the potential to be affected by others' suffering, and with the history of resistance to race-based oppression. Thus, where Evans and Reid saw neoliberal resilience foreclosing on alternatives, Grove saw it shadowed by other, often hidden transcripts. "Resilience is saturated with the potential to be otherwise," he wrote, "and can thus be both radical and reactionary *at one and the same time*. Resilience techniques and rationalities articulate with other cultural, political economic and ecological trajectories in ways that create new possibilities for both power and resistance."[9]

It's starting to look as if resilience can be anything and anything can be resilient. This is close to the truth. Resilience is not a normative concept—it includes no built-in "performance measures," no innate morality.[10] To de-ploy it is to deploy a set of definitions, what ecologists call "defining the resilience of what to what."[11] A city can be resilient to an earthquake, but a dictatorship can also be resilient to dissent. Diseases, invasive weeds, and fire ants are resilient to extermination. Patriarchy and apartheid are resilient to justice. The world's most destructive corporations spend millions of dollars planning their resilience to everything. An international terrorist network is a master class in resilient design.

Resilience is an art as much of the powerful as of the powerless. Of course, the powerful are much better at it, and they seldom get their own feet wet. As Zygmunt Bauman, the sociologist of liquid modernity, wrote, "The prime technique of power is now escape, slippage, elision and avoidance, the effective rejection of any territorial confinement with its cumbersome corollaries of order-building, order-maintenance and the responsibility for the consequences of it all as well as of the necessity to bear their costs. . . . Holding to the ground is not that important if the ground can be reached and abandoned at whim, in a short time or in no time."[12] This is the same dance we expect of the poor, but the rich are its adepts. Regarding them, Bauman warned, "Attempts to anticipate their moves and the unanticipated consequences of their moves (let alone the efforts to avert or arrest the most undesirable among them) have a practical effectivity not unlike that of a League to Prevent Weather Change."[13]

HINGES OF RESILIENCE

If resilience methods entail a decentralization of responsibility for disaster management, then where does power lie within the system? There are clues in the standard definitions of resilience, such as the Sendai Framework's:

> The ability of a system, community or society exposed to hazards to resist, absorb, accommodate to and recover from the effects of a hazard in a timely and efficient manner, including through the preservation and restoration of its essential basic structures and functions.[14]

The phrase "essential basic structures and functions," or something like it, lingers vaguely inside many such definitions. The Intergovernmental Panel on Climate Change replaces it with "essential function, identity, and structure."[15] But what *are* the essential functions of a community? Resilience hinges on the power to define them.

Once we start defining, the questions start coming and don't stop. Is a community's essential function to provide its members with bodily security, a healthy environment, and natural resources? Is it to offer livelihood opportunities—and if so, how good should they be? Is it essential that the community provide something to the outside world as well? If we observe that one essential function of a slum is to absorb risks from surrounding areas, do the residents have to accept this as part of their role? If we observe that one of the essential functions of an American suburb is to be a thirsty market for fossil fuels, do its residents (and the rest of us) have to accept that? To remain itself, must a community maintain its function and structure *in a particular place*—or will it still be the same community if it's thirty or three thousand miles away, out of range of the storm surge, mud volcano, or fire? What if that hazard is also the source of livelihood? Conversely, is it the same community if the settlement stays put but the people change?

In Australia and Russia, can human resilience build on the concept's volatile ecological fundamentals? Do fires threaten us, or do invasive species (like us) threaten fire regimes? Are fires sustaining or destroying the functionality of the landscape? Are the fires of the fossil fuel economy sustaining or destroying the world?

In the Philippines, can cultural mechanisms of solidarity, risk acceptance, and a healthy cynicism continue to sustain the indomitable poor as they strive to live and work in the country's most hazardous zones? Do these tools increase Filipinos' resilience or deepen their vulnerability? Do perpetual cycles of disaster make the country what it is or keep it from becoming something more?

In New York, would the Lower East Side become more resilient with the importation of a different, wealthier group of residents, or is it essential that the neighborhood function as a safe and pleasant home for the people who live there now? When homeowners on Staten Island fight for the right to retreat and to let their former neighborhoods become *Phragmites* wetlands, do these areas cease to become a functional part of the city, or do they gain a new, stronger function?

Has the Lusi mudflow victims' determination to fight the corporation that buried their home been an essential part of their resilience? Did the Bakrie conglomerate prove to be even more resilient in the face of that rebellion, ducking its responsibilities for nine long years and living on to keep creating fresh hazards with its mining and pumping? When the resilient poor come up against the resilient rich, is it a clash of functions or a class war?

Where did the system fail in the L'Aquila earthquake? Was it the function of science to interject lifesaving signals into the city? Or was science just a scapegoat for a government that failed to do its duty? And can a new, more resilient L'Aquila survive the hole in its heart where the old city was?

Is Miami the least resilient city in the United States or the most? In denying the importance of geographic place in favor of a wholly economic resilience—a plan that says to get out of town at the first sight of surging water or falling property values—has this metropolis built on swamp and sand in the hurricane belt become no more than a fragile Ponzi scheme? Or is it the future: a trillion-dollar tent city for a nomadic society?

Will Mumbai, Kampala, and other cities of the global South become even more central to the global economy and ecology than Miami and Rotterdam? Their slums are very much part of the urban machine, but to what end? Do they recycle hazard into innovation or just soak it up? Is their life of risk and repair a service that slum dwellers provide to the city, or has their city failed them?

Can a people living on one-third of a tiny island keep finding silver linings in anything the Earth chooses to hit them with? Should they have to? If Montserrat has outlived its function in a colonial system of exploitation, is it time to call it quits? Or, once the Soufrière Hills have quieted down, will the intense desire to live again in "the Caribbean the way it used to be" draw the community back to the island's twice-destroyed south? And would doing that show resilience or recklessness?

Is the function of a rural Kansas town simply to give people a place to live and work, or is its carbon footprint also part of its (dys)function? Is the new Greensburg's greenness intimately tied to the whole of the atmosphere, as its planners and architects maintained, or is it just an economic development strategy unrelated to climate change? If Greensburg is rebuilt green and the population doesn't come back, has it still proved its resilience? Will it become a frontier town of post-carbon America or just another traffic light on U.S. 54?

Finally, in Uttarakhand, what is more resilient: to rebuild the roads to the four holy temples in order to keep the pilgrimage economy going, or to withdraw, allowing the gods to reclaim their abodes and the people their forests? Does hydroelectric development need to bounce back from the floods and landslides of 2013, or does Uttarakhand need to bounce back from hydroelectric? Do the sources of the Ganges function as a spiritual landscape for all Hindus, a renewable power source for all Indians, a fragile ecosystem, or a homeland?

No model will get us out of answering questions like these. The hinges of resilience—of "resilience of what to what"—are moral questions, political questions. They are decisions about what we should change in the name of adaptation and what we should defend to the bitter end. Who gets the power to answer these questions? Catastrophic events might sound a wake-up call, but they won't naturally, effortlessly lead us to the right answers; they can too easily be hijacked. Fitting events into the existing shape of individual, state, or corporate agendas is something everyone is skilled at.

From this perspective it looks as if it's those in the economic power centers of the neoliberal world, the disciples of creative destruction, who are best placed to master resilience. This is their claim on the future. And it must be contested: their resilience is catastrophic by design, morally unhinged, because it counts on the vulnerable to absorb what the market sheds so that the market's irreparable fragility can be conserved. The vulnerable and the marginalized must have power, but not just the power to adapt; they must recapture the terms of adaptation, recapture the means by which a consensus of the vulnerable is possible. The fundamental choices are rightfully theirs. The answers should not belong to those who float free in the rainbow of chaos, never setting foot in the mud and ash and black water of disaster.

NOTES

(All Internet addresses were accessed on February 19, 2016, unless otherwise noted.)

Introduction

1 There's no known mechanism by which the glacial collapse, flood, or landslides could have produced such a gas, though several of our interviewees had seen or heard of its effects. One possibility is that this was a description of the effects of the extreme stress and disorientation of the survivors, and particularly of the many pilgrims and tourists who were not acclimated to the low oxygen levels at 12,000 feet.

2 Ramala Khumriyal, interview with the authors, Guptkashi, January 29, 2014. Translation by Adarsh Tribal and Priti Gulati Cox.

3 Internal Displacement Monitoring Centre, *Global Estimates 2014: People Displaced by Disasters* (Geneva: IDMC, 2014). From June to October 2013, floods displaced an estimated 1,042,000 people in the Indian states of Bihar, Kerala, Uttarakhand, Assam, Andhra Pradesh, West Bengal, and Uttar Pradesh.

4 The book most often credited with guiding this shift was Kenneth Hewitt, ed., *Interpretations of Calamity from the Viewpoint of Human Ecology* (Boston: Allen & Unwin, 1983). Today, the most widely cited work on this point is probably Ben Wisner, Piers Blaikie, Terry Cannon, and Ian Davis, *At Risk: Natural Hazards, People's Vulnerability and Disasters*, 2nd ed. (London: Routledge, 2004).

5 Anthony Oliver-Smith and Susanna Hoffman, *The Angry Earth: Disaster in Anthropological Perspective* (New York: Routledge, 1999), 29.

6 The events leading up to almost every "natural" disaster have geological, climatic, and meteorological aspects, as well as connections to other disciplines of physical geography and earth science: coastal geography, glaciology, hydrology, landscape ecology, limnology, oceanography, or pedology. Some are more prominent in, say, a landslide and others more prominent in a flood or wildfire, but to us their combined presence justifies the term "geoclimatic." The word is widely used in biological and physical research to denote an array of influences or characteristics that includes both climatic and geographic or geologic factors.

7 We include drought in our analysis only as it relates to more acute hazards, such as fire. Drought is a deadly natural process but not quite an event. Aaron Popp has characterized the difference this way: "Droughts are a parasite to economically unhealthy nations whereas other natural disasters are a shock." Aaron Popp, "The Effects of Natural Disasters on Long Run Growth," *Major Themes in Economics* 1 (2006): 61–81.

8 This literature came out of environmental concerns that long predated the climate wars, and was consolidated in the 1990s by the sociologists Ulrich Beck and Anthony Giddens under their concept of the "risk society," in which manufactured hazards

and risks are considered as primary products of modernity itself. Ulrich Beck, *Risk Society: Towards a New Modernity* (London: SAGE Publications, 1992). The subject of blame was, in the same period, taken up by the anthropologist Mary Douglas as part of the cultural theory of risks she established together with political scientist Aaron Wildavsky. Mary Douglas, *Risk and Blame: Essays in Cultural Theory* (London: Routledge, 1992); Mary Douglas and Aaron B. Wildavsky, *Risk and Culture: An Essay on the Selection of Technical and Environmental Dangers* (Berkeley: University of California Press, 1983).

9 A thorough, if not recent, consensus can be found in the Intergovernmental Panel on Climate Change, *Special Report: Managing the Risks of Extreme Events and Disasters to Advance Climate Change Adaptation (SREX)* (Cambridge: Cambridge University Press, 2012). This was bolstered by new data collected in the IPCC's *Climate Change 2014: Impacts, Adaptation, and Vulnerability. Contribution of Working Group II to the Fifth Assessment Report of the Intergovernmental Panel on Climate Change* (Cambridge: Cambridge University Press, 2014).

10 Naomi Oreskes and Erik M. Conway, *Merchants of Doubt* (London: Bloomsbury Press, 2010).

11 Roger Pielke Jr., "Disasters Cost More than Ever—But Not Because of Climate Change," *FiveThirtyEight*, March 19, 2014, fivethirtyeight.com/features/disasters-cost-more-than-ever-but-not-because-of-climate-change. The article was mostly rewritten from material in Pielke's book *The Climate Fix: What Scientists and Politicians Won't Tell You about Global Warming* (New York: Basic Books, 2011).

12 For research results refuting the protection-through-growth idea, see pages 98–100, 209–11, and 246–48.

13 See, for example, William Nordhaus, "Economic Policy in the Face of Severe Tail Events," *Journal of Public Economic Theory* 14 (2012): 197–219; William Nordhaus, "The Economics of Tail Events with an Application to Climate Change," *Review of Environmental Economics and Policy* 5 (2011): 240–57; Robert Pindyck, "Fat Tails, Thin Tails, and Climate Change Policy," *Review of Environmental Economics and Policy* 5 (2011): 258–74.

14 John Bellamy Foster, "Capitalism and the Accumulation of Catastrophe," *Monthly Review* 63, no. 7 (2011): 1–17.

15 An argument to this effect was made formally by Harvard University economist Martin Weitzman in what he called the "dismal theorem." Martin Weitzman, "On Modeling and Interpreting the Economics of Catastrophic Climate Change," *Review of Economics and Statistics* 91 (2009): 1–19; Martin Weitzman, "Fat-tailed Uncertainty in the Economics of Catastrophic Climate Change," *Review of Environmental Economics and Policy* 5 (2011): 275–92.

16 Nicholas Stern, "The Structure of Economic Modeling of the Potential Impacts of Climate Change: Grafting Gross Underestimation of Risk onto Already Narrow Science Models," *Journal of Economic Literature* 51 (2013): 838–59; Nicholas Stern, *The Economics of Climate Change: The Stern Review* (Cambridge: Cambridge University Press, 2007).

17 Ibid.

18 See Nicholas Georgescu-Roegen, *The Entropy Law and the Economic Process* (Cambridge, MA: Harvard University Press, 1971); Herman Daly, *Steady-State Economics:*

With New Essays (Washington, DC: Island Press, 1991); Wes Jackson and Wendell Berry, *Nature as Measure: The Selected Essays of Wes Jackson* (Berkeley: Counterpoint Press, 2011); John Bellamy Foster, Richard York, and Brett Clark, *The Ecological Rift: Capitalism's War on the Earth* (New York: Monthly Review Press, 2010); Joel Kovel, *The Enemy of Nature: The End of Capitalism or the End of the World?* (London: Zed Books, 2002); Naomi Klein, *This Changes Everything: Capitalism vs. the Climate* (New York: Simon and Schuster, 2014); James Gustave Speth, *The Bridge at the Edge of the World: Capitalism, the Environment, and Crossing from Crisis to Sustainability* (New Haven: Yale University Press, 2008); Richard Heinberg, *The End of Growth: Adapting to Our New Economic Reality* (Gabriola Is., BC, Canada: New Society Publishers, 2011); Ted Trainer, *The Conserver Society: Alternatives for Sustainability* (London: Zed Books, 1995); Tim Jackson, *Prosperity Without Growth: Economics for a Finite Planet* (New York: Routledge, 2011).

19 Klein, *This Changes Everything*, 40.

20 Brian Tokar, *Toward Climate Justice: Perspectives on the Climate Crisis and Social Change* (Porsgrunn, Norway: New Compass Press, 2014), 73.

21 Robert Verchick has developed a legal basis for disaster justice. See Robert Verchick, "Disaster Justice: The Geography of Human Capability," *Duke Environmental Law and Policy Forum* 23 (2013): 23–71.

22 Among the stories we relate are those of two banner-headline storms of recent years: Superstorm Sandy in 2012 and Super Typhoon Yolanda (known as Haiyan internationally) in 2013. In the other stories, we focus on disasters that hold important lessons for us all but that so far have not received the attention they warrant. As for other headline tragedies of the past dozen years—the 2004 Indian Ocean tsunami, 2005's Hurricane Katrina, the 2010 earthquake in Haiti, the 2011 Great East Japan earthquake and tsunami, the 2015 Nepal earthquake, and several big typhoons in the western Pacific—we discuss crucial implications of these events, relying on excellent reporting and research done by others.

23 D.E. Alexander, "Resilience and Disaster Risk Reduction: An Etymological Journey," *Natural Hazards and Earth System Science* 13 (2013): 2707–16.

24 UNISDR, *Sendai Framework for Action 2015–2030* (Geneva: United Nations Office for Disaster Risk Reduction, 2015).

25 UNISDR, *Living with Risk: A Global Review of Disaster Reduction Initiatives, 2004 Edition, Volume 1* (Geneva: United Nations Office for Disaster Risk Reduction, 2004). This report is critically analyzed in Julian Reid, "The Disastrous and Politically Debased Subject of Resilience," *Development Dialogue* 58 (2004): 67–80. Its importance was avowed by UNISDR leader Margareta Wahlström in 2015, after negotiating the Sendai Framework, when she pointed to the review as the beginning of the effort to promote a common understanding of DRR terminology. Denis McClean, "Work Starts on Sendai Indicators," UNISDR, September 29, 2015, unisdr.org/archive/45961.

26 Brad Evans and Julian Reid, *Resilient Life: The Art of Living Dangerously* (Cambridge: Polity, 2014).

27 Kevin Grove, "Agency, Affect, and the Immunological Politics of Disaster Resilience," *Environment and Planning D: Society and Space* 32, no. 2 (2014): 240–56.

28 Writes Grove (ibid.), "Foundational vulnerability studies drawing on Marxist political economy and Freirian pedagogy emphasized that local people possessed their own

knowledge on hazards and vulnerability reduction that technocentric approaches to hazard studies silenced and delegitimized. Here, the 'agency' of local peoples became the foundation for participatory disaster mitigation programs that made vulnerability reduction a matter of changing uneven political economic relations that caused vulnerability in the first place." As examples of such studies he cites Hewitt, *Interpretations of Calamity*; Ben Wisner, Phil O'Keefe, and Ken Westgate, "Global Systems and Local Disasters: The Untapped Power of Peoples' Science," *Disasters* 1 (1977): 47–57; and Andrew Maskrey, *Disaster Mitigation: A Community-Based Approach* (Oxford: Oxfam, 1989). Distinct from this, he continues, "resilience recognizes disaster victims as active agents with inherent self-help capacities that can be strengthened through proper resilience programming, rather than passive victims who require external aid to overcome structural constraints."

29 UNISDR, "International Day for Disaster Reduction," *United Nations Office for Disaster Risk Reduction*, unisdr.org/we/campaign/iddr.

30 Grove, "Agency, Affect."

31 Nietzsche stated this as one of his axioms, attributing it only to "life's school of war," and it has remained axiomatic ever since, even though it is easily falsifiable. Many things that do not kill people—including disasters—frequently do make them weaker. Friedrich Nietzche, *Götzen-Dämmerung, oder, Wie man mit dem Hammer philosophirt* (Leipzig: C.G. Naumann, 1889).

32 Fernanda Santos, "Battling Flames in Forests, with Prison as the Firehouse," *New York Times*, November 26, 2013.

33 Vanessa Barford, "The Prisoners Fighting Wildfires in California," BBC News, September 24, 2015, bbc.com/news/magazine-34285658.

34 Seth Robert Reice, *The Silver Lining: The Benefits of Natural Disasters* (Princeton, NJ: Princeton University Press, 2001), 18.

35 The disutility of labor—the idea that the enjoyment of spending an hour on your paying job is almost always less than the enjoyment you'd derive from your favorite pastime—is fundamental to economics. In his monumental book *The Entropy Law and the Economic Process*, Nicholas Georgescu-Roegen gave enjoyment an even more central role. In his analysis, because no material products of an economy are permanent—they and the materials of which they are made all degrade to the point of uselessness as inputs—the only outputs of an economy are non-usable waste and "enjoyment of life," which can be saved and valued only as a memory. He wrote, "The true 'product' of the economic process is not a material *flow* but a psychic *flux*—the enjoyment of life by every member of the population." Georgescu-Roegen, *The Entropy Law and the Economic Process*, 284.

36 John M. Anderies, Carl Folke, Brian Walker, and Elinor Ostrom, "Aligning Key Concepts for Global Change Policy: Robustness, Resilience, and Sustainability," *Ecology and Society* 18, no. 2 (2013): 8.

37 This term was jointly coined by a biologist and a systems engineer: Marie E. Csete and John C. Doyle, "Reverse Engineering of Biological Complexity," *Science* 295 (2002): 1664–69.

38 Steve Carpenter, Brian Walker, J. Marty Anderies, and Nick Abel, "From Metaphor to Measurement: Resilience of What to What?," *Ecosystems* 4, no. 8 (2001): 765–81.

Chapter 1: Fire Regimes

1 Jennifer Isaacs, *Australian Dreaming: 40,000 Years of Aboriginal History* (Melbourne: Lansdowne Press, 1980). In this book the story is recalled by two Arnhem Land story-tellers, Mandarg and Spider Murululmi of Maningrida Outstation, but the text is also attributed to, and follows closely, Eric Joseph Brandl, *Australian Aboriginal Paintings in Western and Central Arnhem Land: Temporal Sequences and Elements of Style in Cadell River and Deaf Adder Creek Art* (Canberra: Australian Institute of Aboriginal Studies, 1973).

2 Mary-Lou Keating, Jenny Bigelow, and Susan Templeman, *As the Smoke Clears* (Faulconbridge, NSW: Springwood Printing, 2014).

3 Sales of the book raised more than $18,500. The group successfully applied to the Relief Fund committee to use this money to renovate a park in Winmalee that was damaged in the bushfires, to create a native garden at a school in Mount Victoria that will attract small birds, and to run a group to help advocate and plan for animals to be included in disaster management in the Blue Mountains. Jenny Bigelow, e-mail interview with Paul Cox, August 31, 2015.

4 On the videos, see Katherine Fry, *Constructing the Heartland: Television News and Natural Disaster* (Cresskill, NJ: Hampton Press, 2003).

5 Belinda Medlyn, Melanie Zeppel, Niels Brouwers, Kay Howard, Erner O'Gara, Giles Hardy, Thomas Lyons, Li Li, and Bradley Evans, *An Assessment of the Vulnerability of Australian Forests to the Impacts of Climate Change. II. Biophysical Impacts of Climate Change on Australia's Forests* (Southport, QLD: National Climate Change Adaptation Research Facility, 2011).

6 Stephen Pyne, *Burning Bush: A Fire History of Australia* (Seattle: University of Washington Press, 1998), 26.

7 Ibid., 31.

8 Andrew Scott, David Bowman, William Bond, Stephen Pine, and Martin Alexander, *Fire on Earth: An Introduction* (West Sussex: John Wiley and Sons, 2014), 125.

9 *Blue Mountains National Park Fire Management Strategy* (Sydney: NSW National Parks and Wildlife Service, 2004).

10 Excerpt from Arna Radovich, "Mosaic of Loss," in Keating, Bigelow, and Templeman, *As the Smoke Clears*. Reprinted by permission of the author.

11 C.S. Holling, "Resilience and Stability of Ecological Systems," *Annual Review of Ecology and Systematics* 4 (1973): 1–23.

12 On the evolution of Holling's theory, see Jeremy Walker and Melinda Cooper, "Genealogies of Resilience: From Systems Ecology to the Political Economy of Crisis Adaptation," *Security Dialogue* 42 (2011): 143–60.

13 Information provided by Peter Chinn of the Springwood Historical Society, Paul Koen of Blue Mountains City Council, and "The History of Winmalee Rural Fire Brigade," Winmalee Rural Fire Service, winmaleerfs.com.au/wp/history.

14 Phil and Christie Le Breton, interview with Paul Cox, Winmalee, December 7, 2014.

15 Pyne, *Burning Bush*, 83.

16 Rhys Jones, "Fire-Stick Farming," *Australian Natural History* 16, no. 7 (1969): 224–28;

Bill Gammage, *The Biggest Estate on Earth: How Aborigines Made Australia* (Crows Nest, NSW: Allen and Unwin, 2011).

17 For usage of the term "have a go," see Sue Neales, "Prime Minister Tony Abbott Praises Farmers' Drought Resilience," *The Australian*, February 17, 2014.

18 This amenity-led migration is often referred to in the Australian press as the "tree-change," matching a similar "sea-change" migration to the coast.

19 Stephen Pyne, *The Still-Burning Bush* (Melbourne: Scribe, 2006), 93–96.

20 See David Bowman, Jessica O'Brien, and Johann Goldammer, "Pyrogeography and the Global Quest for Sustainable Fire Management," *Annual Review of Environment and Resources* 38 (2013): 57–80. This review calculates the critical threshold of intensity at approximately 2,500 kilowatts per meter, using data from Mark Adams and Peter Attiwill, *Burning Issues: Sustainability and Management of Australia's Southern Forests* (Melbourne: CSIRO, 2011).

21 J.B. Gledhill, "Community Self-Reliance During Bushfires: The Case for Staying at Home," paper presented at the 3rd International Wildland Fire Conference, Sydney, October 2003.

22 On this issue the Royal Commission recommended more community education on fire risk, better local bushfire planning, and greater emphasis on how "the heightened risk on the worst days demands a different response." Out of the sixty-seven recommendations, one of the few rejected by the Victorian premier was a suggestion that "the State develop and implement a retreat and resettlement strategy for existing developments in areas of unacceptably high bushfire risk, including a scheme for non-compulsory acquisition by the State of land in these areas."

23 Katharine Haynes, John Handmer, John McAneney, Amalie Tibbits, and Lucinda Coates, "Australian Bushfire Fatalities 1900–2008: Exploring Trends in Relation to the 'Prepare, Stay and Defend or Leave Early' Policy," *Environmental Science and Policy* 13 (2010): 185–94.

24 Christine Eriksen, Nicholas Gill, and Lesley Head, "The Gendered Dimensions of Bushfire in Changing Rural Landscapes in Australia," *Journal of Rural Studies* 26 (2010): 332–42.

25 Ibid. Quote attributed to "Female tree-changer, Windellama, February 2009."

26 Pyne, *Burning Bush*, 88.

27 Debra Parkinson and Claire Zara, "The Hidden Disaster: Violence in the Aftermath of Natural Disaster," *Australian Journal of Emergency Management* 28, no. 2 (2013); Debra Parkinson and Claire Zara, "Women-Led Sustained Efforts Give Birth to Key Gender and Disaster Body, Many Australian 'Firsts,'" in *Women's Leadership in Risk-Resilient Development*, ed. Michele Cocchiglia (Bangkok: UNISDR, 2015).

28 Pyne, *Burning Bush*, 341–45.

29 David Howell, interview with Paul Cox, Sydney, December 8, 2014.

30 Superintendent David Jones, district manager, and Helen Belshaw, fire mitigation officer, Blue Mountains Rural Fire Service, interviews with Paul Cox, Katoomba, NSW, December 9, 2014. Prescribed burning can only take place, on average, ten days a year, depending on weather conditions and personnel availability. It must be preceded by extensive modeling of fire and smoke behavior. Even so, it remains a preferred strategy for protecting key assets under extreme risk. The problem is, says Belshaw, "most of our assets are listed as extreme" in the fire management plan.

31 Pyne, *Burning Bush*, 405.

32 Ibid., 366–68.

33 David Bowman, interview with Paul Cox, Hobart, December 11, 2014.

34 See David Bowman, "Don't Risk the Homefire Heartache," *The Mercury*, November 1, 2014.

35 Gay Hawkes, *Time and Chance: A Story of the Fires* (Hobart: Monotone Art Printers, 2014), 26.

36 Michael Brown, interview with Paul Cox, Hobart, December 11, 2014. Modeling is used most outside of peak fire season, to find dangerous ignition areas and prioritize them for prescribed burning. As in the Blue Mountains, the Tasmanian Fire Service undertakes burns only after extensive planning and modeling. Sandra Whight, State Fire Management Council, interview with Paul Cox, December 11, 2014.

37 Peter Hannam, "Temperatures off the Charts as Australia Turns Deep Purple," *Sydney Morning Herald*, January 8, 2013.

38 Sandra Whight, interview with Paul Cox, Hobart, December 11, 2014. This trend exploded in January 2016 when 27,000 acres, a full 1.2 percent of the World Heritage Area, burned in lightning fires, destroying rare plant communities and centuries-old trees.

39 Sophie C. Lewis and David J. Karoly, "Anthropogenic Contributions to Australia's Record Summer Temperatures of 2013," *Geophysical Research Letters* 40 (2013): 3705–9.

40 P. Fox-Hughes, R.M.B. Harris, G. Lee, J. Jabour, M.R. Grose, T.A. Remenyi, and N.L. Bindoff, *Climate Futures for Tasmania: Future Fire Danger* (Hobart: Antarctic Climate and Ecosystems Cooperative Research Centre, 2015).

41 *Amanpour*, CNN, October 21, 2013. Figueres was the executive secretary of the UN Framework Convention on Climate Change (UNFCCC).

42 Fairfax Radio 3AW, October 23, 2013. Taking the round of media coverage to its conclusion, Al Gore compared Abbott to American politicians who make spurious health arguments on behalf of tobacco companies (*7.30*, ABC TV, October 23, 2013), and Abbott's environment minister was mocked for turning to Wikipedia for evidence on Australian fire history. Esther Han and Judith Ireland, "Greg Hunt Uses Wikipedia Research to Dismiss Links Between Climate Change and Bushfires," *Sydney Morning Herald*, October 23, 2013.

43 Emma Griffiths, "Tony Abbott Tells National Press Club That Election Will Be Referendum on Carbon Tax," ABC News, September 2, 2013.

44 Investor Group on Climate Change, "Company Emissions Fell 7% in the First Year of the Carbon Price," press release, May 7, 2014.

45 One notable factor affecting the trend was a continuing decline in electricity demand since 2009. See Frontier Economics, "Post Hoc Ergo Propter Hoc," client briefing, July 2013.

46 Peter Hannam, "Australia's Large-Scale Renewable Investment Dives in 2014," *Sydney Morning Herald*, January 12, 2015.

47 John Connor, "Australia's Carbon Reduction Goal Won't Make Enough Difference," *The Australian*, August 13, 2015.

48 Abbott's removal from the prime ministership by his own party in September 2015 signaled only a hint of change. His usurper, Malcolm Turnbull, had previously lost party leadership in 2009 over his support for the carbon tax, and he assured his party

that he would steer clear of the subject this time. Turnbull's personal attendance at the Paris Climate Conference months later was widely seen as a signal of reengagement, but the Abbott government's basic plan of action, and the economic forces behind it, remained in effect.

49 Holling, "Resilience and Stability."

50 Scott et al., *Fire on Earth*, 128. Pyne himself usually prefers a pathogenic analogy: "A biological conception would imagine fire less as a physical process that slams into the living world than as a quasi-organic process propagated within and by a living substrate which manifests itself by physical means, notably heat and light. It more resembles an instantaneous epidemic, a contagion of combustion, like an ecological SARS, than a mudslide or hurricane" (Pyne, *Still-Burning Bush*, 115–16).

51 Scott et al., *Fire on Earth*, 144. It's interesting that the authors use the same evolutionary language in reference to human fire use in prehistoric times: "Landscape burning was probably an emergent property of hunter-gatherer fire management, given that people were moving to different habitats, tracking seasonal variation of resources" (ibid., 176).

52 Ibid., 184–85.

53 David M.J.S. Bowman, Ben J. French, and Lynda D. Prior, "Have Plants Evolved to Self-Immolate?," *Frontiers in Plant Science* 5 (2014): article 590.

54 Guy Bannink, interview with Paul Cox, Hobart, December 12, 2014. Bunkers are also starting to catch on in Victoria, where a company has buried hundreds of precast concrete rooms at $10,000 each, and reported in 2015 that they were becoming popular as wine cellars and "man caves." The builder also stated that most of his customers kept their bunkers a secret from their neighbors; each room was large enough to accommodate only the family that owned it. Olivia Lambert, "Fire Safety Bunkers Being Used as Short-Term Man Caves," News.com.au, November 11, 2015; Belinda Mackowski, "Bushfire Bunker Owners Keeping Them Secret from Neighbours in Melbourne's Outer East," *Melbourne Free Press Leader*, November 9, 2015.

55 Hawkes, *Time and Chance*, 26–27.

56 Krasnoyarsk is the name of the city and the region (Krasnoyarsk Krai), which stretches almost all the way across the middle of Russia from north to south. At 903,400 square miles, Krasnoyarsk region is the third-largest subnational division in the world (after the neighboring Sakha Republic and the State of Western Australia).

57 The Soviet strategy ended up suppressing information as much as fire, so the exact consequences of the resulting fuel buildup are unknown. But one indication was the great Black Dragon Fire of May 1987, the largest recorded fire in history. Between Transbaikal and northern China this cataclysmic fire burned an area larger than all of Tasmania, around 18 million acres or 28,000 square miles. The size of the whole fire complex that tore through the two countries that year is unmeasured, but based on their own satellite images U.S. experts guessed that 30 to 35 million acres had burned—roughly the area of New York State. Stephen Pyne, *Vestal Fire: An Environmental History, Told Through Fire, of Europe and Europe's Encounter with the World* (Seattle: University of Washington Press, 2012), 523–28.

58 Elena Kukavskaya, Sukachev Institute, interview with Paul Cox, Krasnoyarsk, November 15, 2013; Alexey Yaroshenko, Greenpeace Russia, interview with Paul Cox, Moscow, November 19, 2013. The 75 percent figure is from Greenpeace surveys of

forestry employees. Before 2007 the share of time spent on reporting averaged 10 percent.

59 Victor Romasko, interview with Paul Cox, Krasnoyarsk, November 15, 2013.
60 William Lau and Kyu-Myong Kim, "The 2010 Pakistan Flood and Russian Heat Wave: Teleconnection of Hydrometeorological Extremes," *Journal of Hydrometeorology* 13, no. 1 (2012): 392–403.
61 "Overall Picture of Natural Catastrophes in 2010," Munich Re, January 3, 2011.
62 Climate modelers have found, as usual, that climate change is both implicated and not. An occurrence such as the 2010 heat wave "can be both mostly internally-generated in terms of magnitude and mostly externally-driven in terms of occurrence-probability." F.E.L. Otto, N. Massey, G.J. van Oldenborgh, R.G. Jones, and M.R. Allen, "Reconciling Two Approaches to Attribution of the 2010 Russian Heat Wave," *Geophysical Research Letters* 39 (2012): L04702.
63 Daria Mokhnacheva, "Wild Fires in Russia," in *The State of Environmental Migration 2010*, ed. François Gemenne, Pauline Brücker, and Joshua Glasser (Paris: Institute for Sustainable Development and International Relations, 2011), 27.
64 Irina Kurganova, Valentin Lopes de Gerenyu, Johan Six, and Yakov Kuzyakov, "Carbon Cost of Collective Farming Collapse in Russia," *Global Change Biology* 20, no. 3 (2014): 938–47.
65 Grigory Ioffe and Tatyana Nefedova, "Rural Exodus in Eurasia: Consequences for Agriculture and Forestry," presentation at the International Congress on Forest Fire and Climate Change, Novosibirsk, Russia, November 12, 2013.
66 Elena Kukavskaya, interview with Paul Cox, Krasnoyarsk, November 15, 2013.
67 International Congress on Forest Fire and Climate Change: Challenges for Fire Management in Natural and Cultural Landscapes of Eurasia, November 11–12, 2013.
68 Pyne, *Vestal Fire*, 508–9.
69 Furyaev is an equally long-standing proponent of prescribed fire in Krasnoyarsk. He had experimented with burning in the 1960s but could find no official support. When he sought to publish extracts from American experiments in prescribed burning he was denied permission; the authorities suspected that these might be a form of CIA disinformation, presumably aimed at tricking the Soviets into burning their forest down around them (ibid., 529).
70 Today the Bor Forest Island lies just a few miles from the ZOTTO mast, within the observatory's footprint area. More than twenty years on you can still see the blackened oval of the island in online satellite images at 60°45'02.2"N, 89°24'45.7"E.
71 Johann G. Goldammer, ed., *Prescribed Burning in Russia and Neighbouring Temperate-Boreal Eurasia: A Publication of the Global Fire Monitoring Center (GFMC)* (Remagen-Oberwinter: Kessel, 2013).
72 Greenpeace was invited to the conference as a civil society representative, and its presenter, Grigory Kuksin, also spoke at length on this issue: "Right now we have misinformation on the benefits of fire and propaganda of using fire to prevent wildfire. Scientific principles like those presented by Johann are used by authorities to say we don't need to suppress fires or provide the necessary resources. This is the new ideology." Grigory V. Kuksin, "Role of Russian Civil Society in Prevention and Suppression of Forest Fires," presentation at the International Congress on Forest Fire and Climate Change, Novosibirsk, Russia, November 11, 2013.

73   Alexey Yaroshenko, interview with Paul Cox, Moscow, November 19, 2013.

74   Models reviewed in Charles D. Koven, "Boreal Carbon Loss Due to Poleward Shift in Low-Carbon Ecosystems," *Nature Geoscience* 6 (2013): 452–56; E.A.G. Schuur, A.D. McGuire, C. Schädel, G. Grosse, J.W. Harden, D.J. Hayes, G. Hugelius, et al., "Climate Change and the Permafrost Carbon Feedback," *Nature* 520 (2015): 171–79.

75   Koven, "Boreal Carbon Loss." A later review corroborating this outlook was S. Gauthier, P. Bernier, T. Kuuluvainen, A.Z. Shvidenko, and D.G. Schepaschenko, "Boreal Forest Health and Global Change," *Science* 349, no. 6250 (2015): 819–22. It should also be noted that Koven's concerns about the taiga's southern edge have been around for some time; see, for example, Brian Rizzo and Ed Wiken, "Assessing the Sensitivity of Canada's Forest to Climatic Change," *Climate Change* 21 (1992): 37–55.

76   Koven, "Boreal Carbon Loss." To reiterate, boreal forest expansion in the Arctic is likely to follow a mass release of carbon dioxide and methane from thawing permafrost beneath both forest and tundra, which could prove catastrophic enough to make the whole debate irrelevant. Permafrost underlies two-thirds of Russia and more than half of its forests. Its melting would also dry out taiga in low rainfall areas, potentially killing trees and providing more fuel for fires. M. Torre Jorgenson, Vladimir Romanovsky, Jennifer Harden, Yuri Shur, Jonathan O'Donnell, Edward A.G. Schuur, Mikhail Kanevskiy, et al., "Resilience and Vulnerability of Permafrost to Climate Change," *Canadian Journal of Forest Research* 40, no. 7 (2010): 1219–36; Schuur et al., "Climate Change."

77   Elena Kukavskaya, interview with Paul Cox, Krasnoyarsk, November 15, 2013; see E.A. Kukavskaya, L.V. Buryak, G.A. Ivanova, S.G. Conard, O.P. Kalenskaya, S.V. Zhila, and D.J. McRae, "Influence of Logging on the Effects of Wildfire in Siberia," *Environmental Research Letters* 8, no. 4 (2013): 045034. Alongside and interlinked with fire, insect outbreaks are another source of substantial disturbance associated with higher temperatures—most dramatically the mountain pine beetle in North America and Siberian silk moth in Russia.

78   N.M. Tchebakova, E. Parfenova, and A.J. Soja, "The Effects of Climate, Permafrost and Fire on Vegetation Change in Siberia in a Changing Climate," *Environmental Research Letters* 4, no. 4 (2009): 045013.

79   Nadja Tchebakova, interview with Paul Cox, Krasnoyarsk, November 15, 2013.

Chapter 2: Leave It Up to Batman

1   Greg Bankoff, *Cultures of Disaster: Society and Natural Hazard in the Philippines* (London: RoutledgeCurzon, 2003), 11.

2   DJ Yap, "Victims' Toughness, Grace, Generosity Awe Rescuers," *Philippine Daily Inquirer*, December 9, 2012.

3   Tropical cyclones occurring in the western Pacific, where they are called typhoons, are categorized differently in different countries. In most cases, a "super typhoon" corresponds roughly to a category-5 hurricane in the Western Hemisphere. Yolanda was much stronger, garnering an unofficial label of "category 6" (see for instance I.-I. Lin, Iam-Fei Pun, and Chun-Chi Lien, "'Category-6' Supertyphoon Haiyan in Global Warming Hiatus: Contribution from Subsurface Ocean Warming," *Geophysical Research Letters* 41 [2014]: 8547–53). It was only in January 2015, more than a year

after Yolanda, that the Philippines weather agency PAGASA adopted the super typhoon category, which was defined as having sustained winds of 220 kilometers (132 miles) per hour or greater ("PAGASA Adopts 'Super Typhoon' Category," ABS-CBN News, January 10, 2015). In referring to strengths of tropical cyclones in the Western Hemisphere, where they are called hurricanes, we will use the standard Saffir-Simpson scale, which is employed by the U.S. National Weather Service. A category-5 hurricane has sustained winds of 157 statute miles per hour or faster. National Hurricane Center, "Saffir Simpson Hurricane Wind Scale," n.d., nhc.noaa.gov/aboutsshws .php.

4   The authoritarian president Ferdinand Marcos renamed the barrios "barangays" in 1974, reviving a more dignified and traditional Filipino name to promote what he saw as the ideal scale for democratic participation. In a rural setting barangays are villages or groups of settlements; in an urban setting they are wards or neighborhoods, and here to be "in the barangay" can have a ring similar to "in the slums."

5   The most visible of these ships, the M/V *Eva Jocelyn*, stayed on the spot for more than a year, and when it was finally removed the bow was kept by the city and built into the Super Typhoon Yolanda Ship Remnant Monument.

6   Judith Buhay, interview with the authors, Tacloban, January 11, 2014.

7   This is the system that became Tropical Depression Agaton and is suspected of a connection to winter storms in the United States and Europe. See pages 148–49.

8   Michael Lim Ubac and Marlon Ramos, "P-Noy Orders Uprooting of People in Harm's Way," *Philippine Daily Inquirer*, December 14, 2012. In an ambush interview, Aquino told reporters, "We will convince the communities that it will be in their interest to relocate to higher, safer grounds," but "in certain instances, we really can't wait for all the consultations to be over before we transfer them." To date, no communities in the Pablo-affected region have been moved.

9   Greg Bankoff, "A History of Poverty: The Politics of Natural Disasters in the Philippines, 1985–95," *Pacific Review* 12 (1999): 381–420.

10 Bankoff, *Cultures of Disaster*, 177–78.

11 F. Landa Jocano, *Management by Culture: Fine-Tuning Management to Filipino Culture* (Manila: Punlad Research House, 1990), 32. Similar concepts and labor traditions exist in many other countries, such as Indonesia's *gotong royong*—which was, however, seldom mentioned in the course of the Lusi mud eruption (Chapter 5).

12 Greg Bankoff, "Dangers to Going It Alone: Social Capital and the Origins of Community Resilience in the Philippines," *Continuity and Change* 22 (2007): 327–55.

13 In an information age analogue, a government-funded open-source development project to create a free localized operating system went under the name Bayanihan Linux, distrowatch.com/table.php?distribution=bayanihan.

14 Jocano, *Management by Culture*, 98.

15 Bankoff, "Dangers to Going It Alone."

16 Official resilience-building policies reinforce this acute disaster awareness. In 2015, high schools introduced an eighty-hour course called Disaster Readiness and Risk Reduction as part of the core curriculum in the science and technology stream. This gave disaster the same classroom time as courses such as General Mathematics, Reading and Writing, and Understanding Culture, Society, and Politics.

17 On the annual number of typhoons: Philippine Atmospheric, Geophysical and

Astronomical Services Administration (PAGASA), "Member Report to the ESCAP/ WMO Typhoon Committee, 41st Session," January 2009, typhooncommittee.org/41st /docs/TC2_MemberReport2008_PHILIPPINES1.pdf.

18 Juvy Tanio, assistant to Mayor Michelle Rabat of Mati City in Davao Oriental, quoted in Germelina Lacorte, "'So, That's What a Typhoon Is Like,'" *Philippine Daily Inquirer*, December 5, 2012.

19 Statements on December 6, 2012, 18th Conference of Parties to the UNFCCC, Doha, Qatar.

20 Yap, "Victims' Toughness."

21 Ibid.

22 Ayan Mellejor, "National Grid, PSALM, NPC Slammed for Keeping Pablo-Hit Areas Without Power," *Philippine Daily Inquirer*, December 30, 2012.

23 Nikko Dizon and Marlon Ramos, "Police, Military Cancel Christmas Parties in Deference to 'Pablo' Victims," *Philippine Daily Inquirer*, December 12, 2012.

24 Cathy Yamsuan, "Filipinos Urged to Shun Lavish Christmas Celebrations," *Philippine Daily Inquirer*, December 15, 2012.

25 Frinston Lim and Karlos Manlupig, "Christmas Among the Ruins," *Philippine Daily Inquirer*, December 24, 2012.

26 Paolo Montecillo, "Cell Phone Users Text P1M for Victims of Typhoon 'Pablo,'" *Philippine Daily Inquirer*, December 28, 2012.

27 DJ Yap, "Paje Pushes 'Voluntourism,'" *Philippine Daily Inquirer*, February 24, 2013.

28 Kate McGeown, "Philippine President Bans Logging After Deadly Floods," BBC News, February 4, 2011. From the report: "This is not the first time a Philippine leader has tried to rein in the loggers. Almost every recent president has tried some sort of ban, with little success. The last attempt at a ban was in 2004, under the presidency of Gloria Arroyo. The policy had few noticeable results and was lifted after a year."

29 R.D. Lasco, R.G. Visco, and J.M. Pulhin, "Secondary Forests in the Philippines: Formation and Transformation in the 20th Century," *Journal of Tropical Forest Science* 13 (2001): 652.

30 Jeannette I. Andrade, "Illegal Logging, Mining Blamed for High Death Toll in Compostela Valley," *Philippine Daily Inquirer*, December 6, 2012.

31 DJ Yap, "Don't Blame Mining, Says Compostela Valley Governor," *Philippine Daily Inquirer*, December 7, 2012.

32 Neal H. Cruz, "Why So Many People Died in Mindanao," *Philippine Daily Inquirer*, December 10, 2012.

33 Ramon Tulfo, "Illegal Logging Caused Floods," *Philippine Daily Inquirer*, December 13, 2012.

34 DJ Yap, "DENR to Hire Mindanao Typhoon Victims for Reforestation Work," *Philippine Daily Inquirer*, February 26, 2013.

35 Randy David, "Disasters and the Poor," *Philippine Daily Inquirer*, December 13, 2012.

36 Michael Lim Ubac, "Palace Chides Bishop for Linking Tragedies to RH Bill Push," *Philippine Daily Inquirer*, December 8, 2012.

37 "'Liwanag' World Festival Starts Jan. 29 in Davao," *Philippine Daily Inquirer*, January 18, 2013.

38 "Philippines (Mindanao) Humanitarian Action Plan 2013," UN Office for the

Coordination of Humanitarian Affairs (OCHA), December 10, 2012; "Global Aid for PH Launched," *Philippine Daily Inquirer*, December 11, 2012.

39 Tarra Quismundo, "1M 'Pablo' Victims Still Need Food Aid," *Philippine Daily Inquirer*, January 5, 2013.

40 "Philippines (Mindanao) Humanitarian Action Plan 2013, Revision: January 2013," OCHA, January 23, 2013; "Filipino Typhoon Victims Need More Help—UN," *Philippine Daily Inquirer*, January 25, 2013.

41 Jerome Aning, "Former Boxing Champ Evander Holyfield Arrives in Manila," *Philippine Daily Inquirer*, February 14, 2013.

42 Jocelyn R. Uy, "For Bouncing Back from Disaster, PH Receives Major Tourism Award," *Philippine Daily Inquirer*, May 28, 2013. The following year, the award went to India for rebuilding after the 2013 Uttarakhand disaster (see Chapter 14) and Cyclone Phailin.

43 Frinston L. Lim, "Makeshift Classrooms in Typhoon-Hit Village," *Philippine Daily Inquirer*, June 3, 2013.

44 "Philippines (Mindanao) Humanitarian Action Plan 2013, Mid-Year Review," OCHA, June 25, 2013.

45 "WFP Appeals for US$21.6 Million to Assist Victims of Typhoon Bopha," *World Food Programme News*, December 11, 2012, wfp.org/news/news-release/wfp-appeals -us216-million-assist-victims-typhoon-bopha; Judy Quiros, "Int'l Aid Group Concerned by Malnutrition Among Children in Davao Oriental Town," *Philippine Daily Inquirer*, December 4, 2013.

46 Kristine Angeli Sabillo, "DBM to Release P1.06B for Rehab of Areas Damaged by 'Pablo,'" *Philippine Daily Inquirer*, December 5, 2013.

47 "DOST Wind Tunnel to Simulate Storms," *Philippine Daily Inquirer*, December 15, 2012.

48 "DOST to Launch MOSES for Project NOAH," Department of Science and Technology, Republic of the Philippines, July 22, 2013, dost.gov.ph/index.php?option=com _content&view=article&id=1320:dost-to-launch-moses-for-project-noah.

49 Nina P. Calleja, "Rehab of Banana Plantations Set," *Philippine Daily Inquirer*, January 6, 2013.

50 Nico Alconaba, "Tenants Leave Coconut Farms Destroyed by Typhoon 'Pablo,'" *Philippine Daily Inquirer*, December 17, 2012.

51 Allan A. Nawal, "Mindanao Banana Growers' Losses Exceed P1B," *Philippine Daily Inquirer*, August 14, 2013.

52 Riza T. Olchondra, "Economy Headed for 7% GDP Growth," *Philippine Daily Inquirer*, December 17, 2012.

53 Matikas Santos, "Philippines Is Fastest Growing Asian Country for First Quarter of 2013," *Philippine Daily Inquirer*, May 30, 2013.

54 Nancy C. Carvajal, "NBI Probes P10-B Scam," *Philippine Daily Inquirer*, July 12, 2013.

55 Kristine Felisse Mangunay, "Napoles Gets Life Sentence," *Philippine Daily Inquirer*, April 15, 2015; Camille Diola, "Luy's Camp: Napoles Pocketed More than P2 Billion," *Philippine Star*, June 4, 2014.

56 Frinston L. Lim, "Infra Projects in Provinces Left Hanging with 'Pork' Suspension, Say Local Officials," *Philippine Daily Inquirer*, October 21, 2013.

57 Christian V. Esguerra and Jerome Aning, "More Cases Filed Asking SC to Rule DAP

Illegal," *Philippine Daily Inquirer*, October 18, 2013; Artemio V. Panganiban, "DAP Is Not PDAF," *Philippine Daily Inquirer*, November 9, 2013.

58 Christine O. Avendaño, "Supreme Court Slays PDAF," *Philippine Daily Inquirer*, November 20, 2013; Tetch Torres-Tupas, "Gov't Lawyers Tell SC: Gov't Doesn't Need DAP Anymore," *Philippine Daily Inquirer*, January 28, 2014.

59 Bankoff, *Cultures of Disaster*, 177–78.

60 Frances Mangosing, "Army to Honor 21 Heroes of Typhoon 'Pablo,'" *Philippine Daily Inquirer*, December 26, 2012.

61 Edwin O. Fernandez, Frinston Lim, and Nico Alconaba, "Santa Comes to ComVal," *Philippine Daily Inquirer*, December 24, 2012.

62 Nikko Dizon, "We Have Soldier, Cop: NPA," *Philippine Daily Inquirer*, January 20, 2013.

63 Karlos Manlupig, "Agusan Folk Caught in the Crossfire," *Philippine Daily Inquirer*, August 10, 2013.

64 Carlos Isagani T. Zarate, "Rising Up," *Philippine Daily Inquirer*, February 3, 2013.

65 TJ Burgonio and Karlos Manlupig, "Davao Food Protest Over but DSWD Suing Leaders," *Philippine Daily Inquirer*, March 1, 2013.

66 "Storm Survivor Felled by Assassins," *Philippine Daily Inquirer*, March 16, 2013.

67 The eruption buried U.S. air and naval bases in ash. With nationalist sentiment already growing, the legislature took the opportunity to shut the door on the American presence altogether.

68 Germelina Lacorte, "'PH-US Military Exercises a Cover to Open Mining Activities'— Church Group," *Philippine Daily Inquirer*, April 9, 2013.

69 Ambassador Harry K. Thomas Jr., remarks at Balikatan Opening Ceremony, Camp Aguinaldo, April 5, 2013.

70 TJ Burgonio, "Palace Defends US Military's Use of Drones in 'Special Cases,'" *Philippine Daily Inquirer*, August 16, 2013.

71 Germelina Lacorte, "DENR Eyes Drones vs. Illegal Logging; Militants Believe Rebels Are Real Targets," *Philippine Daily Inquirer*, July 16, 2013. A year later, the New People's Army captured five employees of the drone contractor Skyeye and held them for four days for their alleged spying activities. "Rebels Release DENR Surveyors," *Inquirer Mindanao*, June 2, 2014.

72 International names for these 2013 storms, assigned by the Japan Meteorological Agency, were Utor (Labuyo), Trami (Maring), Usagi (Odette), Nari (Santi), Krosa (Vinta), and Podul (Zoraida).

73 Maricar Cinco, "Unstable Garbage Mountain Slows Down Search for 4 in Rizal 'Trash Slide,'" *Inquirer Southern Luzon*, April 20, 2013.

74 Jean-Christophe Gaillard and Jake Rom D. Cadag, "From Marginality to Further Marginalization: Experiences from the Victims of the July 2000 Payatas Trashslide in the Philippines," *Journal of Disaster Risk Studies* 2 (2009): 197.

75 We use the moment-magnitude scale to express the sizes of the earthquakes discussed in this book. U.S. Geological Service, "Measuring the Size of an Earthquake," n.d., earthquake.usgs.gov/learn/topics/measure.php.

76 Momerto Bautista, interview with the authors, Inabanga, Philippines, January 13, 2014.

77 Ernesto C. Torres Jr., *A Success Story of Philippine Counterinsurgency: A Study of Bohol*, master's thesis, U.S. Army Command and General Staff College, 2011.

78 Department of Tourism, "Bohol Tourism Back in Business," *Official Gazette*, October 30, 2013.

79 Jeandie O. Galolo, "Bohol Ruins 'Are Tourist Spots,'" *Sun Star Cebu*, October 29, 2013; "Philippines: Post-Quake Bohol Refocusing the Local Tours," Tourism-Review.com, February 24, 2014.

80 Contributions were flowing to Bohol from the nearby metropolitan island of Cebu, according to Chinkee Sabanpan of the Gawad Kalinga Community Development Foundation, but after Yolanda "it seems like everybody forgot about the earthquake," especially as northern Cebu itself was hit hard by the typhoon (telephone interview with Paul Cox, January 14, 2014). National reconstruction funds did, however, still come to the island, thanks in part to Governor Chatto's strong personal connections in Manila. Eight months after the quake the Department of Interior and Local Government handed over an infrastructure recovery fund worth $54 million. See "Bohol Bankable to Investments: Roxas," *Bohol Chronicle*, June 8, 2014.

81 Carmel Loise Matus, "Bohol Quake Triggers a Phenomenon: Land Rising from Bottom of the Sea," *Philippine Daily Inquirer*, August 9, 2015.

82 Jeannette I. Andrade and Michael Lim Ubac, "PH Braces for a Category 5 Storm," *Philippine Daily Inquirer*, November 8, 2013.

83 Michael Lim Ubac, "Gov't Prepositions P195M Relief in Eastern Visayas," *Philippine Daily Inquirer*, November 8, 2013.

84 Michael Lim Ubac and Nikko Dizon, "Mass Evacuation Ready," *Philippine Daily Inquirer*, November 7, 2013.

85 "Thousands Flee Fury of Supertyphoon 'Yolanda,'" *Philippine Daily Inquirer*, November 8, 2013.

86 Ubac and Dizon, "Mass Evacuation Ready."

87 *Philippine Daily Inquirer*, "Thousands Flee Fury."

88 Statements on November 13, 2013, 19th Conference of Parties to the UNFCCC, Warsaw, Poland. For the following year's conference in Lima, Saño was removed from the national delegation at the last minute. After this he took to organizing civil society, leading a "people's pilgrimage" to the 2015 Paris Climate Change Conference, and in 2016 becoming the executive director of Greenpeace Southeast Asia.

89 "Statement by the President on Super Typhoon Haiyan/Yolanda," White House, November 10, 2013.

90 Juan L. Mercado, "Ignored Crisis," *Philippine Daily Inquirer*, January 10, 2014; Matikas Santos, "How a Forest of Mangroves Saved a Village from 'Yolanda,'" *Philippine Daily Inquirer*, November 6, 2014.

91 Michael Lim Ubac, "P41B for 'Yolanda' Rehabilitation," *Philippine Daily Inquirer*, December 1, 2013; Nestor Corrales, "Aquino Signs 'Yolanda' Rehabilitation Plan," *Philippine Daily Inquirer*, October 30, 2014.

92 Financial Tracking Service, UN Office for the Coordination of Humanitarian Affairs, "Strategic Response Plan(s): Philippines—Typhoon Haiyan Strategic Response Plan (November 2013–November 2014)," http://docs.unocha.org/sites/dms/CAP/SRP _2013-2014_Philippines_Typhoon_Haiyan.pdf.

93 Ronnel W. Domingo, "Coco Oil Exports Down 18% Due to 'Yolanda' Damage," *Philippine Daily Inquirer*, March 11, 2014.

94 *Global Economic Prospects*, World Bank, January 2015.

95 Global Coal Plant Tracker, http://endcoal.org/tracker.

96 Fernando del Mundo, "'Yolanda' Bunkhouses Overpriced," *Philippine Daily Inquirer*, January 6, 2014; Christian V. Esguerra, "35% Bunkhouse Kickbacks Probed," *Philippine Daily Inquirer,* January 7, 2014; Joey A. Gabieta, "20,570 Yolanda Survivors Still Live in Bunk Houses, Says Archbishop," *Philippine Daily Inquirer*, October 28, 2014; People Surge, "Statement on the 2nd Anniversary of Typhoon Yolanda," peoplesurge phils.wordpress.com/ph-disaster-survivors-to-the-world-intensify-our-demands-for-jus tice-our-survival-is-non-negotiable.

97 Jamie Marie Elona, "PNP on NPA Leyte Attack: 'It's a Hoax,'" *Philippine National Inquirer*, November 13, 2013; Allan Nawal and Germelina Lacorte, "NPA Extends Ceasefire in Areas Devastated by 'Yolanda' for 1 Month," *Philippine National Inquirer*, November 24, 2013; Delfin T. Mallari Jr., "Communist Rebels to Launch 'Yolanda' Relief Drive on 45th Anniversary," *Philippine National Inquirer*, December 18, 2013; Bong Lozada, "Bukidnon Town Mayor Dies in NPA Ambush," *Philippine National Inquirer*, July 2, 2014.

98 Katungod-Sinirangan Bisayas cited in Marissa Cabaljao, "Aquino's Vindictiveness Puts Yolanda Survivors Under Siege," *People Surge*, November 3, 2015, peoplesurge phils.wordpress.com/2015/11/03/aquinos-vindictiveness-puts-yolanda-survivors -under-siege-people-surge.

99 Judith Buhay, e-mail interview with Stan Cox, January 21, 2015. Buhay wrote that Tacloban suffered no casualties from Ruby, but countless homes had once again been left roofless.

100 Rehabilitation czar Panfilo Lacson, quoted in "Gov't Eyes No-Dwelling Zones in 'Yolanda' Areas," ABS-CBN News, January 28, 2014, abs-cbnnews.com/nation/01/28 /14/govt-eyes-no-dwelling-zones-yolanda-areas; see also IBON Foundation, *Disaster upon Disaster: Lessons Beyond Yolanda* (Quezon City, Philippines: IBON Foundation, 2015).

101 People Surge, "Statement."

102 DJ Yap, "'No-Build Zones' Marked in 'Yolanda'-Hit Areas," *Philippine Daily Inquirer*, December 20, 2013; Lalaine M. Jimenea, "Solve Water Problems First Before 'Great Wall,'" *The Freeman*, October 6, 2015.

103 U.S. Embassy, Manila, Philippines, "JTF 505 Disestablished," press release, December 1, 2013.

104 "Agreement Between the Government of the United States of America and the Government of the Republic of the Philippines on Enhanced Defense Cooperation," signed at Quezon City, April 28, 2014.

105 Liwanag World Festival website, liwanagworldfest.net.

106 "Homily of Pope Francis," mass with survivors of Super Typhoon Yolanda, Daniel Z. Romualdez International Airport, Tacloban, January 17, 2015 (delivered originally in Spanish); Marc Jayson Cayabyab and Nestor Corrales, "Pope Francis Leaves Leyte Early for Manila," *Philippines Daily Inquirer*, January 17, 2015.

Chapter 3: Neighbors to the Sky

1 Translation of Eli Siegel, *Hail, American Development* (New York: Definition Press, 1968), 118. Originally published in Jean de La Fontaine, *Fables Choisies* (1668).

2 Devin Leonard, "Gary Barnett, Controversial Master of New York City Luxury Real Estate," *Bloomberg Businessweek*, October 2, 2014. The penthouse sale, still in progress at the time of the profile, ultimately closed at $100.5 million—the most expensive apartment in the city's history by a margin of $12.5 million.

3 CityRealty, *CityRealty 100 Semi-Annual Report,* November 4, 2015. These super-tall, super-skinny towers (111 West 57th would be only sixty feet wide) populated the skyline despite the now clear threat from high winds. In 2015, Brad Gair, vice president of emergency management and enterprise resilience at New York University's Langone Medical Center, told writer Eric Klinenberg, "No one in New York City is really prepared for a major wind event, and we could be facing Category 3 or 4 hurricanes. The debris from broken windows, construction sites, and even cranes would be truly dangerous." In other words, none of the post-Sandy measures taken so far—mostly aimed at storm surge defense—would prevent an even more menacing version of the One57 crane accident from occurring. Eric Klinenberg, "Is New York Ready for Another Sandy?," *New Yorker*, October 27, 2015.

4 Ken Belson and Mary Pilon, "Marathon Is Set to Go On, Stirring Debate," *New York Times*, October 31, 2012.

5 Office of Electricity Delivery and Energy Reliability, "Hurricane Sandy Situation Report #6," U.S. Department of Energy, October 31, 2012.

6 Generators were already becoming a symbol of the city's injustices. A Reuters photo circulated of a wine-dark Lower Manhattan with the power switched off, with the exception of a blazing bright but vacant Goldman Sachs building, powered by its own automatic generator.

7 Tara Palmeri, "This Is No Way to Get Us Up and Running," *New York Post*, November 2, 2012.

8 "Giuliani's Hurricane Sandy Marathon Prediction," TEN Eyewitness News, Sydney, Australia, October 31, 2012.

9 Ken Belson and David W. Chen, "Marathon Stumbled Along a Route of Indecision on Its Way Toward Cancellation," *New York Times*, November 4, 2012.

10 Carey Vanderborg, "Hurricane Sandy 2012 Live Update: New York and New Jersey Deal with the Unrelenting Damage," *International Business Times*, October 22, 2012.

11 Both GOLES and the Sixth Street Community Center use the name of the Lower East Side in its larger historic sense. GOLES provides services to residents north of the Brooklyn Bridge, south of Fourteenth Street, and east of Bowery and Third and Fourth Avenues. Both organizations are located in what is now more commonly called the East Village, Alphabet City, or Loisaida (a Hispanicization of "Lower East Side"). The organizations use the name Lower East Side for the same reason real estate agents don't: its associations with a poor but tough immigrant neighborhood of nearly two centuries' standing.

12 Goldi Guerra, interview with Paul Cox, New York City, October 28, 2014.

13 David Noriega, "The Undocumented Immigrants Who Rebuilt New York After Sandy," *BuzzFeed*, October 28, 2014.

14 Ibid.

15 Alan Feuer, "Where FEMA Fell Short, Occupy Sandy Was There," *New York Times*, November 9, 2012.

16 *Hurricane Sandy: Youthful Energy and Idealism Tackles Real-World Disaster Response*, Department of Homeland Security, 2013, https://www.hsdl.org/?view&did=783226.

17 Nick Pinto, "Occupy's Undercover Cop: 'Shady,' Ubiquitous, and Willing to Get Arrested," *Gothamist*, November 10, 2013.

18 Bill Sothern, telephone interview with Paul Cox, October 30, 2014.

19 "Far Rockaway Cough," MyFoxNY.com, November 15, 2012.

20 "Weathering the Storm: Rebuilding a More Resilient New York City Housing Authority Post-Sandy," Alliance for a Just Rebuilding, Alliance in Greater New York (ALIGN), Community Development Project at the Urban Justice Center, Community Voices Heard, Faith in New York, Families United for Racial and Economic Equality (FUREE), Good Old Lower East Side (GOLES), New York Communities for Change (NYCC), and Red Hook Initiative (RHI), 2014.

21 "A Tale of Two Sandys," Superstorm Research Lab, 2013, superstormresearchlab.org /white-paper.

22 Ibid.

23 Max Liboiron and David Wachsmuth, "The Fantasy of Disaster Response: Governance and Social Action During Hurricane Sandy," *Social Text Journal—Periscope Topics*, October 29, 2013.

24 Lee Clarke, *Mission Improbable: Using Fantasy Documents to Tame Disaster* (Chicago: University of Chicago Press, 1999).

25 Justin Elliott, Jesse Eisinger, and Laura Sullivan, "The Red Cross' Secret Disaster," ProPublica, October 29, 2014.

26 Justin Elliott, "Red Cross: How We Spent Sandy Money Is a 'Trade Secret,'" ProPublica, June 26, 2014.

27 Amy Peterson, interview with Paul Cox, New York City, November 5, 2014.

28 Christopher Robbins, "'Build It Back' Sandy Recovery Program Has Built Nothing Back," *Gothamist*, February 24, 2014; Matthew Schuerman, "Zero Homes Repaired, and 19,920 to Go," WNYC News, February 28, 2014.

29 Russ Buettner and David W. Chen, "Broken Pledges and Bottlenecks Hurt Mayor Bloomberg's Build It Back Effort," *New York Times*, September 4, 2014.

30 "One City, Rebuilding Together—Progress Update," Mayor's Office of Housing Recovery Operations, October 2015.

31 Ilya Jalal, interview with Paul Cox, New York City, October 29, 2014.

32 Patricia Kane, RN, treasurer, New York State Nurses Association, interview with Paul Cox, New York City, October 29, 2014.

33 Jeffrey Kluger, "Is Global Warming Fueling Katrina?," TIME.com, August 29, 2005; Eric Klinenberg, "Adaptation: How Can Cities be 'Climate-Proofed'?," *New Yorker*, January 7, 2013; Kate Sheppard, "Flood, Rebuild, Repeat: Are We Ready for a Superstorm Sandy Every Other Year?," *Mother Jones*, July/August 2013.

34 Charles Greene, "The Winters of Our Discontent," *Scientific American* 307, no. 6 (2012): 50–55.

35 Daniel Doctoroff, "Without PlaNYC, Hurricane Sandy's Devastation Would Have Been Much Worse," *New York Observer*, December 11, 2012.

36 "PlaNYC: A Stronger, More Resilient New York," Office of the Mayor, New York City, 2013.

37 "A Tale of Two Sandys," Superstorm Research Lab, 2013.

38 "One City: Built to Last," Office of the Mayor, New York City, 2014.

39 Jonathan Mahler, "How Coastline Became a Place to Put the Poor," *New York Times*, December 4, 2012.

40 BIG had established a New York office in 2011 to design another major residential tower on the Fifty-Seventh Street corridor, the striking pyramidal West 57. This tower and the Dry Line raised the firm's profile so much that they were subsequently hired to design Two World Trade Center, as well as, on the other side of the country, a new corporate campus for Google.

41 Daniel Kidd, interview with Paul Cox, New York City, October 27, 2014.

42 Jane Jacobs was the writer and community organizer who famously fought Moses's plans to build a freeway through Greenwich Village, sparking a reevaluation of urban renewal projects and the role of community.

43 Lilah Mejia, interview with Paul Cox, New York City, October 27, 2014.

44 Oshrat Carmiel, "NYC's Luxury-Condo King Shifts Gears in Bid for Chinese Buyers," *Bloomberg Businessweek*, October 14, 2015; Maggie Livingstone, "Community Accuses Extell of 'Economic Segregation' in LES," *Curbed NY*, June 19, 2014.

45 *Resilience in the Wake of Superstorm Sandy*, Associated Press–NORC Center for Public Affairs Research, 2013.

46 Liz Koslov, "Fighting for Retreat after Sandy: The Ocean Breeze Buyout Tent on Staten Island," *Metropolitics*, April 23, 2014.

47 New York governor Andrew Cuomo, 2013 State of the State Address, January 9, 2013.

48 For New York City's history of expansion and building back bigger, see Ted Steinberg, *Gotham Unbound: The Ecological History of Greater New York* (New York: Simon and Schuster, 2014).

49 The other neighborhoods asking for buyouts were Midland Beach, Great Kills, Tottenville, and South Beach. The governor's spokeswoman said that the state had never intended to remove development from the entire shore and was instead directing money into the city's acquisition for redevelopment program. Matthew Schuerman, "Cuomo Limits Sandy Buyout Program to Three City Nabes," WNYC News, April 11, 2014.

50 It is *Phragmites* that stars in the fable of the oak and the reed cited in this chapter's epigraph, and of which La Fontaine wrote, "You come to be most often / On the wet edges of the kingdoms of the wind" (Siegel, *Hail, American Ingenuity*, 118). *Phragmites* is a much maligned invader in North America but its storm resilience is (literally) fabled. Asked about Sandy's long-term effects on the *Phragmites*-heavy New Jersey Meadowlands, Francisco Artigas, director of the Meadowlands Environmental Research Institute, dismissed any such thing: "There's no smoking gun showing that Sandy did anything positive or negative to the wetlands. Everything was back to normal two weeks later." Francisco Artigas, interview with Paul Cox, Lyndhurst, NJ, October 28, 2014.

51 Ronnie Loesch, interview with Paul Cox, New York City, October 29, 2014.

Chapter 4: Every Silver Lining . . .

1 Translation of Joseph McCabe, *Toleration and Other Essays by Voltaire* (New York: G.P. Putnam's Sons, 1912), 257.

2   Pearl S. Buck, *The Big Wave* (New York, NY: John Day, 1948), 45.

3   J.M. Albala-Bertrand, *The Political Economy of Large Natural Disasters: With Special Reference to Developing Countries* (Oxford: Oxford University Press, 1993), 186, 197. Thanks to the durability of those sociopolitical systems, argues Albala-Bertrand, "it should be clear by now that the social structure of affected localities, let alone of whole countries, is never destroyed by sudden natural disasters and rarely terminated by slow-developing ones." Never destroyed, perhaps, but many have since argued that those social structures can undergo dramatic change.

4   J. Birkmann, P. Buckle, J. Jaeger, M. Pelling, N. Setiadi, M. Garschagen, N. Fernando, and J. Kropp, "Extreme Events and Disasters: A Window of Opportunity for Change?," *Natural Hazards* 55 (2010): 637–55; R.S. Olson and V.T. Gawronski, "Disasters as Critical Junctures? Managua, Nicaragua 1972 and Mexico City 1985," *International Journal of Mass Emergencies and Disasters* 21 (2003): 3–35.

5   Sandra Bulling, "Typhoon Haiyan: An Aid Worker's Diary of a Disaster," *The Guardian*, December 21, 2013.

6   Agustino Fontevecchia, "Despite $50B in Damages, Hurricane Sandy Will Be Good for the Economy, Goldman Says," *Forbes*, November 6, 2012.

7   Jeff Kearns, Susanna Pak, and Noah Buhayar, "Post-Sandy Reconstruction to Boost Economy," *San Francisco Chronicle*, November 26, 2012.

8   Michelle Krebs and Bill Visnic, "November Car Sales Ride Tailwind of Hurricane Recovery," Edmunds.com, December 3, 2012.

9   Frédéric Bastiat, *Selected Essays on Political Economy* (Irvington-on-Hudson, NY: Foundation for Economic Education, 1964), 2–3.

10 The economist Amartya Sen conceived the concept of capability. He explained it thus in 1993: "Capability is not an awfully attractive word. It has a technocratic sound, and to some it might even suggest the image of nuclear war strategists rubbing their hands in pleasure over some contingent plan of heroic barbarity. . . . Perhaps a nicer word could have been chosen when some years ago I tried to explore a particular approach to well-being and advantage in terms of a person's ability to do valuable acts or reach valuable states of being. The expression was picked to represent the alternative combinations of things a person is able to do or be—the various 'functionings' he or she can achieve. The capability approach to a person's advantage is concerned with evaluating it in terms of his or her actual ability to achieve various valuable functionings as a part of living. . . . It differs from other approaches using other informational focuses, for example, personal utility (focusing on pleasures, happiness, or desire fulfilment), absolute or relative opulence (focusing on commodity bundles, real income, or real wealth), assessments of negative freedoms (focusing on procedural fulfilment of libertarian rights and rules of non-interference), comparisons of means of freedom (e.g. focusing on the holdings of 'primary goods,' as in the Rawlsian theory of justice), and comparisons of resource holdings as a basis of just equality (e.g. as in Dworkin's criterion of 'equality of resources')." Amartya Sen, "Capability and Well-Being," in *The Philosophy of Economics: An Anthology, Third Edition*, ed. Daniel Hausman (Cambridge: Cambridge University Press, 2008), 270–93.

11 Mark Skidmore and Hideki Toya, "Do Natural Disasters Promote Long-Run Growth?," *Economic Inquiry* 40 (2002): 664–87.

12 Joseph Schumpeter, *Capitalism, Socialism and Democracy* (New York: Harper, 1975),

82–85; Eduardo Cavallo and Ilan Noy, "Natural Disasters and the Economy—A Survey," *International Review of Environmental and Resource Economics* 5 (2011): 63–102. The authors of this survey write that Skidmore and Toya "explain their somewhat counterintuitive finding by suggesting that disasters may be speeding up the Schumpeterian 'creative destruction' process that is at the heart of the development of market economies."

13 Schumpeter, *Capitalism*, 51.

14 With "innovation," Schumpeter took an old word of negative or suspicious meanings and gave it a new definition within economics that burst into the popular culture only in recent decades. Jill Lepore, "The Disruption Machine: What the Gospel of Innovation Gets Wrong," *New Yorker*, June 23, 2014.

15 Schumpeter, *Capitalism*, 82.

16 Joseph Berliner, *The Innovation Decision in Soviet Industry* (Cambridge, MA: MIT Press, 1976), 527–28. He wrote, "If Adam Smith had taken as his point of departure not the coordinating mechanism but the innovation mechanism of capitalism, he [might] well have designated competition not as an invisible hand but as an invisible foot."

17 Andrea Leiter, Harald Oberhofer, and Paul Raschky, "Creative Disasters? Flooding Effects on Capital, Labour and Productivity Within European Firms," *Environmental Resource Economics* 43 (2009): 333–50.

18 Eric Strobl, "The Economic Growth Impact of Hurricanes: Evidence from U.S. Coastal Counties," *Review of Economics and Statistics* 92 (2011): 575–89.

19 Cavallo and Noy, "Natural Disasters"; Thomas McDermott, Frank Barry, and Richard Tol, "Disasters and Development: Natural Disasters, Credit Constraints, and Economic Growth," *Oxford Economic Papers* 66 (2014): 750–73; Ilan Noy and Aekkanush Nualsri, "What Do Exogenous Shocks Tell Us About Growth Theories?," University of Hawaii working paper, October 29, 2007, www.economics.hawaii.edu/research/workingpapers/WP_07-28.pdf.

20 Ilan Noy and Tam Bang Vu, "The Economics of Natural Disasters in a Developing Country: The Case of Vietnam," *Journal of Asian Economics* 21 (2010): 345–54.

21 Norman Loayza, Eduardo Olaberria, Jamele Rigolini, and Luc Christiaensen, "Natural Disasters and Growth: Going Beyond the Averages," *World Development* (2012): 1317–36. Only medium-size disasters were sometimes followed by increased growth rates, while severe disasters never were.

22 David Strömberg, "Natural Disasters, Economic Development, and Humanitarian Aid," *Journal of Economic Perspectives* 21 (2007): 199–222.

23 J.C. Cuaresma, J. Hlouskova, and M. Obersteiner, "Natural Disasters as Creative Destruction? Evidence from Developing Countries," *Economic Inquiry* 46 (2008): 214–26; Loayza et al., "Natural Disasters"; Ilan Noy, "The Macroeconomic Consequences of Disasters," *Journal of Development Economics* 88 (2009): 221–31; Michael Carter, Peter Little, Tewodaj Mogues, and Workneh Negatu, "Poverty Traps and Natural Disasters in Ethiopia and Honduras," *World Development* 35 (2007): 835–56; Hideki Toya and Mark Skidmore, "Economic Development and the Impacts of Natural Disasters," *Economics Letters* 94 (2007): 20–25.

24 Eduardo Rodriguez-Oreggia, Alejandro de la Fuente, Rodolfo de la Torre, and Hector A. Moreno, "Natural Disasters, Human Development and Poverty at the Municipal Level in Mexico," *Journal of Development Studies* 49 (2013): 442–55.

25 Patrick Premand and Renos Vakis, "Do Shocks Affect Poverty Persistence? Evidence Using Welfare Trajectories from Nicaragua," *Well-Being and Social Policy* 6, no. 1 (2010): 95–129.

26 Tobias Rasmussen, "Macroeconomic Implications of Natural Disasters in the Caribbean," Social Science Research Network Scholarly Paper, December 1, 2004, papers .ssrn.com/abstract=879049; Martin Heger, Alex Julca, and Oliver Paddison, "Analysing the Impact of Natural Hazards in Small Economies: The Caribbean Case," Research Paper, United Nations University, 2008, econstor.eu/handle/10419/63505.

27 Stéphane Hallegatte, Jean-Charles Hourcade, and Patrice Dumas, "Why Economic Dynamics Matter in Assessing Climate Change Damages: Illustration on Extreme Events," *Ecological Economics* 62 (2007): 330–40.

28 George Horwich, "Economic Lessons of the Kobe Earthquake," *Economic Development and Cultural Change* 48 (2000): 521–42. Horwich wrote that the quick restoration of growth was "attributed to the primacy of human capital, which suffered comparatively little loss, over the physical stock of capital in determining the output of a developed economy." With the more than six thousand dead apparently not registering as a major loss of human capital, he continued, "Price-directed market responses were spurred both by the quake and by monetary stimulus in a setting of excess production capacity. These responses appear to have led the recovery."

29 William duPont IV and Ilan Noy, "What Happened to Kobe? A Reassessment of the Impact of the 1995 Earthquake in Japan," University of Wellington Working Paper No. 09/2012, September 2012, researcharchive.vuw.ac.nz/handle/10063/2087; Ilan Noy, "The Enduring Economic Aftermath of Natural Catastrophes," *Vox* (Centre for Economic Policy Research), September 5, 2012, voxeu.org/article/economic-conse quences-natural-catastrophes.

30 William Nordhaus, "Economic Policy in the Face of Severe Tail Events," *Journal of Public Economic Theory* 14 (2012): 197–219; William Nordhaus, "The Economics of Tail Events with an Application to Climate Change," *Environmental Economics and Policy* 5 (2011): 240–57; Richard Tol, "On the Uncertainty About the Total Economic Impact of Climate Change," *Environmental and Resource Economics* 53 (2012): 97–116.

31 Richard Samuels, *3.11: Disaster and Change in Japan* (Ithaca, NY: Cornell University Press, 2013), 35–36.

32 "Government Support Indispensable for Restart of Nuclear Power Plants," *Japan News*, December 19, 2014; Esther Tanquintic-Misa, "Japan Approves Restart of Two Nuclear Reactors," *International Business Times*, December 19, 2014.

33 Stephen Stapczynski. "Japan to Restart 3rd Nuclear Reactor Under Post-Fukushima Rules," *Bloomberg*, January 28, 2016.

34 Osamu Tsukimori, Kiyoshi Takenaka, and Elaine Lies, "Japanese Volcano Erupts, Nearby Nuclear Plant Unaffected," Reuters, February 5, 2016.

35 Alice Fothergill and Lori Peek, "Poverty and Disasters in the United States: A Review of Recent Sociological Findings," *Natural Hazards* 32 (2004): 89–110; Nejat Anbarci, Monica Escaleras, and Charles Register, "Earthquake Fatalities: The Interaction of Nature and Political Economy," *Journal of Public Economics* 89 (2005): 1907–33.

36 Fothergill and Peek, "Poverty"; Eric Kleinenberg, *Heat Wave: A Social Autopsy of Disaster in Chicago* (Chicago: University of Chicago Press, 2002); Michelle Chen, "In Sandy's Wake, New York's Landscape of Inequity Revealed," *In These Times*,

November 1, 2012; Daniel Weiss, Jackie Weidman, and Mackenzie Bronson, *Heavy Weather: How Climate Destruction Harms Middle- and Lower-Income Americans* (Washington, DC: Center for American Progress, 2012); Narayan Sastry and Mark Van Landingham, "One Year Later: Mental Illness Prevalence and Disparities Among New Orleans Residents Displaced by Hurricane Katrina," *American Journal of Public Health* 99 (2009): S725–31; Jacob Vigdor, "The Economic Aftermath of Hurricane Katrina," *Journal of Economic Perspectives* 22 (2008): 135–54; Russell Sobel and Peter Leeson, "Government's Response to Hurricane Katrina: A Public Choice Analysis," *Public Choice* 127 (2006): 55–73.

37 Andrew Whitehead, interview with Stan Cox, Joplin, Missouri, August 23, 2014.

38 Mike McGraw, "Housing Troubles Mount, Especially for Joplin's Poor," *Kansas City Star*, December 17, 2011.

39 Klein, *Shock Doctrine*, 398.

40 *Economist*, "Counting the Cost." This article was written not long before publication of analyses showing that even Kobe, the poster child for quick recovery, has actually suffered very long-term stagnation since its earthquake; see DuPont and Noy, "What Happened."

41 Derek Kellenberg and Ahmed Mushfiq Mobarak, "Does Rising Income Increase or Decrease Damage Risk from Natural Disasters?," *Journal of Urban Economics* 63 (2008): 788–802.

42 Mark Pelling and Kathleen Dill, "Disaster Politics: Tipping Points for Change in the Adaptation of Sociopolitical Regimes," *Progress in Human Geography* 34 (2010): 21–37; R. Nordås and N.P. Gleditsch, "Climate Change and Conflict," *Political Geography* 26, no. 6 (2007): 761–82.

43 Popp, "Effects."

44 Voltaire, *Candide* (Rockville, MD: Wildside Press, 2007).

45 Voltaire, *Poem on the Lisbon Disaster*, in McCabe, *Toleration*, 255.

46 See Susan Neiman, *Evil in Modern Thought: An Alternative History of Philosophy* (Princeton, NJ: Princeton University Press, 2004).

47 The quote is apocryphal, and it has been attributed to other figures in Lisbon—for instance, to the Marquis of Alorno's order to "bury the dead, care for the living and close the ports." Filomena Amador, "The 1755 Lisbon Earthquake: Collections of Eighteenth-Century Texts," *História, Ciências, Saúde-Manguinhos* 14, no. 1 (2007): 285–323.

48 Neiman, *Evil in Modern Thought*, 248.

49 Voltaire, *Poem on the Lisbon Disaster*, in McCabe, *Toleration*, 260; Jean-Jacques Rousseau, "Rousseau to Voltaire, 18 August 1756," in *Correspondance complète de Jean Jacques Rousseau*, ed. Ralph Alexander Leigh (Geneva, 1967), 4:37–50, translated by R. Spang.

50 In this book, we express all tornado strengths according to the Enhanced Fujita (EF) scale. An EF-5 tornado produces three-second gusts of an estimated two hundred miles per hour or more. The National Weather Service notes, "The EF scale still is a set of wind estimates (not measurements) based on damage. Its uses three-second gusts estimated at the point of damage based on a judgment of 8 levels of damage." National Weather Service, "The Enhanced Fujita Scale," n.d., www.srh.noaa.gov /oun/?n=efscale.

51 William Brice, *A City Laid Waste: Tornado Devastation at Gainesville, Ga., April 6, 1936* (Atlanta: Webb and Martin, 1936). Among the 203 dead was Lucille Brewer Cox, grandmother of Stan Cox and great-grandmother of Paul Cox. She had just arrived at her job at the Gallant-Belk department store on the city square when the tornado hit. Stan's father, Tom Cox, five years old at the time, was in his home, located about a thousand feet from the square, and was unhurt. See Carolyn Crist, "Community to Observe 75th Anniversary of Deadly Tornado," *Gainesville Times*, April 6, 2011.

52 Johnny Vardeman, interview with Stan Cox, Gainesville, December 30, 2014.

53 B.M. Sigmon, "The Gainesville Tornado," *American Journal of Nursing* 37 (1937): 131–33.

54 Johnny Vardeman, interview with Stan Cox, Gainesville, December 30, 2014.

55 "The 10 Deadliest U.S. Tornadoes on Record," CNN, April 27, 2014; Samantha Grossman, "10 Deadliest Tornadoes in U.S. History," *Time*, May 21, 2013. One who survived the Tupelo tornado was the one-year-old Elvis Presley.

56 Michele Landis, "Fate, Responsibility, and 'Natural' Disaster Relief: Narrating the American Welfare State," *Law and Society Review* 33 (1999): 257–318. The initial payments were to individual merchants for, essentially, business losses: in 1790, an Act for the Relief of Thomas Jenkins and Company remitted duties on a parcel of goods lost in a ship fire, while an Act for the Relief of John Stewart and John Davidson remitted duties on 1,325 bushels of salt "casually destroyed" by a flood in the port of Annapolis.

57 "Hugh Bennett and the Perfect Storm," WETA, Washington, DC, weta.org/tv/program/dust-bowl/perfectstorm; Wellington Brink, *Big Hugh, the Father of Soil Conservation* (New York: Macmillan, 1951), 6.

58 Debabrata Mohanty, "Cyclone Phailin Super Relief: The Night Govt and IMD Saved the Day," *Financial Express*, October 14, 2013.

59 "Indian Media: Praise for Cyclone Rescue Efforts," BBC News, October 14, 2013; Amitabh Srivastava, "Naveen Patnaik's Record Fourth Straight Win as Odisha CM Turns State Politics into a One-Horse Race," *India Today*, May 17, 2014.

60 John Gasper and Andrew Reeves, "Make It Rain? Retrospection and the Attentive Electorate in the Context of Natural Disasters," *American Journal of Political Science* 55 (2011): 340–55.

61 Andrew Healy and Neil Malhotra, "Myopic Voters and Natural Disaster Policy," *American Political Science Review* 103 (2009): 387–406. The authors found that an increase in federal disaster relief spending in a county amounting to $9 per resident increases that party's vote share in the next election by three-quarters of a percentage point. And the favor is returned: for each additional percentage point in the vote share that a county gives a winning presidential candidate, the county receives an average 1.8 percent increase in disaster spending during the next presidential term.

62 Jowei Chen, "Voter Partisanship and the Effect of Distributive Spending on Political Participation," *American Journal of Political Science* 57 (2013): 200–217.

63 Healy and Malhotra, "Myopic Voters."

64 Gasper and Reeves, "Make It Rain?"

65 Healy and Malhotra, "Myopic Voters."

66 Ilan Noy and Aekkanush Nualsri, "Fiscal Storms: Public Spending and Revenues in the Aftermath of Natural Disasters," *Environment and Development Economics* 16 (2011): 113–28.

67 Healy and Malhotra, "Myopic Voters"; Gasper and Reeves, "Make It Rain?"

68 Thomas Garrett and Russell S. Sobel, "The Political Economy of FEMA Disaster Payments," *Economic Inquiry* 41 (2003): 496–509.

69 Russell Sobel and Peter Leeson, "Government's Response to Hurricane Katrina: A Public Choice Analysis," *Public Choice* 127 (2006): 55–73.

70 FEMA, "Disaster Declarations by Year," n.d., fema.gov/disasters/grid/year.

71 Associated Press, "Mississippi's Palazzo Sought Katrina Aid, Voted Against Sandy Insurance Funds," January 8, 2013.

72 Joseph Straw, "Gulf Coast Republican Rep. Steve Palazzo, Who Initially Opposed the $51-Billion Hurricane Sandy Relief Bill, Visits Wreckage Sites on Staten Island and Announces He Will Vote for Bill After All," New York *Daily News*, January 8, 2013.

73 Office of Jim Bridenstine, "Fiscal Responsibility Is Real Compassion," press release, January 15, 2013.

74 Pete Kasperowicz, "GOP: Predict Storms, Not Climate Change," *The Hill*, March 28, 2014; "Jim Bridenstine, Frank Lucas Push Weather Bill Through House," *Tulsa World*, May 20, 2015.

75 David A. Fahrenthold and Paul Kane, "Conservative Okla. Lawmakers Face Dilemma: Will They Support Tornado Relief Funding?," *Washington Post*, May 21, 2013.

76 Richard Simon, "Oklahoma Lawmakers Who Opposed Disaster Aid Now Face Own Disaster," *Los Angeles Times*, May 21, 2013.

77 Naomi Klein, *This Changes Everything: Capitalism vs. the Climate* (New York: Simon and Schuster, 2014), 10, 25.

78 Ian Angus, "The Myth of Environmental Catastrophism," *Monthly Review*, September, 2013.

79 This portion of Angus's essay is a critical analysis of Eddie Yuen, "The Politics of Failure Have Failed: The Environmental Movement and Catastrophism," in *Catastrophism: The Apocalyptic Politics of Collapse and Rebirth*, ed. Sasha Lilley, David McNally, Eddie Yuen, and James Davis (Oakland: PM Press, 2012), 15–43. Yuen offers a rebuttal and Angus responds in Eddie Yuen and Ian Angus, "Reply to 'The Myth of Environmental Catastrophism,'" *Monthly Review*, December, 2013.

80 Angus, "Myth"; emphasis is his.

81 Yuen and Angus, "Reply." Others on the left have strongly supported Klein's arguments in *This Changes Everything* even as they argue that she does not issue an explicit call for deep enough systemic change or socialism and is overly optimistic about prospects for quickly turning the ecological crisis around. See John Bellamy Foster and Brett Clark, "Crossing the River of Fire," *Monthly Review*, February 2015, and Robert Jensen, "Review: Naomi Klein, This Changes Everything," *Texas Observer*, September 25, 2014.

82 Kristen McQueary, "Chicago, New Orleans, and Rebirth," *Chicago Tribune*, August 13, 2015.

83 See, for example, Sara Jerde, "Chicago Tribune Writer: I'm 'Praying' for a Storm like Katrina in Chicago," *Talking Points Memo*, August 14, 2015.

84 Kristin McQueary, "Hurricane Katrina and What Was in My Heart," *Chicago Tribune*, August 15, 2015.

85 Patrick Rucker, "Obama Gives Unexpected Nod to Climate as Second Term Priority," Reuters, January 21, 2013; Office of the White House Press Secretary,

"Remarks by the President at the GLACIER Conference—Anchorage, AK," September 1, 2015, www.whitehouse.gov/the-press-office/2015/09/01/remarks-president-glacier-conference-anchorage-ak.

86 Samuels, *3.11*, chap. 4.

87 Ibid., 18.

88 Ibid., 92.

89 Barbara Nimri Aziz, "Prospects for a New Nepal: Life and Politics After the Big Quakes," *Counterpunch* 22, no.7 (2015): 17–21; Tom Esslemont, "Global Charities Accused of Misleading Public on Nepal Quake Aid," Reuters India, September 21, 2015; Amie Ferris-Rotman, "The Earthquake-Wrecked Town That the Nepali Government Forgot," *The Atlantic*, May 1, 2015; Andrew Marshall and Ross Adkin, "Nepalese Army Gets Image Boost from Quake Relief Work," Reuters India, April 30, 2015.

90 Pradeep Thakur, "India, Pak Agree for SAARC Disaster Response Force," *Times of India*, December 26, 2015.

91 Jennifer Elsea and R. Chuck Mason, "The Use of Federal Troops for Disaster Assistance: Legal Issues," Report No. RS22266, Congressional Research Service, November 28, 2008, archived at fas.org/sgp/crs/natsec/RS22266.pdf; Kevin Cieply, "Charting a New Role for Title 10 Reserve Forces: A Total Force Response to Natural Disasters," *Military Law Review* 196 (2008): 1–45. In a rare snub to the National Guard, Mayor Bloomberg denied that their aid was needed in Sandy-struck Brooklyn, saying at an October 31, 2012, press conference, "The National Guard has been helpful, but the NYPD is the only people we want on the street with guns. We don't need it." His choice of words caused a furor among gun rights advocates, who claimed that Bloomberg was barring troops from the borough because of his pro-gun-control views. Tom King, president of the New York State Rifle and Pistol Association, accused Bloomberg of "becoming a psychopath" because "if you can't trust the National Guard, who can you trust?" *Cam and Company*, NRA News, November 5, 2012.

92 Erik Auf der Heide, "Common Misconceptions About Disasters: Panic, the 'Disaster Syndrome,' and Looting," in *The First 72 Hours: A Community Approach to Disaster Preparedness*, ed. Margaret O'Leary (Lincoln, NE: iUniverse, 2004), 340–80.

93 Cheryl Harris and Devon Carbado, "Loot or Find: Fact or Frame?," in *After the Storm: Black Intellectuals Explore the Meaning of Hurricane Katrina*, ed. David Dante Troutt (New York: The New Press, 2006), 86–110.

94 Ibid. The debate over the photo captions was triggered initially by Aaron Kinney, "'Looting' or 'Finding'?," *Salon*, September 1, 2005.

95 Jonathan Katz, *The Big Truck That Went By: How the World Came to Save Haiti and Left Behind a Disaster* (New York: Palgrave Macmillan, 2013), 79–84.

96 Ibid., 85.

97 Ibid., 58.

98 Barack Obama, "President Obama Speaks on the Ongoing Response to Hurricane Sandy," White House YouTube channel, December 12, 2012, youtube.com/watch?v=QV2u1YUNK4E.

99 Peter Kropotkin, *Mutual Aid: A Factor of Evolution* (London: William Heinemann, 1902); Rebecca Solnit, *A Paradise Built in Hell: The Extraordinary Communities That Arise in Disasters* (New York: Viking, 2009).

100 Statements in Moonachie, NJ, on October 29, 2013, youtube.com/watch?v=JASOIg K80TE. One week later, Christie referred back to this in his gubernatorial reelection victory speech: "My pledge to you tonight is I will govern with the Spirit of Sandy," speech in Asbury Park, NJ, November 5, 2013.

101 Brian Epstein, "Producer's Notebook: As Reconstruction Crawls, the Spirit of Tacloban Prevails," *PBS NewsHour*, July 18, 2014.

102 Roza Sage, "NSW Bushfires: Clearing a Path to Rebuilding," *Sydney Morning Herald*, October 27, 2013.

103 See Jean-Christophe Gaillard, Elsa Clavé, and Ilan Kelman, "Wave of Peace? Tsunami Disaster Diplomacy in Aceh, Indonesia," *Geoforum* 39 (2008): 511–26. They cautioned against treating the tsunami as the sole vector of peace in Aceh, instead seeing it as a "powerful catalyst" in negotiations that were already favored by pre-disaster political changes. And this secondary role was a good thing, because "it appears that the slow, unequal and often poor reconstruction process is not hindering, or even threatening, the peace process because tsunami disaster related factors are less important for peace than non-tsunami disaster related factors."

104 See Richard Stuart Olson and Vincent T. Gawronski, "Disasters as Critical Junctures? Managua, Nicaragua 1972 and Mexico City 1985," *International Journal of Mass Emergencies and Disasters* 21 (2003): 5–35. This places the earthquake in the larger twenty-year context of the Institutional Revolutionary Party's loss of legitimacy and economic woes.

105 Rebecca Solnit, "Hope in the Face of Disaster," *Occupied Times*, December 19, 2012.

106 Solnit, *Paradise*, 169.

107 Victor W. Turner, *The Ritual Process: Structure and Anti-Structure* (Chicago: Aldine, 1969), 178.

108 Edith L.B. Turner, *Communitas: The Anthropology of Collective Joy* (New York: Palgrave Macmillan, 2011), 73.

109 Turner, *Ritual Process*, 132.

110 On the importance of blame, see Mary Douglas, *Risk and Blame: Essays in Cultural Theory* (London: Routledge, 1992).

111 Turner, *Communitas*, 5.

112 Barbara Bode, "Disaster, Social Structure, and Myth in the Peruvian Andes: The Genesis of an Explanation," *Annals of the New York Academy of Sciences* 293 (1977): 246–74.

113 Ibid.

114 Barbara Bode, *No Bells to Toll: Destruction and Creation in the Andes* (New York: Scribner, 1989), 244.

115 Anthony Oliver-Smith, "Anthropological Research on Hazards and Disasters," *Annual Review of Anthropology* 25 (1996): 303–28.

116 Linda Jencson, "Disastrous Rites: Liminality and Communitas in a Flood Crisis," *Anthropology and Humanism* 26 (2001): 46–58. This sort of magic seems at times to work on the most personal, psychological level. See Ronald Koegler and Shelby Hicks, "The Destruction of a Medical Center by Earthquake: Initial Effects on Patients and Staff," *California Medicine* 116 (1972): 63–67.

117 Susanna M. Hoffman, "The Worst of Times, the Best of Times: Toward a Model of Cultural Response to Disaster," in *The Angry Earth: Disaster in Anthropological Perspective*, ed. Anthony Oliver-Smith and Susanna Hoffman (New York: Routledge, 1999), 143.

118 Ibid., 144–45.

119 Turner, *Ritual Process*, 96, 129. This talk of structure reveals a particularly 1960s view of the social world, but it's very similar to Solnit's sentiments: "To make fellowship, joy and freedom work for a day or a week is far more doable than the permanent transformation of society, and it can inspire people to return to that society in its everyday incarnation with renewed powers and ties" (Solnit, *Paradise*, 169).

120 Anthony Oliver-Smith, "The Brotherhood of Pain: Theoretical and Applied Perspectives on Post-Disaster Solidarity," in *The Angry Earth*, 168.

121 In a conversation sixteen years later, Oliver-Smith said that the now-dominant resilience doctrine too easily descends into "victim blaming," but that he still saw a need to look beyond the vulnerability framing and acknowledge the forms of local agency that it leaves out. Anthony Oliver-Smith, interview with Stan Cox, Copenhagen, Denmark, December 10, 2015.

122 David Graeber, *Debt: The First 5,000 Years* (Brooklyn: Melville House, 2011), 95–96.

Chapter 5: Gray Goo

1 "Lapindo Innuendo," *Jakarta Post*, April 2, 2014.

2 Nur Hadi, "Korban Lumpur Berterima Kasih Kepada Jokowi, Bukan Ke Lapindo," *Tempo Nasional*, October 11, 2015 (original statement in Bahasa).

3 Deborah Stone, "Causal Stories and the Formation of Policy Agendas," *Political Science Quarterly* 104 (1989): 281–300.

4 Nancy Krieger, "Epidemiology and the Web of Causation: Has Anyone Seen the Spider?," *Social Science and Medicine* 39 (1994): 887–903.

5 Richard Stuart Olson, "Toward a Politics of Disaster: Losses, Values, Agendas, and Blame," *International Journal of Mass Emergencies and Disasters* 18 (2000): 265–87.

6 Even the question of what to call the mud volcano is highly contentious. When in Sidoarjo, the most effective way to ask directions is to ask for "Lumpur Lapindo," that is, "Lapindo's mud." Local maps show a blue-gray area that is so labeled. But of course Lapindo and the Bakrie Group don't like that kind of terminology, so they have urged the use of the term "Lusi." (To our knowledge, Lapindo's critics haven't yet suggested the parallel nickname "Lula," for "Lumpur Lapindo.") Those using the name "Lusi" are sometimes viewed as taking sides with the company, but in using the name in this book, we are not implying any kind of advocacy. We use "Lusi" in this book because most experts we interviewed on both sides of the trigger issue use it, and we are citing and quoting them extensively.

7 J.R. Richards, "March 2011 Report into the Past, Present and Future Social Impacts of Lumpur Sidoarjo," Humanitus Sidoarjo Fund, March 2011.

8 *Historical Tragedy Lapindo Hot Mud*, video compact disc produced by Lusi victims, purchased by authors on west levee of Lusi, January 18, 2014. No producer, director, videographer, or any other credits are provided in the video or on the case.

9 Hardi Prasetyo, interview with the authors, Porong district, East Java, Indonesia, January 16, 2014.

10 More famously, "gray goo" refers to a doomsday scenario named by the nanotechnologist Eric Drexler in 1986. In this sci-fi thought experiment, self-replicating

nanomachines convert surrounding matter into microscopic copies of themselves, turning everything into a uniform mass of goo in an uncontrollable process of "ecophagy." If we take ecophagy as a metaphorical description of the global economy, as others have, then standing on the edge of Lusi gives that metaphor a particular viscosity. K. Eric Drexler, *Engines of Creation: The Coming Era of Nanotechnology* (New York: Doubleday, 1986); Robert A. Freitas Jr., "Some Limits to Global Ecophagy by Biovorous Nanoreplicators, with Public Policy Recommendations," 2000, rfreitas.com/Nano/Ecophagy.htm; Robin Stoate, "Gray Goo and You: The Ecophagy of Global Capital," in *Criticism, Crisis, and Contemporary Narrative: Textual Horizons in an Age of Global Risk*, ed. Paul Crosthwaite (New York: Routledge, 2011), 110–26.

11 Rachel Nuwer, "9 Years of Muck, Mud and Debate in Java," *New York Times*, September 21, 2015.

12 Hardi Prasetyo, interview with the authors, Porong subdistrict, East Java, Indonesia, January 17, 2014.

13 Lynette McDonald and Wina Widaningrum, "Muddied Waters: Lapindo Brantas' Response to the Indonesian Mudflow Crisis," in *Communication, Creativity and Global Citizenship*, ed. Terry Flew (Brisbane: Australian and New Zealand Communication Association, 2009), 1101–21.

14 Mark Tingay, e-mail interview with Stan Cox, December 9, 2013.

15 Ibid.

16 Mark Tingay, "Initial Pore Pressures Under the Lusi Mud Volcano, Indonesia," *Interpretation* 3 (2015): SE33–49; Richard Davies, Maria Brumm, Michael Manga, Rudi Rubiandini, Richard Swarbrick, and Mark Tingay, "The East Java Mud Volcano (2006 to Present): An Earthquake or Drilling Trigger?," *Earth and Planetary Science Letters* 272 (2008): 627–38. Richard Davies, a petroleum geologist now serving as a pro-vice-chancellor at the University of Newcastle, says that just before and during the initial mud eruption there were six coincidences, the simultaneous occurrence of which is very hard for those who support the earthquake trigger to explain: "The first coincidence is spatial: the well and the mud volcano occurred in virtually the same place. Then there's the temporal coincidence: they happened at the same time, and these first two coincidences were noted immediately by the villagers themselves. Third, the well operators had failed to provide sufficient casing for the drill, making it vulnerable to the very kind of accident that happened. Fourth, we know the 'kick' from the well accident was large enough to exceed the tolerance of the rock. Fifth, pressure measurements showed pressure dropping within the well after the 'kick,' implying that fluid was escaping from the well—it had to have gone somewhere. As a final coincidence, the records show that when the well operators attempted to plug the well by pumping mud into it, they saw an increase in the quantity of mud coming out of the nearby volcano; the two were clearly connected. It's only if you ignore the convergence of all of those coincidences that you can construct an alternative explanation." Richard Davies, telephone interview with Stan Cox, December 11, 2013.

17 Adriano Mazzini, Giuseppe Etiope, and Henrik Svensen, "A New Hydrothermal Scenario for the 2006 Lusi Eruption, Indonesia. Insights from Gas Geochemistry," *Earth and Planetary Science Letters* 317–318 (2012): 305–18; Nurrochmat Sawolo,

Edi Sutriono, Bambang P. Istadi, and Agung B. Darmoyo, "The LUSI Mud Volcano Triggering Controversy: Was It Caused by Drilling?," *Marine and Petroleum Geology* 26 (2009): 1766–84.

18 Mark Tingay, e-mail interview with Stan Cox, December 9, 2013; M. Manga, "Did an Earthquake Trigger the May 2006 Eruption of the Lusi Mud Volcano?," *Eos Transactions*, American Geophysical Union, 2007. In the summer of 2013, a new analysis (M. Lupi, E.H. Saenger, F. Fuchs, and S.A. Miller, "Lusi Mud Eruption Triggered by Geometric Focusing of Seismic Waves," *Nature Geoscience* 6 [2013]: 642–46) helped revive the earthquake trigger hypothesis. In it, the authors described an extremely dense rock layer in the earth below the drilling site that could have concentrated and focused the seismic waves from the quake. This was viewed by many as the most convincing argument yet that an earthquake could have set off the eruption. However, according to Richard Davies, there was a serious problem with this scenario: "The rock layer doesn't exist." It was the upper part of the drill with its steel casing, not a hypothetical, super-dense rock layer, that was responsible for the wave velocity data, according to Davies and his colleagues. Lupi and colleagues acknowledged the error in a subsequent publication (M. Lupi, E.H. Saenger, F. Fuchs, and S.A. Miller, "Corrigendum: 'Lusi Mud Eruption Triggered by Geometric Focusing of Seismic Waves,'" *Nature Geoscience* 7 [2014]: 687–88) and proposed yet another mechanism by which the earthquake could have triggered the eruption. This second explanation has been refuted in Tingay, "Initial Pore."

19 Davies et al., "East Java." Davies and colleagues' argument for a drilling trigger was followed by a reply: Sawolo et al., "LUSI Mud Volcano," which prompted its own response: Richard Davies, Michael Manga, Mark Tingay, Susila Luciagna, and Richard Swarbrick, "Discussion: Sawolo et al. (2009) the Lusi Mud Volcano Controversy: Was It Caused by Drilling?," *Marine and Petroleum Geology* 27 (2010): 1651–57. In the same issue of the journal, Sawolo and colleagues then published a detailed, highly technical response to Davies and colleagues' response to their response, still concluding that the eruption was not caused by the blowout: Nurrochmat Sawolo, Edi Sutriono, Bambang P. Istadi, and Agung B. Darmoyo, "Was LUSI Caused by Drilling?—Authors Reply to Discussion," *Marine and Petroleum Geology* 27 (2010): 1658–75. Mark Tingay says this final Sawolo response is useful in that it provides the scientific community more basic data with which to analyze the drilling incident. But the authors, he says, could reach the conclusion they did only because, out of the broad range of data available, "they took only the 'best case' values at the edge of data uncertainty." Davies agrees: "The authors of that paper managed to find the *one and only* method that gives a pressure low enough not to crack the rock." When asked why, Davies paused and then said, "What's the best way to say this? There was a conflict of interest. I should leave it at that." All of the Sawolo papers' authors were employees of the parent company of Lapindo.

20 Adriano Mazzini, e-mail interview with Stan Cox, April 15, 2014. Then why, given Mazzini's evidence, is there such a widely held view that the drilling accident was the trigger? He explains it this way: "This is human I think. We all need instinctively to give some sort of closure to things. It is reassuring (in addition to the economic compensations) to blame something or someone that we can define, like, for example, a company or a driller chief. Something or someone tangible. On the other hand,

blaming nature or a combination of domino-like events is certainly not rewarding or reassuring for people."

21 Richard Davies, e-mail interview with Stan Cox, November 24, 2014; Mark Tingay, e-mail interview with Stan Cox, November 24, 2014. Tingay added, "To so simply dismiss this data really does indicate that [Mazzini] does not understand it. . . . I would point out that my offer to go through the drilling data with him, and explain drilling practices, was met with polite refusal." Davies believes the persistence of the earthquake trigger idea can be attributed to the scientific tendency to look to one's own specialty for answers: "When you have geologists studying a problem, they will come up with geological explanations. They can create beautiful ideas about how an earthquake could have caused a disaster. Meanwhile, engineers will tend to come up with engineering explanations." Because the research groups he has worked with on the Lusi problem have included both geologists and drilling experts, he says, they have been forced to weigh the two explanations against one another, and in the end the group came to full consensus that the drilling accident solution had far more weight. Richard Davies, telephone interview with Stan Cox, December 11, 2013.

22 M.R.P. Tingay, Maxwell Rudolph, M. Manga, R.J. Davies, and Chi-Yuen Wang, "Initiation of the Lusi Mudflow Disaster," *Nature Geoscience* 8 (2015): 493–94. Upon the paper's publication, Tingay wrote to us, "Liquefaction of clays is ALWAYS (and I mean that in a scientific way—100% of the time) associated with gas release. Furthermore, liquefaction is caused by the actual physical shaking from the earthquake waves—hence it will start when those waves actually pass through the rock." Mark Tingay, e-mail interview with Stan Cox, July 18, 2015.

23 Mark Tingay, e-mail interview with Stan Cox, July 18, 2015.

24 Nuwer, "9 Years."

25 Ibid.

26 Denise Leith, *The Politics of Power: Freeport in Suharto's Indonesia* (Honolulu: University of Hawaii Press, 2003), 70.

27 Bakrie Metals Industries, "Our Major Clients," 2014, bakrie-metal.com/statis-15 -klienbesarkita.html.

28 B. Lynn Pascoe, "East Java: Increasing Mudflow Threatens Infrastructure and More Homes," U.S. embassy cable, September 7, 2006. https://wikileaks.org/plusd/cables /06JAKARTA11110_a.html.

29 Bagus Saragih and Imanuddin Razak, "People's Longing for New Order a Benefit to Golkar: Aburizal," *Jakarta Post*, February 20, 2014.

30 Richards, "March 2011 Report"; McDonald and Widaningrum, "Muddied Waters."

31 McDonald and Widaningrum, "Muddied Waters."

32 Gunardi, interview with the authors, Renojoyo, Porong subdistrict, January 17, 2014. Translated by Dwinanto Prasetyo.

33 Unidentified resident of Renojoyo, interview with the authors, Renojoyo, Porong district, January 17, 2014. Translated by Dwinanto Prasetyo.

34 Dwinanto Prasetyo, interview with the authors, Porong district, January 17, 2014.

35 Agence France-Presse, "Crisis Sets Back Payouts for Indonesian Mud Volcano Victims," November 27, 2008.

36 Bosman Batubara, interview with the authors, Yogyakarta, Indonesia, January 20, 2014.

37 Mujtaba Hamdi, Wardah Hafidz, and Gabriela Sauter, "Uplink Porong: Supporting Community-Driven Responses to the Mud Volcano Disaster in Sidoarjo, Indonesia," International Institute for Environment and Development, August 2009, pubs.iied .org/14581IIED.html?a=N.

38 Batubara has helped organize efforts to rebuild livelihoods in the area, with some success. He told us, "In 2009 we started with thirty-two women in three villages, including Renokenongo. Our idea was that the men were absorbed with the compensation issue, but the women needed to do something. They were used to labor jobs with fixed hours and regular paychecks. They tried to apply for existing jobs outside the area, but those companies wanted only new laborers who would work for less than experienced people. It's hard to switch to being an entrepreneur, so we had to convince them to believe in themselves. The women divided into six groups. They had meetings that rotated among their houses. There was an attempt at a microfinance project; then they switched to a cooperative arrangement." The project was still in place, but "I am not satisfied with the growth—it is not becoming a mass movement." Bosman Batubara, interview with the authors, Yogyakarta, Indonesia, January 20, 2014.

39 Hasyim Widhiarto, "Aburizal Could Be Forced to Settle Lapindo Mudflow," *Jakarta Post*, September 30, 2014; Hasyim Widhiarto, "Govt to Take Over Lapindo's Liabilities," *Jakarta Post*, December 9, 2014.

40 Felix Utama Kosasih, "Sidoarjo Mud Settlement After Nine Long Years," *Global Indonesian Voices*, July 16, 2015.

41 M.L. Rudolph, M. Shirzaei, M. Manga, and Y. Fukushima, "Evolution and Future of the Lusi Mud Eruption Inferred from Ground Deformation," *Geophysical Research Letters* 40 (2013): 1089–92.

42 On the projections about future mudflow, see Tingay, "Initial."

43 Nuwer, "9 Years."

44 Berry wrote words to this effect in a typewritten manuscript that was never published. It was read by Stan Cox but has since apparently been lost. Asked by Cox on September 28, 2012, in Salina, Kansas, if he recalled saying or writing these words, Berry responded, "I sure could have written that."

Chapter 6: How to Booby-Trap a Planet

1 Max Horkheimer and Theodor W. Adorno, *Dialectic of Enlightenment* (New York: Herder and Herder, 1972), 1.

2 Cézanne's line "Nous sommes un chaos irisé" more directly translates as "We are an iridescent chaos," but the quote has entered English-language circulation as "We live in a rainbow of chaos." This is not groundless, as both *irisé* and *iridescent* come from the Latin *iris*, meaning rainbow. Other translations in this passage are those of Julie Lawrence Cochran, in *Conversations with Cézanne*, ed. Michael Doran (Berkeley: University of California Press, 2001), 114–15, originally published in Joachim Gasquet, *Cézanne* (Paris: Bernheim-Jeune, 1921). In any event, Gasquet was known for his poetic license, and some or all of the conversation may have been his invention anyway.

3 Doran, *Conversations with Cézanne*, 113.

4 "La Chapelle Saint-Ser," Syndicat d'Initiative de Puyloubier, www.si-puyloubier.13.fr /Docs/r-stser1.pdf.

5 "La Montagne Sainte-Victoire Ravagée par un Incendie," *France 3*, August 28, 1989. This TV broadcast is archived on Repères Méditerranéens, http://fresques.ina.fr /reperes-mediterraneens.

6 In the 2014 consensus of the Intergovernmental Panel on Climate Change, "Changes in many extreme weather and climate events have been observed since about 1950. Some of these changes have been linked to human influences, including a decrease in cold temperature extremes, an increase in warm temperature extremes, an increase in extreme high sea levels and an increase in the number of heavy precipitation events in a number of regions. . . . It is *very likely* that human influence has contributed to the observed global scale changes in the frequency and intensity of daily temperature extremes since the mid-20th century. It is *likely* that human influence has more than doubled the probability of occurrence of heat waves in some locations. . . . There are *likely* more land regions where the number of heavy precipitation events has increased than where it has decreased. Recent detection of increasing trends in extreme precipitation and discharge in some catchments imply greater risks of flooding at regional scale (*medium confidence*). It is *likely* that extreme sea levels (for example, as experienced in storm surges) have increased since 1970, being mainly a result of rising mean sea level." IPCC, *Climate Change 2014: Synthesis Report* (Geneva: Intergovernmental Panel on Climate Change, 2014), 7–8, ipcc.ch/report/ar5/syr; emphasis in original.

7 Amato T. Evan, James P. Kossin, Chul Chung, and V. Ramanathan, "Arabian Sea Tropical Cyclones Intensified by Emissions of Black Carbon and Other Aerosols," *Nature* 479 (2011): 94–97; Ben B.B. Booth, Nick J. Dunstone, Paul R. Halloran, Timothy Andrews, and Nicolas Bellouin, "Aerosols Implicated as a Prime Driver of Twentieth-Century North Atlantic Climate Variability," *Nature* 484 (2012): 228–33; Daniel Rosenfeld, Ulrike Lohmann, Graciela Raga, Colin O'Dowd, Markku Kulmala, Sandro Fuzzi, Anni Reissell, and Meinrat Andreae, "Flood or Drought: How Do Aerosols Affect Precipitation?," *Science* 321 (2008): 1309–13.

8 Ted Steinberg, *Acts of God: The Unnatural History of Natural Disaster in America* (New York: Oxford University Press, 2000), xxiii.

9 Mark Fischetti, "Was Typhoon Haiyan a Record Storm?," *Scientific American*, November 12, 2013; Peter Hamman, "Typhoon Haiyan Influenced by Climate Change, Scientists Say," *Sydney Morning Herald*, November 11, 2013.

10 The IPCC's Fifth Assessment Report concludes, "Based on process understanding and agreement in 21st century projections, it is *likely* that the global frequency of occurrence of tropical cyclones will either decrease or remain essentially unchanged, concurrent with a *likely* increase in both global mean tropical cyclone maximum wind speed and precipitation rates. . . . Improvements in model resolution and downscaling techniques increase confidence in projections of intense storms, and the frequency of the most intense storms will *more likely than not* increase substantially in some basins." Jens Hesselbjerg Christensen, Krishna Kumar Kanikicharla, Gareth Marshall, and John Turner, "Climate Phenomena and Their Relevance for Future Regional Climate Change," in *Climate Change 2013: The Physical Science Basis. Contribution of Working Group I to the Fifth Assessment Report of the Intergovernmental*

Panel on Climate Change (Cambridge: Cambridge University Press, 2013); emphasis in original.

11 Aslak Grinsted, John Moore, and Svetlana Jevrejeva, "Homogeneous Record of Atlantic Hurricane Surge Threat Since 1923," *Proceedings of the National Academy of Sciences* 109 (2012): 19601–5.

12 Kevin E. Trenberth, John T. Fasullo, and Theodore G. Shepherd, "Attribution of Climate Extreme Events," *Nature Climate Change* 5, no. 8 (2015): 725–30.

13 Chris Mooney, "Why Record-Breaking Hurricanes like Patricia Are Expected on a Warmer Planet," *Washington Post*, October 23, 2015.

14 Noah Diffenbaugh, Martin Scherer, and Robert Trappvol, "Robust Increases in Severe Thunderstorm Environments in Response to Greenhouse Forcing," *Proceedings of the National Academy of Sciences* 110 (2013): 16361–66.

15 Julia Sander, Jan Eichner, Eberhard Faust, and Markus Steuer, "Climate-Driven Increase in the Variability and Multi-Year Mean Level of Severe Thunderstorm-Related Losses and Thunderstorm Forcing Environments in the U.S. Since 1970," *Geophysical Research Abstracts* 15 (2013): EGU2013-5155-1,

16 Eric Holthaus, "The Scariest Part of This Season's Weird Weather Is Coming Soon," *Slate*, December 29, 2015; Bob Henson and Jeff Masters, "U.S. Reeling from Violent Tornadoes, Epic Flooding, Winter Weather, and Weird Heat," Weather Underground, December 28, 2015; Bill Hanna, "North Texas Tornadoes Were Deadliest in Decades," *Fort Worth Star-Telegram*, December 27, 2015; Shannon Jones, "Winter Storms Kill Dozens Across US South and Midwest," *World Socialist Website*, December 29, 2015; Jonah Bromwich, "A Fitting End for the Hottest Year on Record," *New York Times*, December 23, 2015.

17 Chris Mooney, "As California Fires Rage, the Forest Service Sounds the Alarm About Sharply Rising Wildfire Costs," *Washington Post*, August 4, 2015.

18 A. Park Williams, Craig Allen, Alison Macalady, Daniel Griffin, Connie Woodhouse, David Meko, Thomas Swetnam, et al., "Temperature as a Potent Driver of Regional Forest Drought Stress and Tree Mortality," *Nature Climate Change* 3 (2013): 292297.

19 Liu Yongqiang, Scott Goodrick, and John Stanturf, "Future U.S. Wildfire Potential Trends Projected Using a Dynamically Downscaled Climate Change Scenario," *Forest Ecology and Management* 294 (2013): 120–35.

20 Joe Romm, "Historic Blizzard Poised to Strike New England: What Role Is Climate Change Playing?," *ThinkProgress*, February 8, 2015; Jianping Li and Zhiwei Wu, "Importance of Autumn Arctic Sea Ice to Northern Winter Snowfall," *Proceedings of the National Academy of Sciences* 109 (2012): E1898; Zhiwei Wu, Jianping Li, Zhihong Jiang, and Jinhai He, "Predictable Climate Dynamics of Abnormal East Asian Winter Monsoon: Once-in-a-Century Snowstorms in 2007/2008 Winter," *Climate Dynamics* 37 (2010): 1661–69; Jiping Liu, Judith Curry, Huijun Wang, Mirong Song, and Radley Horton, "Impact of Declining Arctic Sea Ice on Winter Snowfall," *Proceedings of the National Academy of Sciences* 109 (2012): 4074–79.

21 U.K. Meteorological Office, "A Global Perspective on the Recent Storms and Floods in the UK," February 9, 2014, metoffice.gov.uk/research/news/2014/uk-storms-and-floods.

22 William K.M. Lau and Kyu-Myong Kim, "The 2010 Pakistan Flood and Russian Heat Wave: Teleconnection of Hydrometeorological Extremes," *Journal of Hydrometeorology* 13 (2012): 392–403.

23 M. Monirul Qader Mirza, "Climate Change, Flooding in South Asia and Implications," *Regional Environmental Change* 11 (2011): S95–S107.

24 P.C.D. Milly, R.T. Wetherald, K.A. Dunne, and T.L. Delworth, "Increasing Risk of Great Floods in a Changing Climate," *Nature* 415 (2002): 514–17.

25 Hirabayashi Yukiko, Roobavannan Mahendran, Sujan Koirala, Lisako Konoshima, Dai Yamazaki, Satoshi Watanabe, Hyungjun Kim, and Shinjiro Kanae, "Global Flood Risk Under Climate Change," *Nature Climate Change* 3 (2013): 816–21.

26 Wolfgang Kron, "Coasts: The High-Risk Areas of the World," *Natural Hazards* 66 (2013): 1363–82.

27 Dushmanta Dutta, "An Integrated Tool for Assessment of Flood Vulnerability of Coastal Cities to Sea-Level Rise and Potential Socio-Economic Impacts: A Case Study in Bangkok, Thailand," *Hydrological Sciences Journal* 56 (2011): 805–23.

28 Laura Carbognin, Pietro Teatini, Alberto Tomasin, and Luigi Tosi, "Global Change and Relative Sea Level Rise at Venice: What Impact in Terms of Flooding," *Climate Dynamics* 35 (2010): 1039–47.

29 Robert Nicholls, Natasha Marinova, Jason Lowe, Sally Brown, Pier Vellinga, Diogo de Gusmão, Jochen Hinkel, and Richard Tol, "Sea-Level Rise and Its Possible Impacts Given a 'Beyond 4°C World' in the Twenty-First Century," *Philosophical Transactions of the Royal Society of London A* 369 (2010): 161–81.

30 Jason Ericson, Charles Vörösmarty, S. Lawrence Dingman, Larry Ward, and Michel Meybeck, "Effective Sea-Level Rise and Deltas: Causes of Change and Human Dimension Implications," *Global and Planetary Change* 50 (2006): 63–82; Robert Nicholls, "Planning for the Impacts of Sea Level Rise," *Oceanography* 24 (2011): 144–57.

31 S. Hanson, R. Nicholls, N. Ranger, S. Hallegatte, J. Corfee-Morlot, C. Herweijer, and J. Chateau, "A Global Ranking of Port Cities with High Exposure to Climate Extremes," *Climatic Change* (2011): 104, 89–111.

32 Congxian Li, Daidu Fan, Bing Deng, and Vlasdilav Korotaev, "The Coasts of China and Issues of Sea Level Rise," *Journal of Coastal Research* 43 (2004): 36–49.

33 David Petley, "Global Losses from Landslides Associated with Dams and Reservoirs," *Italian Journal of Engineering Geology and Environment—Book Series* 6 (2013): 63-72; David Petley, "Global Patterns of Loss of Life from Landslides," *Geology* 40 (2012): 927–30.

34 Dalia Kirschbaum, Robert Adler, David Adler, Christa Peters-Lidard, and George Huffman, "Global Distribution of Extreme Precipitation and High-Impact Landslides in 2010 Relative to Previous Years," *Journal of Hydrometeorology* 13 (2012): 1536–51.

35 Marten Geertsema, John Clague, James Schwab, and Stephen Evans, "An Overview of Recent Large Catastrophic Landslides in Northern British Columbia, Canada," *Engineering Geology* 83 (2006): 120–43. Also see Christian Huggel, John Clague, and Oliver Korup, "Is Climate Change Responsible for Changing Landslide Activity in High Mountains?," *Earth Surface Processes and Landforms* 37 (2012): 77–91.

36 Vladimir Kotlyakov, O. Rototaeva, and G. Nosenko, "The September 2002 Kolka Glacier Catastrophe in North Ossetia, Russian Federation: Evidence and Analysis," *Mountain Research and Development* 24 (2004): 78–83.

37 S.S. Chernomorets, O.V. Tutubalina, I.B. Seinova, D.A. Petrakov, K.N. Nosov, and E.V. Zaporozhchenko, "Glacier and Debris Flow Disasters Around Mt. Kazbek, Russia/

Georgia," in *Debris-Flow Hazards Mitigation: Mechanics, Prediction, and Assessment,* ed. Cheng-Lung Chen (Rotterdam: Millpress, 2007), 691–702.

38 Yueping Yin, "Recent Catastrophic Landslides and Mitigation in China," *Journal of Rock Mechanics and Geotechnical Engineering* 3 (2011): 10–18.

39 Christian Huggel, "Recent Extreme Slope Failures in Glacial Environments: Effects of Thermal Perturbation," *Quaternary Science Reviews* 28 (2009): 1119–30.

40 Rinaldo Genevois and Monica Ghirotti, "The 1963 Vaiont Landslide," *Giornale di Geologia Applicata* 1 (2005): 41–52; E. Semenza and M. Ghirotti, "History of the 1963 Vaiont Slide: The Importance of Geological Factors," *Bulletin of Engineering Geology and Environment* (2000): 59, 87–97.

41 Genevois and Ghirotti, "1963 Vaiont"; E.L. Quarantelli, "The Vaiont Dam Overflow: A Case Study of Extra-Community Responses in Massive Disasters," *Disasters* 3 (1979): 199–212.

42 Genevois and Ghirotti, "1963 Vaiont."

43 Andrew Revkin, "One-Two Punch of Earthquakes and Landslides Exposes Hydropower Vulnerability in Nepal," *New York Times,* May 25, 2015.

44 Kathleen Nicoll, "Geomorphic and Hazard Vulnerability Assessment of Recent Residential Developments on Landslide-Prone Terrain: The Case of the Traverse Mountains, Utah, USA," *Journal of Geographical and Regional Planning* 3 (2010): 126–41.

45 J.J. Bommer, M.B. Benito, M. Ciudad-Real, A. Lemoine, M.A. López-Menjívar, R. Madariaga, J. Mankelow, et al., "The El Salvador Earthquakes of January and February 2001: Context, Characteristics and Implications for Seismic Risk," *Soil Dynamics and Earthquake Engineering* 22 (2002): 389–418.

46 Mark Berman, "After a Record Number of Earthquakes in Oklahoma, Concerns That a Damaging Quake Is Coming," *Washington Post,* May 6, 2014.

47 Dana Branham, "Oklahoma Reports Surge in Earthquakes During 2015," *Oklahoma Daily,* December 14, 2015; Darla Cameron and Dan Keating, "Maps: Oklahoma's Earthquake Problem Is Getting Worse," *Washington Post,* February 3, 2015; Mike Soraghan, "Shaken More than 560 Times, Okla. Is Top State for Quakes in 2014," *EnergyWire,* January 5, 2015; Corey Jones, "Oklahoma on Record Earthquake Pace, Up 90 Percent from Over a Year Ago," *Tulsa World,* June 22, 2015.

48 Ziva Branstetter, "Quake Debate: Scientists Warn of Potential for 'Large Earthquake' as Injection Well Discussion Continues," *Tulsa World,* January 5, 2015.

49 Steve Olafson, "Five Homes Damaged in Record Earthquake in Oklahoma," *Reuters,* November 6, 2011.

50 Branstetter, "Quake Debate."

51 Katie Keranen, Heather Savage, Geoffrey Abers, and Elizabeth Cochran, "Potentially Induced Earthquakes in Oklahoma, USA: Links Between Wastewater Injection and the 2011 Mw 5.7 Earthquake Sequence," *Geology* 41 (2013): 699–702.

52 F. Rall Walsh and Mark Zoback, "Oklahoma's Recent Earthquakes and Saltwater Disposal," *Science Advances* 1 (2015): e1500195.

53 M. Weingarten, S. Ge, J.W. Godt, B.A. Bekins, and J.L. Rubinstein, "High-Rate Injection Is Associated with the Increase in U.S. Mid-Continent Seismicity," *Science* 348 (2015): 1336–40.

54 K.M. Keranen, M. Weingarten, G.A. Abers, B.A. Bekins, and S. Ge, "Sharp Increase in

Central Oklahoma Seismicity Since 2008 Induced by Massive Wastewater Injection," *Science* 345 (2014): 448–51.

55 Eric Hand, "Injection Wells Blamed in Oklahoma Earthquakes," *Science* 345 (2014): 13–14; Keranen et al., "Sharp Increase." The orientation of faults in the vicinity of the Nemaha is said to make such a large quake unlikely, but the precise probability is anyone's guess. For one thing, there could be an "unmapped offshoot" of the Nemaha that could make just the right, or rather wrong, connection. The Prague quake itself was totally unexpected because it occurred in the Wilzetta Fault, which had previously been considered completely dormant and incapable of producing a moderate-sized earthquake.

56 Oklahoma City spokesperson Kristy Yager told *National Geographic* that if wastewater injection is found to be responsible for the quake swarms, it will mean big trouble for the local economy. She said, "We hope it's not a byproduct of oil and gas because so many of our jobs depend on it. It's a difficult position to be in." Joe Eaton, "Oklahoma Grapples with Earthquake Spike—and Evidence of Industry's Role," *National Geographic*, July 31, 2014.

57 Benjamin Elgin, "Oil CEO Wanted University Quake Scientists Dismissed: Dean's E-Mail," *Bloomberg*, May 15, 2015.

58 Adam Wilmoth, "Oklahoma Commissioner Warns Insurers on Denial of Earthquake Claims," *The Oklahoman*, March 4, 2015.

59 Yeganeh Torbati, "Oklahoma Drilling Regulator Calls Spike in Quakes a 'Game Changer,'" Reuters, June 24, 2015.

60 D.E. McNamara, J.L. Rubinstein, E. Myers, G. Smoczyk, H.M. Benz, R.A. Williams, G. Hayes, D. Wilson, R. Herrmann, N.D. McMahon, R.C. Aster, E. Bergman, A. Holland, and P. Earle, "Efforts to Monitor and Characterize the Recent Increasing Seismicity in Central Oklahoma," *Leading Edge* 34 (2015): 628–39.

61 Joe Wertz, "Regulation Accelerates as Officials Move from Hesitation to 'Direct Correlation' on Oil-Linked Earthquakes," *StateImpact*, August 13, 2015.

62 Joe Wertz, "Regulators Issue Tougher Disposal Well Directives as Oklahoma's Quake Risk Rises," *StateImpact*, March 25, 2015; Joe Wertz, "After Spate of Earthquakes, Oklahoma Oil Regulator Slashes Disposal Well Activity in Shaky Region," *StateImpact*, August 3, 2015; Wertz, "Regulation Accelerates"; Katie Valentine, "What's Behind the Spike In Earthquake Activity Oklahoma Has Seen This Year?," *ThinkProgress*, August 20, 2015; "Oklahoma Agency Reveals New Well Plans in the Cushing Area," Associated Press, September 18, 2015.

63 The two January earthquakes were, respectively, tied for fourth and tied for seventh in magnitude. Corey Jones and Paighten Harkins, "20 Quakes—One Tied for Fourth Largest in Oklahoma History—Rattle State Since Wednesday Night," *Tulsa World*, January 7, 2016: "5.1 Magnitude Earthquake Among Several to Shake Oklahoma," Associated Press, February 14, 2016.

64 William Ellsworth, "Injection-Induced Earthquakes," *Science* 341 (2013): 142.

65 Daniel Trugman, Adrian Borsa, and David Sandwell, "Did Stresses from the Cerro Prieto Geothermal Field Influence the El Mayor–Cucapah Rupture Sequence?," *Geophysical Research Letters* 41 (2014): 8767–774.

66 Armand Vervaeck and James Daniell, "Strong Earthquake Near Lorca and Murcia,

Spain," *Earthquake Report*, July 1, 2011; "Damage from Lorca Earthquake Put at €70 million," *New Europe*, May 22, 2011.

67 ABB Communications, "Growing More Crops with Less Water," press release, n.d., abb.us/cawp/seitp202/22ea6ffcc5ac1f79c12576330040c62e.aspx.

68 Pablo González, Kristy Tiampo, Mimmo Palano, Flavio Cannavó, and José Fernández, "The 2011 Lorca Earthquake Slip Distribution Controlled by Groundwater Crustal Unloading," *Nature Geoscience* 5 (2012): 821–25.

69 Jean-Philippe Avouac, "Earthquakes: Human-Induced Shaking," *Nature Geoscience* 5 (2012): 763–64.

70 Harsh Gupta, "A Review of Recent Studies of Triggered Earthquakes by Artificial Water Reservoirs with Special Emphasis on Earthquakes in Koyna, India," *Earth-Science Reviews* 58 (2002): 279–310.

71 Christian Klose, "Evidence for Anthropogenic Surface Loading as Trigger Mechanism of the 2008 Wenchuan Earthquake," *Environmental Earth Sciences* 66 (2012): 1439–47.

72 Fan Xiao, "Did the Zipingpu Dam Trigger China's 2008 Earthquake? The Scientific Case," *Probe International*, December 2012. Fan's report, which has the relentless rhythm of a prosecutor's closing statement, methodically indicts Zipingpu for creating insupportable stresses through its sheer mass while lubricating the surrounding fissures through water seepage. That combination of factors, Fan wrote, set off a seismic chain reaction: a ninety-second sequence of ruptures that started below the reservoir and ended miles away in the north section of a highly stressed fault zone that was "close to the critical state of rupture." In retrospect, he argued, "the rapid rate and dramatic scale of filling and drawdown of Zipingpu's reservoir was unprecedented and dangerous."

73 Fan Xiao, "Chinese Geologist Says Zipingpu Dam Reservoir May Have Triggered China's Deadly Quake, Calls for Investigation," *Probe International*, January 26, 2009.

74 Sharon Lafraniere, "Possible Link Between Dam and China Quake," *New York Times*, February 6, 2009.

Chapter 7: Foreshock, Shock, Aftershock

1 Nicola Nosengo, "Scientists Face Trial over Earthquake Deaths," *Nature News*, May 26, 2011.

2 Thomas Jordan, Y.-T. Chen, Paolo Gasparini, Raul Madariaga, Ian Main, Warner Marzocchi, Gerassimos Papadopoulos, K. Yamaoka, and J. Zschau, "Operational Earthquake Forecasting: State of Knowledge and Guidelines for Utilization," *Annals of Geophysics* 54, no. 4 (2011): 316–91.

3 Kazuhiko Kawashima, Ömer Aydan, Takayoshi Aoki, Ichizo Kishimoto, Kazuo Konagai, Tomoya Matsui, Joji Sakuta, Noriyuki Takahashi, Sven-Peter Teodori, and Atsushi Yashima, "Reconnaissance Investigation on the Damage of the 2009 L'Aquila, Central Italy Earthquake," *Journal of Earthquake Engineering* 14 (2010): 817–41.

4 Ibid.; C.A. Chiarabba et al., "The 2009 L'Aquila (Central Italy) MW6.3 Earthquake: Main Shock and Aftershocks," *Geophysical Research Letters* 36 (2009): L18308.

5 Daryoush Shoyajee, interview with Stan Cox, L'Aquila, September 1, 2013.

6 Mauro Dolce, e-mail interview with Stan Cox, September 21, 2013.

7   In addition to Dolce, the group included Enzo Boschi of the National Institute of Geo-
     physics and Volcanology (INGV), Claudio Eva of the University of Genova, Giulio
     Selvaggi of INGV's National Earthquake Centre, Franco Barberi of the University
     of Rome III, Gian Michele Calvi of the European Centre for Training and Research
     in Earthquake Engineering, and Bernardo De Bernardinis, then vice director of the
     Department of Civil Protection. All except De Bernardinis were earth scientists.

8   Daryoush Shoyajee, interview with Stan Cox, L'Aquila, September 1, 2013.

9   Warner Marzocchi, interview with Stan Cox, Rome, August 26, 2013; David Alex-
     ander, "Communicating Earthquake Risk to the Public: The Trial of the 'L'Aquila
     Seven,'" *Natural Hazards* 72 (2014): 1159–73.

10 Stephen Hall, "Scientists on Trial: At Fault?," *Nature News* 477 (2011): 264–69;
     V. Koschatsky, K. Haynes, P. Somerville, J. McAneney, and D. McAneney, "Guilty?,"
     Risk Frontiers, November 2012; Warner Marzocchi and Massimo Cocco, "Guilty? Or
     Not?," *Risk Frontiers*, June 2013; Warner Marzocchi, "Putting Science on Trial," *Phys-
     ics World*, December 2012; Alexander, "Communicating"; Warner Marzocchi, e-mail
     interview with Stan Cox, April 7, 2014.

11 Ibid.

12 Hall, "Scientists."

13 Koschatsky et al., "Guilty?"

14 Marzocchi and Cocco, "Guilty? Or Not?"; Marzocchi, "Putting."

15 Jordan et al., "Operational Earthquake Forecasting."

16 Warner Marzocchi, interview with Stan Cox, Rome, August 26, 2013.

17 Hall, "Scientists."

18 Thomas Van Stiphout, Stefan Wiemer, and Warner Marzocchi, "Are Short-Term
     Evacuations Warranted? Case of the 2009 L'Aquila Earthquake," *Geophysical Research
     Letters* 37 (2010): L06306; Warner Marzocchi, interview with Stan Cox, Rome, Au-
     gust 26, 2013.

19 Warner Marzocchi, interview with Stan Cox, Rome, August 26, 2013.

20 Hall, "Scientists."

21 Alexander, "Communicating."

22 Warner Marzocchi, e-mail interview with Stan Cox, April 7, 2014.

23 Richard Stuart Olson, Bruno Podesta, and Joanne Nigg, *The Politics of Earthquake
     Prevention* (Princeton, NJ: Princeton University Press, 1989), 54.

24 Warner Marzocchi, interview with Stan Cox, Rome, August 26, 2013.

25 T.H. Jordan, W. Marzocchi, A.J. Michael, and M.C. Gerstenberger, "Operational
     Earthquake Forecasting Can Enhance Earthquake Preparedness," *Seismological Re-
     search Letters* 85 (2014): 955–59.

26 Warner Marzocchi, e-mail interview with Stan Cox, May 8, 2014.

27 Jordan et al., "Operational Earthquake Forecasting."

28 David Alexander and Michele Magni, "Mortality in the L'Aquila (Central Italy)
     Earthquake of 6 April 2009," *PLOS Currents Disasters*, January 7, 2013, doi:10.1371/
     50585b8e6efd1. In the death toll across central L'Aquila, 147 perished in reinforced
     concrete structures, while there were 39 deaths in unreinforced buildings.

29 "Replies of the Commission to the Special Report of the European Court of Auditors:
     The European Union Solidarity Fund's Response to the 2009 Abruzzi Earthquake:
     The Relevance and Cost of Operations," Report No. COM(2012) 783, European

Union, December 13, 2012; Søren Bo Søndergaard, "The European Union Solidarity Fund's Response to the 2009 Abruzzi Earthquake: The Relevance and Cost of the Operations," Special Report No. 24/2012–CONT/7/12526, Committee on Budgetary Control, European Union, October 23, 2013.

30 Søndergaard, "European Union"; Giulia Segreti, "European Report Slams Misuse of Quake Funds," *Financial Times*, November 5, 2013; Mauro Dolce, e-mail interview with Stan Cox, April 9, 2014. The Søndergaard report also raised questions about the quality of the building material used in the CASE project, as well as the many problems that occurred with the electrical, heating, and sanitation systems. One project, the Pagliare di Sassa CASE, "caught fire due to a faulty electrical system and because it was constructed with flammable materials." Meanwhile, the provisional accommodation modules, which provided temporary shelter as per EUSF rules, turned out to be flimsily constructed, with flammable plaster and a host of problems with electrical wiring, water pipes, humidity damage, broken walls and floors, and malfunctioning sewage systems. Three complexes were evacuated by court order for safety reasons, and one set of modules burned because of bad wiring.

31 Daryoush Shoyajee, interview with Stan Cox, L'Aquila, September 1, 2013.

32 Mauro Dolce, e-mail interview with Stan Cox, September 19, 2013.

33 Søndergaard, "European Union."

34 "L'Aquila Quake Verdicts Quashed," BBC News, November 10, 2014.

35 Edwin Cartlidge, "Why Italian Earthquake Scientists Were Exonerated," *Science*, February 10, 2015.

Chapter 8: Atlantis of the Americas

1 Lucas Lechuga, "Rainbow over Biscayne Boulevard," *Miami Condo Investments*, August 17, 2010, accessed February 25, 2015, photos.miamicondoinvestments.com/rainbow-over-biscayne-boulevard.

2 Observed by authors, January 2015. We immediately visualized a "rainbow of chaos" as a silver-lined cloud turned inside out. But for other interpretations, see an article in *Dwntwn Miami Arts and Culture*, "Mural Sites Tour," n.d., downtownartdays.com/mural-sitestours.php. The article explains that "the artists five and Kemo revive the quote [from Cézanne], infusing it with contemporary meaning by suggesting we consider history from today's vantage point. Looking at the words independently from the source, it could also imply that sunny Miami is bubbling with activity." Indeed, rainbows are almost inherently positive phenomena in art and culture. Consider the promise made to Noah in Genesis 9: "I have set my rainbow in the clouds, and it will be the sign of the covenant between me and the Earth. Whenever I bring clouds over the Earth and the rainbow appears in the clouds, I will remember my covenant between me and you and all living creatures of every kind. Never again will the waters become a flood to destroy all life."

3 Bruce Mowry, interview with Stan Cox, Miami Beach, January 16, 2015.

4 Alfonso Chardy, "With Rising Waters in South Beach, FDOT Busy on Alton Road Drainage," *Miami Herald*, April 27, 2014.

5 Stephane Hallegatte, Colin Green, Robert J. Nicholls, and Jan Corfee-Morlot, "Future Flood Losses in Major Coastal Cities," *Nature Climate Change* 3 (2013): 802–6;

S. Hanson, R. Nicholls, N. Ranger, S. Hallegatte, J. Corfee-Morlot, C. Herweijer, and J. Chateau, "A Global Ranking of Port Cities with High Exposure to Climate Extremes," *Climatic Change* 104 (2011): 89–111.

6   L. Barry, M. Arockiasamy, F. Bloetscher, E. Kaisar, J. Rodriguez-Seda, P. Scarlatos, R. Teegavarapu, and N.M. Hammer, *Development of a Methodology for the Assessment of Sea Level Rise Impacts on Florida's Transportation Modes and Infrastructure* (Boca Raton, FL: Florida Atlantic University, 2012), 5–23.

7   John Edward Hoffmeister, *Land from the Sea: The Geologic Story of South Florida* (Coral Gables, FL: University of Miami Press, 1974), 19–26; Daniel Muhs, Kathleen Simmons, R. Randall Schumann, and Robert Halley, "Sea-Level History of the Past Two Interglacial Periods: New Evidence from U-Series Dating of Reef Corals from South Florida," *Quaternary Science Reviews* 30 (2011): 570–90; Michael Faught, "Submerged Paleoindian and Archaic Sites of the Big Bend, Florida," *Journal of Field Archaeology* 29 (2004): 273–90.

8   Hoffmeister, *Land from the Sea*.

9   Paul George, "Brokers, Binders, and Builders: Greater Miami's Boom of the Mid-1920s," *Florida Historical Quarterly* 65 (1986): 27–51.

10   Ibid.

11   Frank Sessa, "Miami in 1926," *Tequesta* 16 (1956): 15–16.

12   Ibid.

13   Ibid.

14   Michael Grunwald, *The Swamp: The Everglades, Florida, and the Politics of Paradise* (New York: Simon and Schuster, 2006), 5.

15   Sessa, "Miami in 1926."

16   Aslak Grinsted, John Moore, and Svetlana Jevrejeva, "Homogeneous Record of Atlantic Hurricane Surge Threat Since 1923," *Proceedings of the National Academy of Sciences* 109 (2012): 19601–5.

17   R. Pielke, J. Gratz, C. Landsea, D. Collins, M. Saunders, and R. Musulin, "Normalized Hurricane Damage in the United States: 1900–2005," *Natural Hazards Review* 9 (2008): 29–42.

18   Ibid.; Ken Kaye, "Could Another Hurricane Andrew Sneak Up on Us 20 Years Later?," *Sun Sentinel*, May 26, 2012.

19   Hoffmeister, *Land from the Sea*, 27–53; McPherson and Halley, "South Florida Environment."

20   Harold Wanless, interview with the authors, Coral Gables, Florida, January 12, 2015.

21   Peter Harlem, interview with Stan Cox, Miami, January 20, 2015.

22   Henry Briceño, interview with the authors, Miami, January 12, 2015.

23   Jeff Goodell, "Goodbye, Miami," *Rolling Stone*, June 2013; Robin McKie, "Miami, the Great World City, Is Drowning While the Powers That Be Look Away," *The Guardian*, July 11, 2014.

24   Christina Viega, "Miami Beach to Spend up to $400 Million to Deal with Flooding Issues," *Miami Herald*, February 12, 2014.

25   Joey Flechas, "Miami Beach Streets Stay Dry During King Tide Peak," *Miami Herald*, October 9, 2014.

26   Nicole Hernandez Hammer, interview with the authors, Gumbo Limbo Nature

Center, Boca Raton, FL, January 8, 2015. Hammer was to make national news reports twelve days after this interview when she was invited to sit next to First Lady Michelle Obama at the State of the Union address.

27 Randall Parkinson, Peter Harlem, and John Meeder, "Managing the Anthropocene Marine Transgression to the Year 2100 and Beyond in the State of Florida U.S.A.," *Climatic Change* 128 (2014): 85–98.

28 "Margaritaville" was a hit song written and performed by troubadour of the semi-tropics Jimmy Buffett on his 1977 album *Changes in Latitudes, Changes in Attitudes.* Two years later, Buffett would record the more explicitly prophetic album *Volcano* at Montserrat's AIR Studios (see page 261).

29 Michael Gerrity, "Miami Bucks National Trend as U.S. Home Prices Decelerate," *World Property Journal*, November 25, 2014; Christopher Rugaber, "US Home Price Gains Slow for 6th Straight Month," Associated Press, July 29, 2014. For one explanation of continued high real estate prices, see Devin Bunten and Matthew Kahn, "The Impact of Emerging Climate Risks on Urban Real Estate Price Dynamics," National Bureau of Economic Research, 2014, nber.org/papers/w20018.homeprices.pdf.

30 Martha Brannigan, "Miami's Condo Boom Redux," *Miami Herald*, July 6, 2014.

31 John Dorschner, "Rising Sea Levels, Falling Real Estate Values," *Miami Herald*, November 9, 2013; Tom Hudson, "Underwater Real Estate," WLRN, November 14, 2013, wlrn.org/post/underwater-real-estate.

32 Cathy Booth, "Miami: The Capital of Latin America," *Time*, June 24, 2001.

33 Hudson, "Underwater."

34 Philip Stoddard, interview with the authors, South Miami, January 6, 2015.

35 Tristram Korten, "In Florida, Officials Ban Term 'Climate Change,'" *Miami Herald*, March 8, 2015.

36 Miami ranks twenty-ninth among sixty-six metropolitan areas nationwide in per capita carbon emissions. Emissions caused by heating are small, of course, but Miami is number one in emissions from electricity generation to run air-conditioning systems. Emissions from private motor vehicles are close to average, but for emissions from public transportation, Miami trails only New York and Washington, DC. Edward Glaeser and Matthew Kahn, "The Greenness of Cities: Carbon Dioxide Emissions and Urban Development," *Journal of Urban Economics* 67 (2010): 404–18.

37 Jenny Staletovich, "Miami-Dade County Takes Reins on Climate Change," *Miami Herald*, January 21, 2015.

38 Celeste De Palma, e-mail interview with Stan Cox, March 9, 2015. At the time of the January 21 Miami-Dade County Board of Commissioners meeting, De Palma was representing the Tropical Audubon Society.

39 Douglas Hanks, "Climate Change Dominates Miami-Dade Budget Hearing," *Miami Herald*, September 3, 2015.

40 The project was called HighWaterLine Miami; see highwaterline.org/miami.

41 U.S. Census Bureau, "Household Income Inequality Within U.S. Counties: 2006–2010," American Community Survey, February 2012, lostcoastoutpost.com/media/uploads/post/2387/acsbr10-18.pdf.

42 J. Chakraborty, T. Collins, M. Montgomery, and S. Grineski, "Social and Spatial Inequities in Exposure to Flood Risk in Miami, Florida," *Natural Hazards Review* 15 (2014): 04014006.

43 Goodell, "Goodbye, Miami."

44 Patricia Sagastume, "Researchers Aim to Resolve Inequity in Miami's Flood Preparation," Al Jazeera America, March 5, 2014.

45 Bob Bolin, "Race, Class, Ethnicity, and Disaster Vulnerability," in *Handbook of Disaster Research*, ed. Havidan Rodriguez, Enrico L. Quarantelli, and Russell Dynes (New York: Springer, 2007), 113–29.

46 Hudson, "Underwater." Asked by Hudson where her own home was located, Plater-Zyberk said, "I have property on the ridge, my house, a historic Coral Gables house, and we have an apartment in Miami Beach as well. And so we do talk about it and say, 'What about that apartment in Miami Beach? Should we . . . ?' I believe that Miami Beach could protect itself."

47 Ross Palombo, "South Florida Towns Prepare for Unexpected with Military Equipment," *Local10 News*, October 1, 2014.

48 Gary Nelson, "Global Warming Blamed for South Florida Severe Flooding," CBS Miami, September 28, 2015; Jenny Staletovich, "King Tide Sets Stage for Climate Talks in South Florida," *Miami Herald*, September 28, 2015; "High Tides Cause Persistent Flooding in South Florida," WSVN-TV, September 30, 2015.

49 Jessica Weiss, "Dutch Sea Level Rise Expert: Miami Will Be 'the New Atlantis,' a City in the Sea," *Miami New Times*, May 21, 2015.

Chapter 9: Engineer, Defend, Insure, Absorb, Leave

1 Deborah Stone, "Causal Stories and the Formation of Policy Agendas," *Political Science Quarterly* 104 (1989): 281–300. Note that if Stone is correct, her insight should reach policy circles in about 2039.

2 James Morgan, "Great Walls of America 'Could Stop Tornadoes,'" BBC News, March 7, 2014; Sam Frizell, "Physicist Claims Giant Walls Could Stop Tornadoes in Midwest," *Time*, March 8, 2014.

3 John McPhee, *The Control of Nature* (New York: Farrar, Straus and Giroux, 1989).

4 J.E. Dugan, L. Airoldi, M.G. Chapman, S.J. Walker, and T. Schlacher, "Estuarine and Coastal Structures: Environmental Effects, a Focus on Shore and Nearshore Structures," in *Treatise on Estuarine and Coastal Science: 8. Human-Induced Problems (Uses and Abuses)*, ed. M.J. Kennish and M. Elliott (Amsterdam: Elsevier, 2011), 17–41.

5 McPhee, *Control*.

6 Jean-Christophe Gaillard, "Alternative Paradigms of Volcanic Risk Perception: The Case of Mt. Pinatubo in the Philippines," *Journal of Volcanology and Geothermal Research* 172 (2008), 315–28.

7 Ted Steinberg, *Acts of God: The Unnatural History of Natural Disasters in America* (New York: Oxford University Press, 2000), 119.

8 Bob Marshall, "New Orleans' Flood Protection System: Stronger than Ever, Weaker than It Was Supposed to Be," *The Lens*, May 15, 2014.

9 Manhattan's Dry Line proposal was led by the Danish company Bjarke Ingels Group, but engineering and costing work was handled by Arcadis, the same Dutch company that designed the harbor barrier. Arcadis does the majority of its business in the United States. Bjarke Ingels himself started his career working for the legendary Dutch architect Rem Koolhaas at Rotterdam's Office for Metropolitan Architecture.

10 Russell Shorto, "How to Think Like the Dutch in a Post-Sandy World," *New York Times Magazine*, April 9, 2014. According to the profile, "For his part, Ovink said it dawned on him during [Secretary] Donovan's visit that the post-Sandy turmoil in the U.S. was an opportunity. Dutch water-management experts have done such a good job of protecting their country that they rarely get to practice with water crises—whereas America was facing something monumental that as a culture it didn't yet grasp."

11 KuiperCompagnons, "The Great Garuda to Save Jakarta," accessed February 24, 2015, kuiper.nl/en/news/the-great-garuda-to-save-jakarta.

12 Herman Gerritsen, "What Happened in 1953? The Big Flood in the Netherlands in Retrospect," *Philosophical Transactions of the Royal Society A* 363 (2005): 1271–91.

13 Arnoud Molenaar, interview with Paul Cox, Rotterdam, August 21, 2014.

14 In ZUS's presentation, the Room for the River project (discussed below) was named Delta Works 2.0.

15 Kristian Koreman and Florian Boer, interview with Paul Cox, Rotterdam, August 21, 2014.

16 Rotterdam Delta City app, rotterdamdeltacity.nl.

17 Project Atelier Rotterdam exhibition at the International Architecture Biennale Rotterdam, Kunsthal Rotterdam, May 29–August 24, 2014. These are rough conversions to short tons; the original figures were 691,860 kilotons of rainwater and 220,000 kilotons of cargo.

18 Daniel Hoornweg, Lorraine Sugar, and Claudia Lorena Trejos Gomez, "Cities and Greenhouse Gas Emissions: Moving Forward," *Environment and Urbanization* 23, no. 1 (2011): 207–27.

19 The unfeasibility of the carbon capture and storage plan was reported to the RCI by a member of its own board, sustainability professor Jan Rotmans, in 2007. Rachel Keeton, "After Nixing Plans to Bury CO2 Underground, Rotterdam Shifts Its Climate Gaze Skyward," *Next City*, November 13, 2013.

20 Rachel Keeton, "Europe's Dysfunctional Carbon-Pricing System Is Keeping Coal Dirt Cheap," *Next City*, May 20, 2014. However, in the summer of 2015 the Dutch government's flagging commitment to mitigation was challenged in court by a citizens' platform called Urgenda Foundation, which sued the government on behalf of the country's climate-vulnerable people. In the first ruling of its kind anywhere, the Hague District Court ruled that the state must take more action to ensure that Dutch emissions in the year 2020 will be at least 25 percent lower than in 1990. "State Ordered to Further Limit Greenhouse Gas Emissions," *De Rechtspraak*, June 24, 2015.

21 Ben Schiller, "The 10 Most Sustainable Cities That Will Thrive as the World Crumbles," *Fast Company Co.Exist*, February 20, 2015.

22 Maarten Nijpels, *Rijnhaven—Metropolitan Delta Innovation*, Stadshavens Rotterdam, July 2013.

23 Harold Wanless, interview with the authors, Coral Gables, Florida, January 12, 2015.

24 Duncan Vlag, interview with Paul Cox, Rotterdam, August 23, 2014. A furniture designer who rents desks to the pavilion, he was manning the open house while the regular staff were on holiday.

25 Jenny Staletovich, "Dutch Solution to Miami's Rising Seas? Floating Islands," *Miami Herald*, August 23, 2014.

26 Zygmunt Bauman, *Liquid Modernity* (Cambridge: Polity Press, 2000).

27 All of the detention facilities are gone and the pair in Zaandam no longer appear in Waterstudio.NL's portfolio; however, the foreign press, including the *Miami Herald* article (Staletovich, "Dutch Solution"), continue to list the "floating prison" among Olthius's feats, overlooking the political context. In 2008, Olthius showed off the facility to a web video team from National Public Radio. John Poole, "Floating Architecture for a Changing Climate," *NPR Special Series: Adaptation*, April 21, 2008, npr.org/templates/story/story.php?storyId=89771102; *Report to the Authorities of the Kingdom of the Netherlands on the Visits Carried Out to the Kingdom in Europe, Aruba and the Netherlands Antilles by the European Committee for the Prevention of Torture and Inhuman or Degrading Treatment or Punishment in June 2007*, CPT, 2008.

28 *Report to the Authorities*. The first pair of boats, the *Reno* and *Bibby Stockholm* in Rotterdam's Merwe-Vierhavens harbor, became notorious thanks to a journalist for *Vrij Nederland* who worked undercover as a security guard on board. Robert van de Griend, "Undercover op de Illegalenboot," *Vrij Nederland*, March 25, 2006. Amnesty International later condemned the boats as the worst blight on the spotty Dutch human rights record with regard to migrants. AI Netherlands, "The Netherlands: The Detention of Irregular Migrants and Asylum-Seekers," Amnesty International (Index No. EUR 35/002/2008), June 27, 2008.

29 Boats4People, "Bajesboot Verrast met Concert op het Wat," IndyMedia–Vrij Media Centrum Nederland, December 18, 2011, indymedia.nl/node/1408.

30 The Seasteading Institute commissioned DeltaSync, the designers of Rotterdam's Floating Pavilion, to create their crowdfunded Floating City design.

31 Rachel Keeton, "Has Floating Architecture's Moment Finally Arrived?," *Next City*, October 1, 2014.

32 Mothership, enterthemothership.com/en/projects/bobbing-forest.

33 Hoornweg et al., "Cities."

34 Prasanth Divakaran, Vaso Kapnopoulou, Erin McMurtry, Min-Guk Seo, and Liwei Yu, *Towards an Integrated Framework for Coastal Eco-Cities: EU-Asia Perspectives* (Southampton: University of Southampton, 2013).

35 Susan Hanson, Robert Nicholls, Nicola Ranger, Stéphane Hallegatte, Jan Corfee-Morlot, Celine Herweijer, and Jean Chateau, "A Global Ranking of Port Cities with High Exposure to Climate Extremes," *Climatic Change* 104 (2011): 89–111.

36 Divakaran et al., *Towards an Integrated*.

37 Jonathan Kaiman, "China's 'Eco-Cities': Empty of Hospitals, Shopping Centres and People," *The Guardian*, April 13, 2014.

38 Federico Caprotti, "Eco-Urbanism and the Eco-City, or, Denying the Right to the City?," *Antipode*, March 1, 2014. As Caprotti puts it, eco-cities are being looked to as "a site where the problematic of industrialisation and environmental degradation can be reconciled with the imperative for sustained and rapid economic growth."

39 Andrew Jacobs, "As Fire Smolders in Tianjin, Officials Rush to Stanch Criticism," *New York Times*, August 13, 2015.

40 "天津爆炸: 中新天津生态城 一名新加坡员工受轻伤 [Tianjin Explosion: In Sino-Singapore Tianjin Eco-City an Employee Suffers Minor Injuries]," Channel 8 News, August 13, 2015, channel8news.sg/news8/singapore/20150813-sg-tianjin/2048298.html. The gases included sodium cyanide, seven hundred tons of which, it was later learned, had been

stored illegally at the site. Beyond the nearly two hundred dead and missing, many hundreds more were injured or sickened. The main explosion was recorded as having the power of an earthquake of magnitude 2 to 3 by USGS seismographs a hundred miles away in Beijing. (Scientists there said this was very unusual for a surface explosion, which transfers only a small portion of its total energy down and away through the earth.) Tianjin's citizens had already been holding street protests against the warehousing of vast chemical stockpiles in a major urban area, and the catastrophic explosion drew all of China, and much of the world, into an argument over blame for the tragedy. Amanda Holpuch, "China's Tianjin Blast Sets Off Earthquake-Recording Instruments 100 Miles Away," *The Guardian*, August 12, 2015; Fergus Ryan, "Tianjin Explosions: Sodium Cyanide on Site May Have Been 70 Times Allowed Amount," *The Guardian*, August 16, 2015.

41 Ken Caldeira, Govindasamy Bala, and Long Cao, "The Science of Geoengineering," *Annual Review of Earth and Planetary Sciences* 41 (2013): 231–56. They write, "The term geoengineering as applied in its current context was introduced into the scientific literature by Victor Marchetti in the title of his classic paper describing deep-sea disposal of carbon dioxide (CO_2). . . . Such efforts include both solar geoengineering (also known as solar radiation management, or SRM) and carbon dioxide removal (CDR). SRM aims to diminish the amount of climate change produced by high greenhouse gas concentrations, whereas CDR involves removing CO_2 and other greenhouse gases from the atmosphere."

42 Paul Crutzen, "Albedo Enhancement by Stratospheric Sulfur Injections: A Contribution to Resolve a Policy Dilemma?," *Climatic Change* 77 (2006): 211–20.

43 Caldeira et al., "Science of Geoengineering."

44 Ibid. This form of geoengineering, it is claimed, not only would make much of the world more comfortable and presumably lower the risk of climatic disasters; by moderating temperatures while allowing carbon dioxide concentrations to continue rising, solar radiation management could increase the Earth's crop production capacity, rendering the world, as one analysis put it, "hazy, cool, and well fed." Michael Roderick and Graham Farquhar, "Geoengineering: Hazy, Cool and Well Fed?," *Nature Climate Change* 2 (2012): 76–77.

45 William Nordhaus, "An Optimal Transition Path for Controlling Greenhouse Gases," *Science* 258 (1992): 1315–19.

46 Dale Jamieson, "Some Whats, Whys and Worries of Geoengineering," *Climatic Change* 121 (2013): 527–37.

47 Naomi Klein, *This Changes Everything: Capitalism vs. the Climate* (New York: Simon & Schuster, 2014), 277.

48 Ben Booth, Nick Dunstone, Paul Halloran, Timothy Andrews, and Nicolas Bellouin, "Aerosols Implicated as a Prime Driver of Twentieth-Century North Atlantic Climate Variability," *Nature* 484 (2012): 228–33.

49 Scott Barrett, "The Incredible Economics of Geoengineering," *Environmental and Resource Economics* 39 (2008): 45–54; Alan Robock, "20 Reasons Why Geoengineering May Be a Bad Idea," *Bulletin of the Atomic Scientists*, May 30, 2013; Lynn Russell, Philip Rasch, Georgina Mace, Robert Jackson, John Shepherd, Peter Liss, Margaret Leinen, et al., "Ecosystem Impacts of Geoengineering: A Review for Developing a Science Plan," *AMBIO* 41 (2012): 350–69; Hans Joachim Schellnhuber, "Geoengineering: The

Good, the MAD, and the Sensible," *Proceedings of the National Academy of Sciences* 108 (2011): 20277–78; Ben Kravitz, Douglas MacMartin, and Ken Caldeira, "Geoengineering: Whiter Skies?," *Geophysical Research Letters* 39 (2012): L11801; Caldeira et al., "The Science of Geoengineering"; Nathan Gillett, Vivek Arora, Kirsten Zickfeld, Shawn Marshall, and William Merryfield, "Ongoing Climate Change Following a Complete Cessation of Carbon Dioxide Emissions," *Nature Geoscience* 4 (2011): 83–87.

50 Scott Barrett, Timothy M. Lenton, Antony Millner, Alessandro Tavoni, Stephen Carpenter, John Anderies, F. Stuart Chapin III, et al., "Climate Engineering Reconsidered," *Nature Climate Change* 4 (2014): 527–29.

51 Jamieson, "Some Whats." Jamieson uses the term "moral hazard" in this context while conceding that it may be problematic. He cites Benjamin Hale, "The World That Would Have Been: Moral Hazard Arguments Against Geoengineering," in *Reflecting Sunlight: The Ethics of Solar Radiation Management*, ed. Christopher Preston (Lanham, MD: Rowman and Littlefield, 2012). Naomi Klein points out the absurdity inherent in any attempt to pilot-test solar radiation management: to determine its effectiveness in lowering global temperatures, the experiment must be done globally, and because climatic averages and extremes fluctuate naturally year to year, the experiment would need to generate many years of data before we could know whether it had either desirable effects or undesirable side effects like higher risk of climatic disasters. By then, we would be committed to hosing down the stratosphere for eternity. Klein, *This Changes*, 269.

52 Stanley Changnon, David Changnon, E. Ray Fosse, Donald Hoganson, Richard Roth, and James Totsch, "Effects of Recent Weather Extremes on the Insurance Industry: Major Implications for the Atmospheric Sciences," *Bulletin of the American Meteorological Society* 78 (1997): 425–35.

53 Swenji Surminski and Delioma Oramas-Dorta, "Do Flood Insurance Schemes in Developing Countries Provide Incentives to Reduce Physical Risks?," Grantham Research Institute on Climate Change and the Environment, July 2013; Pierre Picard, "Natural Disaster Insurance and the Equity-Efficiency Trade-Off," *Journal of Risk and Insurance* 75 (2008): 17–38.

54 Surminski et al., "Do Flood Insurance."

55 Picard, "Natural Disaster Insurance."

56 Ibid.

57 Surminski et al., "Do Flood Insurance"; Picard, "Natural Disaster Insurance."

58 Bruce Alpert, "FEMA Out with New Flood Insurance Rates, Reflecting March Law's Changes," *Times-Picayune*, May 30, 2014. An evaluation of America's flood-insurance system by Australia's National Climate Change Adaptation Research Facility concluded, "The NFIP is a scheme that can be regarded as not very successful in terms of insurance, but very successful in terms of acting as an incentive for ensuring national acceptance of the FEMA flood management guidelines—a rare phenomenon in the United States where local authorities still control most forms of building standards." Picard, "Natural Disaster Insurance."

59 "Roundtable: What Is the Future for Insurance-Linked Securities?," *Artemis*, May 21, 2014.

60 Anthony Harrington, "Catastrophe Bonds, Part 1: Bet on a Hurricane, Anyone?,"

QFinance, January 8, 2010. His comment "Hurricanes don't care what the markets are doing" raises a seemingly important question: with the climate going haywire, will the market as a whole start caring about what hurricanes are doing? If so, the correlation between financial risk and disaster risk will rise.

61 Ibid.

62 "Everglades Re 2014 Completes, Officially the Largest Catastrophe Bond Ever," *Artemis*, May 2, 2014.

63 Rawle King, "Financing Natural Catastrophe Exposure: Issues and Options for Improving Risk Transfer Markets," Congressional Research Service Report No. 7-5700, August 15, 2013; "Text of the Homeowners' Defense Act of 2013," govtrack.us, govtrack.us/congress/bills/113/hr737/text; John McAneney, Ryan Crompton, Delphine McAneney, Rade Musulin, George Walker, and Roger Pielke, *Market-Based Mechanisms for Climate Change Adaptation: Final Report* (Southport, QLD, Australia: National Climate Change Adaptation Research Facility, 2013).

64 Leigh Johnson, "Geographies of Securitized Catastrophe Risk and the Implications of Climate Change," *Economic Geography* 90 (2014): 155–85. Johnson says that although they stand to lose their principal completely if the wrong disaster strikes, investors in the disaster market have high confidence that the third-party modeling firms from whose models critical data such as trigger points and estimated losses are generated have an excellent understanding of exposures. She characterizes the investors' view this way: "We understand much better how an extreme event might wipe out Lower Manhattan than we do the human behavior that leads to mortgage defaults." The specialized investors who inhabit the catastrophe market also have either in-house experts or relationships with firms that provide what they see as state-of-the-art projections. Leigh Johnson, Internet phone interview with Stan Cox, November 3, 2014.

65 Johnson, "Geographies."

66 Ibid.; Leigh Johnson, Internet phone interview with Stan Cox, November 3, 2014.

67 Johnson, "Geographies."

68 Ibid.; R.O. King, *Financing Natural Catastrophe Exposure: Issues and Options for Improving Risk Transfer Markets*, Congressional Research Service, Library of Congress, August 15, 2013, digital.library.unt.edu/ark:/67531/metadc227907.

69 Howard Kunreuther and Erwann Michel-Kerjan, "Enhancing Post-Disaster Economic Recovery: How Improved Flood Insurance Mechanisms Can Help," *Current Research Project Narratives*, Paper No. 71, January 1, 2014, research.create.usc.edu/current_synopses/71; Erwann Michel-Kerjan and Howard Kunreuther, "Redesigning Flood Insurance," *Science* 333 (2011): 408–9.

70 McAneney, "Market-Based."

71 Ibid.

72 Lisi Krall, economics professor, SUNY Cortland, e-mail interview with Stan Cox, June 5, 2014; Caroline Kousky, Resources for the Future, telephone interview with Stan Cox, August 6, 2014; Leigh Johnson, Internet phone interview with Stan Cox, November 3, 2014.

73 Kathryn Schulz, "The Really Big One," *New Yorker*, July 20, 2015.

Chapter 10: The Absorbers

1 Sachin Kadam, interview with Stan Cox, Mumbai, February 19, 2014.

2 Allan Rogers Kibaya, "When Your Car Is Submerged in Floods," *Daily Monitor*, November 19, 2015.

3 Jesse Ribot, "Vulnerability Does Not Fall from the Sky: Toward Multiscale, Pro-Poor Climate Policy," in *Social Dimensions of Climate Change: Equity and Vulnerability in a Warming World,* ed. Robin Mearns and Andrew Morton (Washington, DC: World Bank Publications, 2010), 47–74.

4 Claire Marshall, "Honduras Struggles 10 Years After Mitch," BBC, October 30, 2008.

5 "Deadliest Hurricanes Since 1492," *USA Today*, May 20, 2005.

6 Bradley Ensor and Marisa Ensor, "Hurricane Mitch: Root Causes and Responses to the Disaster," in *The Legacy of Hurricane Mitch: Lessons from Post-Disaster Reconstruction in Honduras*, ed. Marisa Ensor (Tucson: University of Arizona Press, 2009), 34–35.

7 Ibid., 37.

8 Mike Davis, *Planet of Slums* (London: Verso, 2006), 121–27.

9 Dilip Bobb, Malini Bhupta, and Shankkar Aiyar, "Mumbai's Collapse," *India Today*, August 8, 2005.

10 Conservation Action Trust and Concerned Citizens' Commission, *Mumbai Marooned: An Enquiry into Mumbai Floods 2005: Final Report* (Mumbai: Conservation Action Trust, 2006).

11 Sachin Kadam, interview with Stan Cox, Mumbai, February 19, 2014.

12 Aromar Revi, "Lessons from the Deluge: Priorities for Multi-Hazard Risk Mitigation," *Economic and Political Weekly* 40 (2005): 3911–16.

13 Nicole Ranger, Stéphane Hallegatte, Sumana Bhattacharya, Murthy Bachu, Satya Priya, K. Dhore, Farhat Rafique, et al., "An Assessment of the Potential Impact of Climate Change on Flood Risk in Mumbai," *Climatic Change* 104 (2011): 139–67.

14 Shamshaad Sheik, Zahida Qureshi, Mumtaz Qureshi, Halema Sheik, and Aabeda Qureshi, interview with Stan Cox, Dharavi, Mumbai, February 19, 2014. Translation by Sachin Kadam and Priti Gulati Cox. Interviewees with the same surnames were not related to one another.

15 Subhajyoti Samaddar, Roshni Chatterjee, Bijay Anand Misra, and Hirokazu Tatano, "Participatory Risk Mapping for Identifying Spatial Risks in Flood Prone Slum Areas, Mumbai," *Disaster Prevention Research Institute Annals B* 54 (2011): 137–46.

16 Starting around 2010, most of the families in Annanagar had taps installed in or in front of their own rooms, at their own expense.

17 In his book *Mumbai Fables*, Gyan Prakash discusses the origins of this style of long, two-story, single-room-per-unit housing block, known in the local Marathi language as a *chawl*. In nineteenth-century Mumbai under British rule, the *chawl*, he writes, "was the defining emblem of overcrowded working-class space." Then as now, "the government recognized the wretchedness of working-class housing. In fact, colonial officials themselves drew dark pictures of the poor, packed in dense clusters of overcrowded and poorly ventilated slums that were set between open drains and narrow lanes, stables, and warehouses. . . . There is a Dickensian impulse in their focus on squalor and misery but none of the English novelist's insight that industrialization

was directly responsible for the wretched, inhuman conditions." Gyan Prakash, *Mumbai Fables* (Noida: HarperCollins India, 2011), 64–65.

18 The women we spoke with knew of no such predatory activity in the neighborhood during or just after the 2005 flood.

19 Monalisa Chatterjee and James Mitchell, "The Scope for Broadening Climate-Related Disaster Risk Reduction Policies in Mumbai," *Professional Geographer* 66 (2014): 363–71.

20 Daniel Brook, "Slumming It," *The Baffler* 25 (2014), thebaffler.com/salvos/slumming-it.

21 Dan McDougall, "Waste Not, Want Not in the £700m Slum," *The Observer*, March 4, 2007.

22 UN-Habitat, *State of the World's Cities Report 2012/2013: Prosperity of Cities* (Nairobi: United Nations Human Settlements Programme, 2013).

23 Davis, *Planet of Slums*, 126.

24 Ibid., 122.

25 Ibid., 127.

26 Ibid., 121–22.

27 This has likely been a common state of affairs through the history of urbanism. It's notable that archaeologists interested in water infrastructures have turned most of their attention to ancient tropical cities rather than desert metropolises. They recognize that too much water is as much a problem as too little—and that failures of water-removing infrastructure can be among the most catastrophic. One such ancient urban center that declined from a Central Eurasian trading hub to an empty swamp was the Afghan city of Balkh, where all of the inhabitants were ultimately driven out by cholera and malaria. Arash Khazeni, "The City of Balkh and the Central Eurasian Caravan Trade in the Early Nineteenth Century," *Comparative Studies of South Asia, Africa and the Middle East* 30, no. 3 (2010): 463–472; Monica L. Smith, "The Archaeology of Urban Landscapes," *Annual Review of Anthropology* 43, no. 1 (2014): 307–23.

28 The stone was laid as part of the city's Drainage Master Plan; paved drains demarcate the boundaries of "real" neighborhoods. The path described here also passes through parts of Mulimira. The boundaries between this valley's many slums are not easily defined.

29 "Celebrities Living in Slums," *Kampala Sun*, August 23, 2013. Bobi Wine styles himself as a voice of the people, but with Kampala dancehall stars there's always a tension between their high-flying celebrity personas and their roots here in the valleys. In one typical tabloid spread titled "Celebrities Living in Slums!," the *Kampala Sun* devoted four pages to tracking musicians to their homes and taking ambush shots of them jumping over ditches and fleeing through the mud in their BMWs.

30 Madelena Iko, interview with Stan Cox, translated by Pheonah Nabukalu, Kamwokya, Kampala, August 27, 2015.

31 Janet Kankwasa, interview with Stan Cox, translated by Pheonah Nabukalu, Kamwokya, Kampala, August 27, 2015.

32 Shuaib Lwasa, interview with Paul Cox, Kampala, August 31, 2013.

33 Janet Kankwasa, interview with Stan Cox, translated by Pheonah Nabukalu, Kamwokya, Kampala, August 27, 2015.

34 Roland White, Huang Chyi-Yun, Herbert Oule, Martin Onyach-Olaa, John Bach-
    mann, Diane Dale, Brian Goldberg, Maritza Pechin, and Jane Turpie, *Promoting
    Green Urban Development in African Cities: Kampala, Uganda—Urban Environmen-
    tal Profile* (Washington, DC: World Bank Group, 2015).

35 IPCC, *Climate Change 2014: Impacts, Adaptation, and Vulnerability. Contribution of
    Working Group II to the Fifth Assessment Report of the Intergovernmental Panel on
    Climate Change* (Cambridge: Cambridge University Press, 2014), 20.

Chapter 11: Vulnerability Seeps in Everywhere

1   "'It's Time to Stop This Madness'—Philippines Plea at UN Climate Talks," RTCC
    [Responding to Climate Change], November 13, 2013, www.rtcc.org/2013/11/11
    /its-time-to-stop-this-madness-philippines-plea-at-un-climate-talks.

2   Pope Francis, *Laudato Si': On Care for Our Common Home*, encyclical letter, May 24,
    2015.

3   Richard Cash, Shantana Halder, Mushtuq Husain, Mohammed Sirajul Islam, Fuad
    Mallick, Maria May, Mahmudur Rahman, and M. Aminur Rahman, "Reducing the
    Health Effect of Natural Hazards in Bangladesh," *The Lancet* 382 (2013): 2094–103.

4   Ibid.; Ubydul Haque, Masahiro Hashizume, Korine N. Kolivras, Hans J. Overgaard,
    Bivash Das, and Taro Yamamoto, "Reduced Death Rates from Cyclones in Bangla-
    desh: What More Needs to Be Done?," *Bulletin of the World Health Organization* 90
    (2012): 150–56.

5   Cash et al., "Reducing"; Haque et al., "Reduced Death."

6   Haque et al., "Reduced Death."

7   Tania López-Marrero and Ben Wisner, "Not in the Same Boat: Disasters and Differ-
    ential Vulnerability in the Insular Caribbean," *Caribbean Studies* 40 (2012): 129–68;
    A. Pichler and E. Striessnig, "Differential Vulnerability to Hurricanes in Cuba, Haiti,
    and the Dominican Republic: The Contribution of Education," *Ecology and Society*
    18 (2013): 31; Holly Sims and Kevin Vogelmann, "Popular Mobilization and Disaster
    Management in Cuba," *Public Administration and Development* 22 (2002): 389–400;
    Martha Thompson, "How Cuba Turns Early Warning into Joined-up Action," *SciDev
    .Net*, November 21, 2012; "Lessons in Risk Reduction from Cuba," unpublished case
    study prepared for UN-Habitat, *Global Report on Human Settlements 2007,* mirror
    .unhabitat.org/downloads/docs/GRHS.2007.CaseStudy.Cuba.pdf. In the years before
    late 2014, when the relationship between the United States and Cuba began to thaw,
    the approach of a hurricane offered one of only a very few opportunities for cooper-
    ation between the two governments. Most hurricanes that strike the U.S. mainland
    pass through Cuba first—the institute's official in charge of prediction, José Rubiera,
    once told the *New York Times*, "A hurricane that hits Cuba doesn't ask for a visa before
    entering the United States." Jean Friedman-Rudovsky, "Hurricane Tips from Cuba,"
    New York Times, July 29, 2013.

8   López-Marrero and Wisner, "Not in the Same Boat."

9   Pichler and Striessnig, "Differential Vulnerability."

10 López-Marrero and Wisner, "Not in the Same Boat."

11 Pichler and Striessnig, "Differential Vulnerability."

12 Jeff Franks, "Cuba: Hurricane Sandy Leaves Destruction in Its Wake," *Christian*

Science Monitor, October 25, 2012; "A Year After Hurricane Sandy Hit Cuba," *Havana Times*, October 24, 2013.

13 "La Recuperación de Sandy Se Adentra en Su Etapa Más Compleja," *Granma*, October 24. The contrast between disaster impacts in Cuba and the rest of the Caribbean is dramatic, largely because of differences in preexisting socioeconomic conditions. A study comparing disaster preparedness in Cuba and the Dominican Republic found that a majority of Dominicans would refuse to evacuate if ordered because gangs would loot their homes. Others, mostly women, said they would not feel safe in shelters, where sleeping areas and toilets are not separate for males and females. In Haiti following the 2010 earthquake, Amnesty International reported that women had found it necessary to exchange sex for water, food, and safety. Pichler and Striessnig, "Differential Vulnerability."

14 Anil Gupta and Sreeja Nair, *Ecosystem Approach to Disaster Risk Reduction* (Delhi: National Institute of Disaster Management, 2012); W. Neil Adger, Terry Hughes, Carl Folke, Stephen Carpenter, and Johan Rockström, "Social-Ecological Resilience to Coastal Disasters," *Science* 309 (2005): 1036–39.

15 Andrew Cooper and John McKenna, "Working with Natural Processes: The Challenge for Coastal Protection Strategies," *Geographical Journal* 174 (2008): 315–31.

16 Roland Cochard, Senaratne Ranamukhaarachchi, Ganesh Shivakoti, Oleg Shipin, Peter Edwards, and Klaus Seeland, "The 2004 Tsunami in Aceh and Southern Thailand: A Review on Coastal Ecosystems, Wave Hazards and Vulnerability," *Perspectives in Plant Ecology, Evolution and Systematics* 10 (2008): 3–40; Dahdouh Guebas, "How Effective Were Mangroves as a Defence Against the Recent Tsunami?," *Current Biology* 15 (2005): R443–47; Roland Cochard, "On the Strengths and Drawbacks of Tsunami-Buffer Forests," *Proceedings of the National Academy of Sciences* 108 (2011): 18571–72; Nibedita Mukherjee, Farid Dahdouh-Guebas, Vena Kapoor, Rohan Arthur, Nico Koedam, Aarthi Sridhar, and Kartik Shanker, "From Bathymetry to Bioshields: A Review of Post-Tsunami Ecological Research in India and Its Implications for Policy," *Environmental Management* 46 (2010): 329–39; Adger et al., "Social-Ecological."

17 Edward Barbier, Ioannis Georgiou, Brian Enchelmeyer, and Denise Reed, "The Value of Wetlands in Protecting Southeast Louisiana from Hurricane Storm Surges," *PLoS ONE* 8 (2013): e58715.

18 Cochard, "On the Strengths"; Mukherjee et al., "From Bathymetry."

19 Rusty Feagin, Nibedita Mukherjee, Kartik Shanker, Andrew Baird, Joshua Cinner, Alexander Kerr, Nico Koedam, et al., "Shelter from the Storm? Use and Misuse of Coastal Vegetation Bioshields for Managing Natural Disasters," *Conservation Letters* 3 (2010): 1–11.

20 Shing Yip Lee, Jurgene Primavera, Farid Dahdouh-Guebas, Karen McKee, Jared Bosire, Stefano Cannicci, Karen Diele, et al., "Ecological Role and Services of Tropical Mangrove Ecosystems: A Reassessment," *Global Ecology and Biogeography* 23 (2014): 726–43.

21 Juan Carlos Laso Bayas, Carsten Marohn, Gerd Dercon, Sonya Dewi, Hans Peter Piepho, Laxman Joshi, Meine van Noordwijk, and Georg Cadisch, "Influence of Coastal Vegetation on the 2004 Tsunami Wave Impact in West Aceh," *Proceedings of the National Academy of Sciences* 108 (2011): 18612–17.

22 Lee et al., "Ecological Role."

23 Feagin et al., "Shelter from the Storm?"

24 Ibid.

25 Cooper and McKenna, "Working"; Andrew Cooper and John McKenna, "Social Justice in Coastal Erosion Management: The Temporal and Spatial Dimensions," *Geoforum* 39 (2008): 294–306.

26 Often the interest in tourism is to be expected; after all, coastlines and mountains are not only especially disaster-prone but also good places for a vacation. But the shore or the mountain peak is not always at the heart of the tourism quest; the Porong and Greensburg disasters demonstrate that. See Chapters 5 and 13.

27 All of these problems and more are summarized by the UN Environment Program, "Impacts of Tourism," unep.org/resourceefficiency/Business/SectoralActivities/Tourism/FactsandFiguresaboutTourism/ImpactsofTourism/tabid/78774/Default.aspx.

28 Zhiwei Wu, Jianping Li, Zhihong Jiang, and Jinhai He, "Predictable Climate Dynamics of Abnormal East Asian Winter Monsoon: Once-in-a-Century Snowstorms in 2007/2008 Winter," *Climate Dynamics* 37 (2010): 1661–69.

29 Benzhi Zhou, Lianhong Gu, Yihui Ding, Lan Shao, Zhongmin Wu, Xiaosheng Yang, Changzhu Li, et al., "The Great 2008 Chinese Ice Storm: Its Socioeconomic–Ecological Impact and Sustainability Lessons Learned," *Bulletin of the American Meteorological Society* 92 (2011): 47–60; Guizhen He, Yonglong Lu, and Lei Zhang, "Risk Management: Lessons Learned from the Snow Crisis in China," *China Environment Series* 10 (2008/2009): 143–50; Kron, "Coasts."

30 Zhou et al., "Great 2008."

31 Ying Sun, Lianhong Gu, Robert E Dickinson and Benzhi Zhou, "Forest Greenness After the Massive 2008 Chinese Ice Storm: Integrated Effects of Natural Processes and Human Intervention," *Environmental Research Letters* 7 (2012): 35702–8.

32 K. Warner and A. Spiegel, "Climate Change and Emerging Markets: The Role of the Insurance Industry in Climate Risk Management," in *The Insurance Industry and Climate Change—Contribution to the Global Debate* (Geneva: Geneva Association, 2009), 83–96.

33 On the costliest events, see "Counting the Cost of Calamities," *The Economist*, January 14, 2012.

34 Harold Alderman and Trina Haque, "Countercyclical Safety Nets for the Poor and Vulnerable," Workshop on Food Price Risk Management in Low Income Countries, February 28–March 1, 2005.

35 Lauren Brooks, "The Caribbean Catastrophe Risk Insurance Facility: Parametric Insurance Payouts Without Proper Parameters," *Arizona Journal of Environmental Law and Policy* 2 (2012): 135.

36 Ibid.

37 Kevin Grove, "Agency, Affect, and the Immunological Politics of Disaster Resilience," *Environment and Planning D: Society and Space* 32 (2014): 240–56. Indeed, notes Grove, the CCRIF was created in response to the near-collapse of the government of Grenada after the island nation was devastated by Hurricane Ivan in 2004. After the storm—causing damages whose costs reached 200 percent of Grenada's GDP—the government had to operate for a time from a British ship offshore. Grenada did manage, slowly, to rebuild and did not become a failed state, but its near-miss shocked governments around the Caribbean and the world. The incident prompted the World

Bank to take seriously the idea that disasters have the potential to destabilize the region. And in the world of big finance that was an idea that would sell.

38 Ibid.

39 Ibid.

40 Anne Kuriakose, Rasmus Heltberg, William Wiseman, Cecilia Costella, Rachel Cipryk, and Sabine Cornelius, "Climate-Responsive Social Protection," *Development Policy Review* 31 (2013): 19–34; Caribbean Catastrophe Risk Insurance Facility, "Livelihood Protection Policy Launched January 22 2014 in Grenada," January 22, 2014, ccrif.org/news/livelihood-protection-policy-launched-january-22-2014-grenada.

41 Barry Barnett, Christopher Barrett, and Jerry Skees, "Poverty Traps and Index-Based Risk Transfer Products," *World Development* 36 (2008): 1766–85.

42 Kuriakose et al., "Climate-Responsive."

43 Ibid.

44 Robin Stott, "Contraction and Convergence: The Best Possible Solution to the Twin Problems of Climate Change and Inequity," *BMJ* 344 (2012): e1765.

45 Naomi Klein, *This Changes Everything: Capitalism vs. the Climate* (New York: Simon and Schuster, 2014), 409–13.

46 George Kent, "The Human Right to Disaster Mitigation and Relief," *Environmental Hazards* 3 (2001): 137–38; Robert Verchick, "Disaster Justice: The Geography of Human Capability, " *Duke Environmental Law and Policy Forum* 23 (2013): 23–71.

47 United Nations Framework Convention on Climate Change, "Adoption of the Paris Agreement: Draft Decision CP.21," December 12, 2015.

48 Jonathan Katz, *The Big Truck That Went By: How the World Came to Save Haiti and Left Behind a Disaster* (New York: Palgrave Macmillan, 2013), 112–13, 123–33, 203–8.

49 Jonathan Katz, "The Ugly Truth Lurking Behind the Climate Talks," *New Republic*, December 10, 2015.

50 Alex Pashley, "G7 Nods to Loss and Damage Claims with Climate Insurance Pledge," *RTCC*, September 3, 2015.

51 Jesse Ribot, "Vulnerability Does Not Fall from the Sky: Toward Multiscale, Pro-Poor Climate Policy," in *Social Dimensions of Climate Change: Equity and Vulnerability in a Warming World,* ed. Robin Mearns and Andrew Morton (Washington, DC: World Bank Publications, 2010), 47–74.

Chapter 12: Keeping the Lights On

1 Sharon Williams, "Hot Law: Montserrat Lawyers Twenty Years After Volcanic Eruption," West Indian Lawyers, September 14, 2015, westindianlawyers.com/news--articles/hot-law-montserrat-lawyers-twenty-years-after-volcanic-eruption.

2 Donaldson Romeo, interview with Stan Cox, Montserrat, January 31, 2015.

3 B.P. Kokelaar, "Setting, Chronology and Consequences of the Eruption of Soufrière Hills Volcano, Montserrat (1995–1999)," *Geological Society, London Memoirs* 21 (2002): 1–43.

4 Polly Pattullo, *Fire from the Mountain: The Tragedy of Montserrat and the Betrayal of its People* (Dominica: Papillote Press, 2012), 48, 58–59; Katharine Haynes, "Volcanic Island in Crisis: Investigating Environmental Uncertainty and the Complexities It Brings," *Australian Journal of Emergency Management* 21, no. 4 (2006): 21–27.

5   Mark Kurlansky, "Winter in the Sun: Lazy Days in Montserrat," *New York Times*, December 9, 1990.

6   Kokelaar, "Setting."

7   G.B. Wadge, R. Voight, S.J. Sparks, P.D. Cole, S.C. Loughlin, and R.E.A. Robertson, "An Overview of the Eruption of Soufrière Hills Volcano, Montserrat from 2000 to 2010," *Geological Society, London Memoirs* 39 (2014): 1–40.

8   Donaldson Romeo, interview with Stan Cox, Montserrat, January 31, 2015; Bennette Roach, "Montserrat on Pause," *Montserrat Reporter*, January 23, 2015.

9   Haynes, "Volcanic Island"; Katharine Haynes, Jenni Barclay, and Nick Pidgeon, "The Issue of Trust and Its Influence on Risk Communication During a Volcanic Crisis," *Bulletin of Volcanology* 70 (2007): 605–21; Kokelaar, "Setting"; Amy Donovan, Clive Oppenheimer, and Michael Bravo, "Contested Boundaries: Delineating the 'Safe Zone' on Montserrat," *Applied Geography* 35 (2012): 508–14; Amy Donovan and Clive Oppenheimer, "Science, Policy and Place in Volcanic Disasters: Insights from Montserrat," *Environmental Science and Policy* 39 (2014): 150–61.

10 Rod Stewart, interview with Stan Cox, Montserrat, January 29, 2015.

11 Donaldson Romeo, interview with Stan Cox, Montserrat, January 29, 2015.

12 Kokelaar, "Setting"; Willy Aspinall, former director of MVO, e-mail interview with Stan Cox, February 11, 2015.

13 Donovan, "Contested."

14 Pattullo, *Fire on the Mountain*, 21–27.

15 "UK Blamed over Volcano Deaths," BBC, January 12, 1999; "Inquest Verdict on 19 Deaths in Volcanic Eruption," *Montserrat Reporter*, December 31, 1998. Verdict excerpts archived at http://truthsetsusfree.blogspot.in.

16 Donaldson Romeo, interview with Stan Cox, Montserrat, January 31, 2015.

17 Pattullo, *Fire on the Mountain*, 5.

18 John Grattan, "Aspects of Armageddon: An Exploration of the Role of Volcanic Eruptions in Human History and Civilization," *Quaternary International* 151 (2006): 10–18.

19 Ilan Kelman and Tamsin Mather, "Living with Volcanoes: The Sustainable Livelihoods Approach for Volcano-Related Opportunities," *Journal of Volcanology and Geothermal Research* 172 (2008): 189–98.

20 We learned from a confidential source in Montserrat that UK government officials were in fact familiar with the report in the years between Hugo and the eruption.

21 International Development Committee, House of Commons, "First Report: Montserrat," November 18, 1997, http://www.publications.parliament.uk/pa/cm199798/cmselect/cmintdev/267i/id0102.htm.

22 Michael Jarvis, interview with Stan Cox, Montserrat, January 29, 2015.

23 Pattullo, *Fire on the Mountain*, 111–15.

24 Ibid., 116; Donaldson Romeo, interview with Stan Cox, Montserrat, January 31, 2015.

25 David Lea, interview with Stan Cox, Montserrat, January 31, 2015.

26 Pattullo, *Fire on the Mountain*, 117.

27 Warren Cassell Sr., interview with Stan Cox, Montserrat, January 28, 2015.

28 Pattullo, *Fire on the Mountain*, 93, 119.

29 Yvonne Weekes, *Volcano: A Memoir* (Leeds, UK: Peepal Tree, 2006), 24, 48–49.

30 Ibid., 94.

31 Ibid., 94–95.

32 International Development Committee, "First Report."

33 Pattullo, *Fire on the Mountain*, 108–9.

34 International Development Committee, "First Report."

35 "Inquest Verdict on 19 Deaths in Volcanic Eruption," *Montserrat Reporter*, December 31, 1998.

36 Janeen Lester, interview with Stan Cox, Montserrat, January 28, 2015.

37 Rod Stewart, interview with Stan Cox, Montserrat, January 28, 2015.

38 Ibid.

39 Anthony Carrigan, "(Eco)Catastrophe, Reconstruction, and Representation: Montserrat and the Limits of Sustainability," Special Issue: Postcolonial Islands, *New Literatures Review* 47–48 (2011): 111–28.

40 Nerissa Greenaway, former director of information and communications, e-mail exchange via Michael Jarvis, February 2015.

41 Rolston Johnson, interview with Stan Cox, Montserrat, January 30, 2015.

42 Carrigan, "(Eco)Catastrophe"; Polly Pattullo, "After the Volcano," *The Guardian*, July 18, 2005.

43 Rolston Patterson, interview with Stan Cox, Montserrat, January 30, 2015.

44 "Annual Review: Unlocking the Geothermal Potential of Montserrat," document dated October 2014, provided to Stan Cox by Allan Clarkin, UK Department for International Development, on January 20, 2015, archived at iati.dfid.gov.uk/iati_doc uments/4692469.docx.

45 Donaldson Romeo, interview with Stan Cox, Montserrat, January 31, 2015.

46 David Lea, e-mail interview with Stan Cox, February 13, 2015.

Chapter 13: "We Do Things Big Here"

1 Ruth Ann Wedel, interview with Stan Cox, Greensburg, May 10, 2014.

2 The authors watched radar images of the storm's frightening progress that night on TV in Stan's basement in Salina, Kansas. Like many in the state, we were under warnings associated with other tornadoes for much of the evening.

3 Thomas Fox, *Green Town U.S.A.: The Handbook for America's Sustainable Future* (Hobart, NY: Hatherleigh, 2013), 174.

4 Bob Dixson, interview with Stan Cox, Greensburg, May 10, 2014.

5 Keith Schneider, "After a Tornado, a Kansas Town Rebuilds Green," *New York Times*, September 23, 2009.

6 Green was induced to move his stagecoach line's headquarters to the town in the 1880s and was rewarded by naming the town after him. That business, along with a post office poached from a nearby town, cemented Greensburg's claim to become the seat of Kiowa County. Fox, *Green Town U.S.A.*, 101–3.

7 Naomi Klein, *The Shock Doctrine: The Rise of Disaster Capitalism* (New York: Metropolitan, 2007), 3–6. Klein notes that after Katrina, "the Louisiana State Legislature in Baton Rouge had been crawling with corporate lobbyists helping to lock in those big opportunities: lower taxes, fewer regulations, cheaper workers and a 'smaller, safer city'—which in practice meant plans to level the public housing projects and replace

them with condos. Hearing all the talk of 'fresh starts' and 'clean sheets,' you could almost forget the toxic stew of rubble, chemical outflows and human remains just a few miles down the highway."

8 Ibid.; Fox, *Green Town U.S.A.*, 130.

9 Fox, *Green Town U.S.A.*, 32.

10 Robert Berkebile and Stephen Hardy, "Moving Beyond Recovery: Sustainability in Rural America," *National Civic Review* 99 (2010): 36–40.

11 Fox, *Green Town U.S.A.*, 26.

12 Stacey Swearingen White, "Out of the Rubble and Towards a Sustainable Future: The 'Greening' of Greensburg, Kansas," *Sustainability* 2 (2010): 2302–19.

13 City of Greensburg, Kansas, "Resolution No. 2007-17: Pertaining to LEED Building Standards," December 17, 2007, usgbc.org/ShowFile.aspx?DocumentID=691.

14 Fox, *Green Town U.S.A.*, 38–40.

15 Berkebile and Hardy, "Moving Beyond Recovery."

16 Ruth Ann Wedel, interview with Stan Cox, Greensburg, May 10, 2014.

17 Bimal Kanti Paul and Deborah Che, "Opportunities and Challenges in Rebuilding Tornado-Impacted Greensburg, Kansas as 'Stronger, Better, and Greener,' " *GeoJournal* 76 (2011): 93–108.

18 Ibid. Those who remained in Greensburg faced other frustrations: "Conversations with current and former residents of Greensburg revealed that a considerable proportion of them wanted to rebuild within a few weeks after the tornado, but felt zoning issues and the need for building permits slowed progress. Many of them believed the city seemed more interested in rezoning for industrial and business enterprises than for residential development. New easements and strict new zoning regulations established by city officials prohibited some former residents of Greensburg from rebuilding where their homes had once stood. For example, due to the widening of Highway 54, which runs through northern part of the town, several Greensburg residents were not able to rebuild their houses where they had been located before the tornado. Some former and current long-time residents of Greensburg maintain that they could not recall their hometown ever enforcing such strict zoning regulations nor had they seen so much change at one time."

19 Paul and Che, "Opportunities"; Bob Dixson, interview with Stan Cox, Greensburg, May 10, 2014. Some former renters were able to take advantage of a program for low-income first-time home buyers that provided $25,000 toward a down payment on a new house. Programs run by the U.S. Department of Agriculture, the Mennonite Church, Habitat for Humanity, and other groups brought down building costs by using volunteer and homeowner labor.

20 Fox, *Green Town U.S.A.*, 47, 51, 54, 64, 97, 130–35.

21 Bob Dixson, interview with Stan Cox, Greensburg, May 10, 2014.

22 Jana Schwartz, interview with Stan Cox, Greensburg, May 10, 2014.

23 N.S. Diffenbaugh, R.J. Trapp, and H. Brooks, "Does Global Warming Influence Tornado Activity?," *Eos* 89 (2008): 553–54.

24 White, "Out of the Rubble."

25 Matthew Wald, "Citing Global Warming, Kansas Denies Plant Permit," *New York Times*, October 20, 2007.

26 Council on Environmental Quality, "State, Local, and Tribal Leaders Task Force on Climate Preparedness and Resilience," n.d., whitehouse.gov/administration/eop /ceq/initiatives/resilience/taskforce.

27 Fox, *Green Town U.S.A.*, 29–30.

28 Stacy Barnes, interview with Stan Cox, Greensburg, May 10, 2014.

29 White, "Out of the Rubble," emphasis White's.

30 Ruth Ann Wedel, interview with Stan Cox, Greensburg, May 10, 2014.

31 Jana Schwartz, "A Visit to Greensburg Will Help in the Philippines," Greensburg GreenTown, February 26, 2014, greensburggreentown.org.

32 Bimal Paul, telephone interview with Stan Cox, May 23, 2014.

33 White, "Out of the Rubble."

34 Amy Bickel, "Dust 'Bowl' Blows Across Western Kansas," *Kansas Agland*, April 30, 2014.

35 Aiguo Dai, "Increasing Drought Under Global Warming in Observations and Models," *Nature Climate Change* 3 (2013): 52–58; Dim Coumou and Stefan Rahmstorf, "A Decade of Weather Extremes," *Nature Climate Change* 2 (2012): 491–96; Diffenbaugh et al., "Does Global."

36 Fox, *Green Town U.S.A.*, 159.

37 Bimal Paul, telephone interview with Stan Cox, May 23, 2014.

38 Deborah Popper and Frank Popper, "The Great Plains: From Dust to Dust," *Planning* 53 (1987): 12–18.

39 Stan Cox, "The Folly of Turning Water into Fuel," *AlterNet*, March 21, 2008.

40 Donald Worster, *Dust Bowl: The Southern Plains in the 1930s* (New York: Oxford University Press, 1979).

41 In a 2012 Saybrook University Ph.D. dissertation, L. Shuli Rose Goodman compared Greensburg with Parkersburg, Iowa, which was hit by a tornado the year after Greensburg. Being much more affluent on average than Greensburgers, "the people of Parkersburg seemed to have had a very particular idea about what they were willing to accept or not to accept. It seems that very early on in the recovery, there was a decision to return to the identical footprint of their previous world, as quickly as possible. As [resident] Chris says, they rebuilt in the same place, only bigger." L. Shuli Rose Goodman, "Organizational and Community Experiences of Transformation After a Catastrophic Event," Ph.D. diss., Saybrook University, 2012, gradworks.umi .com/34/90/3490705.html.

42 Mike McGraw, "Housing Troubles Mount, Especially for Joplin's Poor," *Kansas City Star*, December 17, 2011.

43 The Joplin tragedy was unusual for another reason: the infection of seventeen people by a "flesh-eating" fungus. *Apophysomyces trapeziformis* normally lives quietly in the soil and almost never infects people, but the storm sucked up spores and spread them through the city. All of those infected had been in the zone of maximum destruction, scattered randomly along a three-mile path, and all had multiple deep wounds from flying debris. Researchers concluded that the tornado picked up dirty water with fungal spores in it soon after touching down and carried the water "swirling around in it" all the way through town. Five of the victims died from the infection. Robyn Neblett Fanfair, Kaitlin Benedict, John Bos, Sarah D. Bennett, Yi-Chun Lo, Tolu Adebanjo, Kizee Etienne, et al., "Necrotizing Cutaneous Mucormycosis After a Tornado in Joplin,

Missouri, in 2011," *New England Journal of Medicine* 367 (2012): 2214–25; Wally Kennedy, "Patients Continue Fight with Rare Fungus Contracted During Tornado," *Joplin Globe*, February 2, 2013.

44 Daniel Sutter and Kevin Simmons, "Tornado Fatalities and Mobile Homes in the United States," *Natural Hazards* 53 (2010): 125–37; Bimal Kanti Paul and Mitchel Stimers, "Exploring Probable Reasons for Record Fatalities: The Case of 2011 Joplin, Missouri, Tornado," *Natural Hazards* 64 (2012): 1511–26. Nationwide, low-income families are more likely than middle- or high-income ones to live in mobile homes, which are notorious for their association with high death rates in tornadoes. From 1985 to 2007, more than 43 percent of tornado deaths occurred in mobile homes, even though they make up less than 8 percent of U.S. housing stock. Mobile homes are more dangerous than fixed houses or apartments only in the event of weaker EF-1 through EF-3–rated storms, however; EF-4 and EF-5 tornadoes, which are capable of destroying almost any building, pose a grave threat to everyone in their path, without differentiating between types of housing. In fact, despite the fact that many of the victims in Greensburg and Joplin were low-income, not a single death in either EF-5 tornado happened in a mobile home.

45 McGraw, "Housing Troubles."

46 Andrew Whitehead, chair, Joplin GreenTown, interview with Stan Cox, Joplin, August 23, 2014, and e-mail interview with Stan Cox, February 18, 2015.

47 Ibid.; McGraw, "Housing Troubles."

48 McGraw, "Housing Troubles."

49 Joplin Area Citizens Advisory Recovery Team, "Vision and Goals," joplinareacart.com/vision-and-goals.

50 Bob Dixson, interview with Stan Cox, Greensburg, May 10, 2014.

51 Barack Obama, "Remarks by the President at the Joplin High School Commencement," Joplin, MO, May 21, 2012, whitehouse.gov/the-press-office/2012/05/21/remarks-president-joplin-high-school-commencement.

Chapter 14: When Mountains Fall

1 Ravi Chopra, "The Untold Story from Uttarakhand," *The Hindu*, June 25, 2013.

2 Nandini, interview with the authors, Guptkashi, January 28, 2014. Translation by Adarsh Tribal and Priti Gulati Cox.

3 S.P. Sati and V.K. Gahalaut, "The Fury of the Floods in the North-West Himalayan Region: The Kedarnath Tragedy," *Geomatics, Natural Hazards and Risk* 4 (2013): 193–201.

4 Alan Ziegler, Robert J. Wasson, Alok Bhardwaj, Yas P. Sundriyal, S.P. Sati, Navin Juyal, Vinod Nautiyal, Pradeep Srivastava, Jamie Gillen, and Udisha Saklani, "Pilgrims, Progress, and the Political Economy of Disaster Preparedness—The Example of the 2013 Uttarakhand Flood and Kedarnath Disaster," *Hydrological Processes* 28 (2014): 5985–90.

5 D. P. Dobhal, Anil K. Gupta, Manish Mehta, and D. D. Khandelwal, "Kedarnath Disaster: Facts and Plausible Causes," *Current Science* 105 (2013): 171–74; Sati and V.K. Gahalaut, "Fury".

6 Adarsh Tribal, interview with the authors, Guptkashi, January 27, 2014.

7 In India, what Americans refer to as the first and second floors are called the ground floor and first floor, respectively. So the first floor in Indian terms had become the American version of a first floor.

8 M. Balasubramanian and P.J. Dilip Kumar, "Climate Change, Uttarakhand, and the World Bank's Message," *Economic and Political Weekly* 49 (2013): 65–68.

9 Amateur Seismic Center, "Earthquakes in Uttarakhand, India," May 6, 2014, asc-india .org/seismi/seis-uttaranchal.htm.

10 Radhika Nagrath, "Nepal Quake Haunts Uttarakhand's Char Dham Revival Plans," *Hindustan Times*, April 30, 2015.

11 Subhajyoti Das, "Uttarakhand Tragedy," *Journal of the Geological Society of India* 82 (2013): 201.

12 Patrick Barnard, Lewis Owen, Milap Sharma, and Robert Finkel, "Natural and Human-Induced Landsliding in the Garhwal Himalaya of Northern India," *Geomorphology* 40 (2001): 21–35.

13 Vineet Gahalaut, interview with Stan Cox, Hyderabad, India, September 8, 2014.

14 Richard Marston, telephone interview with Stan Cox, April 30, 2014; Richard Marston, "Land, Life, and Environmental Change in Mountains," *Annals of the Association of American Geographers* 98 (2008): 507–20; R.A. Marston, M.M. Miller, and L.P. Devkota, "Geoecology and Mass Movement in the Manaslu-Ganesh and Langtang-Jugal Himals, Nepal," *Geomorphology* 26 (1998): 139–50; Patrick Barnard, Lewis Owen, Milap Sharma, and Robert Finkel, "Natural and Human-Induced Landsliding in the Garhwal Himalaya of Northern India," *Geomorphology* 40 (2001): 21–35.

15 Balasubramanian and Kumar, "Climate Change."

16 Anupam Chakravartty, "Tourism in Uttarakhand Needs Regulation," *Down to Earth*, April 13, 2014.

17 Chandra Prakash Kala, "Deluge, Disaster and Development in Uttarakhand Himalayan Region of India: Challenges and Lessons for Disaster Management," *International Journal of Disaster Risk Reduction* 8 (2014): 143–52.

18 Nilanjana Bhowmick, "Devastating North India Floods Likely Worsened by Tourist Boom," *Time*, June 21, 2013.

19 Ziegler et al., "Pilgrims."

20 Not to be confused with the better-known Srinagar that is the capital city of the state of Jammu and Kashmir—and that was totally inundated during an unrelated flooding disaster the following year that killed more than five hundred people.

21 S.P. Sati, e-mail interview with Stan Cox, April 15, 2014.

22 Anil Gupta, Swati Singh, and Sreeja Nair, "Hydrometeorological Hazards in Uttarakhand India, Himalaya—Forensic Assessment of 2013 Flash Flood Disaster: Need of Integrated Planning for Sustainable Development," *International Journal of Geography and Environment Sciences* 1 (2014): 9–20.

23 Maharaj Pandit and R. Edward Grumbine, "Potential Effects of Ongoing and Proposed Hydropower Development on Terrestrial Biological Diversity in the Indian Himalaya," *Conservation Biology* 26 (2012): 1061–71.

24 Naresh Rana, S.P. Sati, Y.P. Sundriyal, Madan Mohan Doval, and Navin Juyal, "Socio-Economic and Environmental Implications of the Hydroelectric Projects in Uttarakhand Himalaya, India," *Journal of Mountain Science* 4 (2007): 344–53.

25 Pandit and Grumbine, "Potential Effects."

26 Vineet Gahalaut, e-mail interviews with Stan Cox, April 12, 2014, and October 14, 2015.

27 It was with this epochal reference point that the media widely referred to the Uttarakhand cataclysm as the "Himalayan Tsunami." See, for example, *Himalayan Tsunami*, Discovery Channel India, December 23, 2013.

28 Ravi Chopra, "The Untold Story from Uttarakhand," *The Hindu*, June 25, 2013.

29 Local resident Gudi Devi, interview with authors, Guptkashi, January 29, 2014; translated by Priti Gulati Cox.

30 Adarsh Tribal, interview with the authors, Guptkashi, January 28, 2014.

31 Chopra, "Untold Story."

32 Chakravartty, "Tourism in Uttarakhand."

33 Nandini, interview with the authors, Guptkashi, January 28, 2014. Translation by Adarsh Tribal and Priti Gulati Cox.

34 M.N. Parth, "After India's 2013 Floods, Key Hindu Pilgrimage Towns Deserted," *Los Angeles Times*, October 31, 2014; Kavita Upadhyay, "Roadmap for Kedarnath Reconstruction Approved," *The Hindu*, October 21, 2014.

35 "Rains Once Again Become Bane for Uttarakhand," Uttarakhand News Network, June 26, 2015.

36 Abhishek Angad, "After the Flood: Kedarnath, a Ghost Town," *Indian Express*, November 15, 2015.

Epilogue: Rainbow of Chaos

1 Russell Banks, *Continental Drift* (New York: HarperCollins, 1985), 46.

2 Harjeet Singh, "Climate Politics Waters Down Ambition of UN Disaster Risk Deal," RTCC [Responding to Climate Change], March 17, 2015. rtcc.org/2015/03/16/climate-politics-waters-down-ambition-of-un-disaster-risk-deal.

3 UNISDR, Twitter post, March 18, 2015, 12:55 p.m., http://twitter.com/unisdr.

4 The Hyogo Framework's crucial but often ignored Priority 4, "Reduce the underlying risk factors," all but disappeared in the Sendai Framework. It was downgraded to a single piece of advice under "guiding principles," section III, item (j): "Addressing underlying disaster risk factors through disaster risk-informed public and private investments is more cost-effective than primary reliance on post-disaster response and recovery, and contributes to sustainable development." *Proceedings: Third UN World Conference on Disaster Risk Reduction* (New York: United Nations, 2015).

5 Brad Evans and Julian Reid, "Dangerously Exposed: The Life and Death of the Resilient Subject," *Resilience* 1, no. 2 (2013): 83–98.

6 Brad Evans and Julian Reid, "Exhausted by Resilience: Response to the Commentaries," *Resilience: International Policies, Practices and Discourses* 3, no. 2 (2015), 4. However, it may turn out that the subject is not so easy to escape even for these writers. Julian Reid soon went on to co-author a book with *Resilience* editor David Chandler, titled *The Neoliberal Subject: Resilience, Adaptation and Vulnerability* (London: Rowman and Littlefield, 2016).

7 Lance H. Gunderson and C.S. Holling, *Panarchy: Understanding Transformations in*

Human and Natural Systems (Washington, DC: Island Press, 2002), 75, 405. Holling calls these nested cycles a "panarchy," after the Greek god Pan: "the all-pervasive, spiritual power of nature. In addition to a creative role, Pan could have a destabilizing, creatively destructive role that is reflected in the word *panic*" (ibid., 74).

8 Mark Pelling, interview with Paul Cox, Geneva, May 19, 2013; Mark Pelling and David Manuel-Navarrete, "From Resilience to Transformation: The Adaptive Cycle in Two Mexican Urban Centers," *Ecology and Society* 16, no. 2 (2011), 11. In IPCC, *Climate Change 2014*, transformative adaptation appears on page 29 of the Summary for Policymakers and in chapter 8, "Urban Areas." The IPCC definition is the following: "Transformation: A change in the fundamental attributes of natural and human systems. Within this summary, transformation could reflect strengthened, altered, or aligned paradigms, goals, or values towards promoting adaptation for sustainable development, including poverty reduction."

9 Kevin Grove, "Hidden Transcripts of Resilience: Power and Politics in Jamaican Disaster Management," *Resilience* 1, no. 3 (2013): 193–209.

10 John M. Anderies, Carl Folke, Brian Walker, and Elinor Ostrom, "Aligning Key Concepts for Global Change Policy: Robustness, Resilience, and Sustainability," *Ecology and Society* 18, no. 2 (2013): 8.

11 Steve Carpenter, Brian Walker, J. Marty Anderies, and Nick Abel, "From Metaphor to Measurement: Resilience of What to What?," *Ecosystems* 4, no. 8 (2001): 765–81.

12 Zygmunt Bauman, *Liquid Modernity* (Cambridge: Polity Press, 2000), 11.

13 Ibid., 33.

14 As originally defined in the "2009 UNISDR Terminology on Disaster Risk Reduction," United Nations Office for Disaster Risk Reduction, May 2009, unisdr.org/we/inform/terminology.

15 IPCC, *Climate Change 2014: Impacts, Adaptation, and Vulnerability. Contribution of Working Group II to the Fifth Assessment Report of the Intergovernmental Panel on Climate Change* (Cambridge: Cambridge University Press, 2014), 5. In sociology and anthropology these concepts are indelibly tainted by associations with so-called structural functionalism, an approach that guided the social sciences into the twentieth century but became fiercely contested in the 1960s. It was almost wholly discarded as a questionable—and generally colonial—means of describing a society in terms of a complex system of interlocking functional parts, working together to maintain equilibrium. Functionalism's apparent revival in resilience literature seems to suffer the same lack of attention to conflict, agency, and change within societies.

INDEX

Abbott, Tony, 28–29, 331n42, 331n48
Ablon, Antonio (bishop in Philippines), 56
"adaptation machine" (Grove), 11
Adorno, Theodor, 143
aerosols of sulfuric acid, 211–12
    problems of, 212–13
AIR Studios, 261
Albala-Bertrand, J.M.
    The Political Economy of Large Natural Disasters, 89
Alexander, Thomas, 169
Altis, Elegio, 61
Ancash earthquakes, in Peru, 117–18
Angus, Ian, 108–9
Aquino, Benigno, III, 43, 47–49, 51, 53, 60–61, 104
Artigas, Francisco, 343n50
Auf der Heide, Erik, 111
Australia
    "Angry Summer," 27–28
    bunkers in, 332n54
    bushfire volunteers in, 23–24
    carbon tax and, 28
    gendered bush culture in, 23
    leave-or-stay fire policy in, 22–23
    prescribed burning in, 330n30
    wildfire management in, 24
Australian fire deaths, demographic history of, 22–23
Australian landscapes
    fire and resilience in, 16–17
    fire regime in, 17
Avouac, Jean-Philippe, 158

bahala na, 44–45, 48, 53
Bakker, Jorge, 208
Bakrie, Aburizal, 133–34, 136–37, 139
Bakrie, Achmad, 133
Bakrie, Nirwana, 136
Bakrie Group, 132–33

Balasubramanian, M., 306
Balikatan (joint U.S.-Philippines military exercises), 56
Balsa Mindanao (disaster relief mobilization in Philippines), 55
Bangladesh, disaster preparedness in, 238–40
Bankoff, Greg, 43–44, 53
    Cultures of Disaster, 38, 43
Banks, Russell, 317
Bannink, Guy, 30
Barangay Andap, Philippines, 42–43, 46, 48, 54
Barberi, Franco, 166, 168
Barnes, Stacy, 288–89
Barnett, Gary, 66, 83–84
Barog Katawhan (People Rise Up), in Philippines, 55
Bastiat, Fréderic, 91–92
Batubara, Bosman, 137–39
Bauman, Zygmunt, 207, 321
Baumohl, Bernard, 91
Bautista, Momerto, 58
bayanihan, 44, 46–48, 50, 53, 61, 62
Beck, Ulrich, 325n8
Beckham, David, 63
Bennett, Hugh Hammond, 103–4
Berkebile, Robert, 281, 283
Berliner, Joseph, 94
Berlusconi, Silvio, 165
Berry, Wendell, 142
Bieber, Justin, 63
Biggert-Waters Act, 215
Billi, Marco, 167, 168
Bimal, Paul, 290–91
bioshields, 243–46
Birkmann, Joern, 89
Biscayne Aquifer, 182
Biscayne Bay, 184
Bjarke Ingels Group (BIG), 82–83, 343n40

Black Dragon Fire, 332n57
"Black Saturday" fires, in Victoria,
    Australia, 22
Bloomberg, Michael, 66–67, 73, 75–76,
    79, 84–85
    National Guard and, 350n91
    "tough" vision of, 80–81
Blue Mountains (Australia), 22, 115
    burning of, 15, 28
Boatswain, Rueben, 266
Bobbing Forest (Rotterdam), 208
Bode, Barbara, 117
Boer, Florian, 203
Bohol, Philippines, earthquake,
    58–59
Bolin, Bob, 190
Bonds, Barry, 146
boreal forest expansion, 334n76
Bor Forest Island Experiment, 35
Boston Consulting Group, 75
Bowman, David, 25–26, 182
    Fire on Earth, 29
BPLS. *See* Sidoarjo Mud Mitigation
    Agency
Brady, Brian, 170–71
Bramble, William, 260
Brandt, David, 268–69
Bremby, Rod, 286
Brice, William, 101
Briceño, Henry, 183, 196
Bridenstine, Jim, 107
Brink, Willington, 102
Brook, Daniel, 232
Brooks, Lauren, 249
Brown, Michael, 27
Browne, Benjamin Joseph, 266
Buck, Pearl S., 88
Buffalo Commons, 292
Buffett, Jimmy, 366n28
    Volcano, 261
Buhay, Judith, 28–29, 41–43, 63
"build back better," 101
Build It Back (NYC recovery program),
    75–76, 80–81, 85
Bulling, Sandra, 90
Burriss, Rhys, 271
Bush, George H.W., 106
Bush, George W., 104, 106
Bush, Jeb, 105

Cage, Jane, 295
Cameron, James
    Titanic, 234
capability, concept of, 344n10
Capasso, Olga, 173
Caprotti, Federico, 210
Carbado, Devon, 112
Caribbean, disaster insurance and,
    249
Caribbean Catastrophe Risk Insurance
    Facility (CCRIF), 249–50
Carrigan, Anthony, 273
Carter, Joe, 114
CASE program, in Italy, 172
Cassell, Warren, 269, 276
Cassell-Sealy, Roselyn, 273
catastrophe bonds, 217–18
    bundling of, 218
    risk attraction and, 219
"catastrophism," 108–9
Cézanne, Paul, 143–44, 319
Char Dham pilgrimage, 298
Char Dham region, India
    earthquakes and, 311
    flooding in, 300–307
    lack of aid to, 305–6
Chatto, Edgar, 59
chawl, 373n17
China
    tree plantations, problems of, 248
    Tianjin Eco-City, 209–11
    2008 earthquake, and Zipingpu dam,
        362n72
    2008 ice storm, 247–48
Chipko Andolan movement, 314
Chopra, Ravi, 298, 313
Chorabari Lake, 300–301, 307
Christie, Chris, 104, 114–15, 296
A City Laid Waste, 103
Clarke, Lee, 73
climate change
    capitalism and, 8
    coastal areas and, 149–50
    cyclones and, 145–46
    economic growth and, 7–8
    fire seasons and, 147–48
    flooding and, 149–50
    hurricanes and, 150
    poverty and, 237

property insurance and, 214–15
reinsurance industry and, 216
taiga and, 36–37
tornadoes and, 146–47
wealthy economies and, 7
climate justice, 9
Clinton, Bill, 106
Cole, Tom, 107
Coloma, Herminio, 61
Columbus, Christopher, 260
communism, in everyday life, 120
"communitas" (Turner), 116–19
demystification of, 120–21
labor and, 121
resilience and, 120
social policy and, 119–20
community functions, questions about,
    321–23
Comprehensive Disaster Management
    Plan (CDMP), of Bangladesh,
    239
conservation of fragility, 13–14
Conway, Erik, 6
Cooper, Andrew, 246
COP18 (18th Conference of the Parties
    to the 1992 UN Framework
    Convention on Climate Change),
    45
COP19 (19th Conference of the Parties
    to the 1992 UN Framework
    Convention on Climate Change),
    62
Cox, Lucille Brewer, v, 348n51
Cox, Tom, 348n51
"creative destruction," 93
Crutzen, Paul, 211
Cruz, Henry, 13
Cruz, Neal H., 48
Cuba, hurricane preparedness in,
    240–42
Cuomo, Andrew, 85
Cyclone Pam, 317
Cyclone Phailin, 104
Cyclone Preparedness Programme (CPP),
    of Bangladesh, 239
cyclones, climate change and, 145–46

Daipan, Dominga, 47
Daly, Herman, 8

dams
in Char Dham region, Uttarakhand,
    299–300, 303, 307, 312
in China, 159
landslides and, 153, 158, 310
river deltas and, 150
Vaiont disaster, 152–53
Zipingpu, 152, 362n72
David, Randy, 49
Davies, Richard, 130, 353n16, 354n18,
    354n19, 355n4
Davis, Mike, 232–33
    Planet of Slums, 226, 232
De Bernardinis, Bernardo, 166, 168, 173,
    174
de Blasio, Bill, 76, 80–81
Delta Works, 201–2, 206
De Palma, Celeste, 187–88, 192
Department for International
    Development (DFID), of UK,
    263–64, 268, 275
Department of Science and Technology
    (DDST), of Philippines, 51
De Urbanisten, 203
Dharavi slum (Mumbai), 229–32
DiCaprio, Leonardo, 285
Dill, Katherine, 99
Dixson, Bob, 285–88, 290, 295
disaster anthropology, 117–18
disaster insurance, ideas for
international, 252–54
universal (for U.S.), 219–23,
    252–53
disaster investors, 217, 372n64
disaster justice, 9, 251, 327n21
disaster preparedness, as cost effective,
    106
disaster prevention, ecological approaches
    to, 243
disaster protection, value of government
    role in, 219–20
disaster readiness course in high schools,
    335n16
disaster relief aid
elections and, 105–6
government spending and, 249
political incentives and, 106–7
Disaster Relief Appropriations Act of 2013,
    107

disasters
  abandoning of emergency plans in,
    73
  blame and, 174
  causal stories of, 122
  downward spirals and, 95
  Dutch experts and, 201–8
  ecological degradations and, 99
  economic growth and, 247
  elections and, 105
  government spending and, 249
  human influence on, 144–45
  human trauma and, 99–100
  inequitable growth and, 99
  insurance and
    social limits of, 219
    universal, 219–23, 252–53
  measuring impact of, 90
  military action and, 110–13
  mutual aid in, 113–15
  opportunities and, 89
  policy lags and, 198
  poverty traps and, 249, 251,
    255–56
  prevention efforts and, drawbacks of,
    199–200
  racial framing and, 112
  risk reduction and, 198
  "silver linings" view of, 90–94, 102–3,
    105
    critique of, 94–95, 97
  socioeconomics and, 97–98
  tourism and, 246–47
  wealth distribution and, 94–95
  See also natural disasters
Disbursement Acceleration Program
    (DAP), of Philippines, 53
disutility of labor, 328n35
Doctoroff, Daniel, 79
Dolce, Mauro, 164–65, 173
Down to Earth, 309
Dresch, Patricia, 78
Drexler, Eric, 352n10
drought, 325n7
Dry Line plan (NYC), 82–84
Dunalley, Australia
  2013 fire, 26
DuPont, William, 95–96
Durkheim, Émile, 117

earthquakes, 4, 6, 7, 91, 101, 122, 145, 151,
    155, 164, 221, 222, 249, 321
  in Baja California, 157
  Char Dham region, India, 311
  damming of rivers and, 158–59
  in El Salvador, 154
  geothermal energy and, 157
  Great East Japan earthquake of 2011, 96,
    99, 110, 170, 317
  groundwater extraction and, 158
  in Haiti, 112–13, 254
  in the Himalaya, 307–8, 311
  human causes of, 155–59
  human structures and, 3, 226
  Indonesia and, 123, 128–30, 138,
    274
  in Kobe, Japan, 95
  in L'Aquila, Italy, 159–63, 166, 172, 259,
    322
  in Lisbon, 100–101
  in Lorca, Spain, 158
  Lusi mud volcano and, 129–32, 138,
    354n18, 355n21
  in Mexico City, 115
  in Nepal, 110, 153, 253, 307, 311, 315
  in Oklahoma, 154–57
  in Peru, 117
  in Philippines, 40, 45, 58–59, 61, 91
  preventing injury from, 171–72
  problems in predicting, 161, 165–66,
    168–71
  at Traverse Mountains, Utah, 153
  tsunami of 2004 and, 244
  in Uttarakhand, 307–8, 311, 315
  wastewater injection and, 155–57
  Wenchuan, 158–59, 247
eco-cities
  as affluent enclaves, 210
  in China, 210–11
economic growth, ecological degradation
    and, 99
ecophagy, 353n10
ecotourism, 247
18th Conference of the Parties to the 1992
    UN Framework Convention on
    Climate Change (COP18), 45
enjoyment, as fundamental to economics,
    328n35
El Salvador, landslides in, 154

Ensor, Bradley, 225
Ensor, Marisa, 225
European Union Solidarity Fund (EUSF), 172–73
Evans, Brad, 319–20
 Resilient Life, 10, 319
Everglades, 180, 182–84
exotic trees in coastal areas, negative impacts of, 245

Fallin, Mary, 157
Fan Xiao, 159, 362n72
Feagin, Rusty, 244–45
federal disaster relief spending, and votes, 348n61
FEMA (Federal Emergency Management Agency), 70, 77, 282
Feuer, Alan, 69
Figueres, Christina, 28
fire, ecology of, 29–30
fire seasons, warming climate and, 147–48
FiveThirtyEight (website), 6
floating cities idea, 208
floating city app, 208
Floating Pavilion, 202–5, 207
Florida, geologic history of, 179
Foster, John Bellamy, 8
Fox, Thomas
 Green Town U.S.A., 279, 291
Framework Convention on Climate Change, 252
Francis (pope), 63, 238
"fracking," 155
Freud, Sigmund, 78
Fukushima crisis, 96
Furyaev, Valentin V., 35, 333n69

Gahalaut, Vineet, 300, 311
Gainesville, Georgia, 1936 tornado, 101–2
Gair, Brad, 341n3
Ganges River, 298
Garrett, Thomas, 106
Gasquet, Joachim, 143
Gaurikund, India, 305
 lack of aid to, 305–6
Gennep, Arnold van
 The Rites of Passage, 116

geoclimatic hazards, 4, 325n6
 insurance and, 214, 218, 249
"geoengineering," 370n41, 370n44
geoengineering projects, for greenhouse gases, 211–12
 problems of, 213
Georgescu-Roegen, Nicholas, 8
 The Entropy Law and the Economic Process, 328n35
Giddens, Anthony, 325n8
Gimenez, Carlos, 187
Giuliani, Gioacchino Giampaolo, 165, 170
Giuliani, Rudy, 67
Goldammer, Johann Georg, 35
Good Old Lower East Side (GOLES), 68–69, 83, 341n11
Gore, Al, 193, 331n42
gotong royong, 335n11
Graeber, David
 Debt, 120
"gray goo," 352n10
Great East Japan earthquake and tsunami of 2011, 96, 99, 110, 170, 317
Great Mangrove Rush, 244
Great Miami Hurricane, 180–81
Great Wall of Bohol, 59–60
Great Wall of Tacloban, 63
"Great Walls of America," 199
Green, Donald "Cannonball," 281
Greene, Charles, 79
Greensburg, Kansas, 256, 278–79, 290–91
 building back green, 279, 281–84, 289
 climate change issue and, 286–87
 LEED resolution of, 282
 obstacles to rebuilding, 283–84
 resilience and, 288
 2007 tornado, 279
 2008 financial crisis, 290
 zoning issues, 381n18
Greensburg GreenTown Organization, 281, 287–88
Grenada, 377n37
Grimm, Michael, 78, 104, 216
Grove, Kevin, 11, 250, 320
Grunwald, Michael
 The Swamp, 180
Guerra, Goldi, 68–71, 73, 77
Gunardi (farmer in Indonesia), 136–37

Haiti, 2010 earthquake, 112–13, 117, 254
Hallegatte, Stéphane, 95
Hamm, Harold, 156
Hammer, Nicole Hernandez, 183, 186, 188, 190, 192, 194
Hardy, Stephen, 281, 283
Harlem, Peter, 182–84, 194–95
Harris, Cheryl, 112
Hart, Catherine, 279, 281
Hawkes, Gay, 31
  Time and Chance, 26
Heimaey, Iceland, 200
Heinberg, Richard, 8
Henson, Bob, 147
Hernkind, Joe, 86
Hewitt, Steve, 279
Hobart, Australia, 1967 fire, 25
Hoffman, Susanna, 3, 118–19
Holling, Crawford Stanley, 18–20, 29, 89, 243, 319
Holyfield, Evander, 50
Honduras, mountain slope farming in, 225
Horkheimer, Max, 143
Horwich, George, 346n28
Howell, David, 24
Hubbard, Barbara Marx, 50
Huggel, Christian, 152
Hurricane Andrew, 181, 186, 190, 192
Hurricane Flora, 240
Hurricane Hugo, 261, 267
Hurricane Ivan, 377n37
Hurricane Katrina, 75, 78, 111–12, 146, 180–81, 201, 243, 284
Hurricane Luis, 262
Hurricane Mitch, 95, 225–26
  deforestation and, 226
Hurricane Patricia, 60, 146
hurricanes, and greenhouse warming, 146
Hurricane Sandy, 242. See also Superstorm Sandy
Hurricane Sandy Rebuilding Task Force, 82
Hyogo Framework (UN), 385n4

Iko, Madelena, 234–35
India. See Kedarnath Temple; Mumbai, India; Orissa, India; Sitapur, India
Indian Ocean, 2004 tsunami, 243–44

Indonesia. See Lusi mud volcano
Inhofe, Jim, 107
"innovation" (Schumpeter), 345n14
insurance industry
  climate risks and, 214–15
  disaster coverage, ideas for, 219–23, 252–54
Intergovernmental Panel on Climate Change (IPCC), 31, 237, 357n6, 357n10
Internal Displacement Monitoring Centre, 2
International Day for Disaster Reduction, 11
international disaster insurance program, ideas for, 252–54
Isaacs, Jennifer
  Australian Dreaming, 15
Italy. See L'Aquila, Italy
iVolunteer, 2, 313

Jackson, Tim, 8
Jackson, Wes, 8
Jacobs, Jane, 83
Japan
  nuclear reactor restarts in, 96–97
  2011 earthquake and tsunami, 96, 110
Jarvis, Michael, 268
Jencson, Linda, 118
Jimenez, Ramon, 50
Jocano, F. Landa, 44
Johnsen, Bill, 87
Johnson, Leigh, 218, 221
Joplin, Missouri
  flesh-eating fungus and, 382n43
  housing problems in, 294–95
  post-disaster goals, 295
  rebuilding in, 293–94, 296–97
  2011 tornado, 97, 292–93
Jordan, Thomas, 169, 171
Jose, Christina, 55
Joseph, Alicia, 266
Joseph, Alister, 266
Joseph I (king), 100

Kadam, Sachin, 224, 229
Kahuripan Nirwana Village (KNV), Indonesia, 136
Kalibeng Formation, 129

Kampala, Uganda
  drainage channel in, 235–36
  flooding and, 233–36
Kamwokya slum (Uganda), 233–35
  flooding in, 234–35
Kankwasa, Janet (Mama Allan), 235–36
Katz, Jonathan, 112–13, 254
Keating, Mary-Lou
  As the Smoke Clears, 15–17
Kedarnath Temple
  glacial dam melting at, 1
  flooding at, 300–301
  hazards of location, 307
  landslides at, 1, 301, 315
  tourism decrease, 315
Kennedy, John F., 88
Keranen, Katie, 155–57
Khumriyal, Ramala, 1–2, 151, 301
Kibaya, Allan Rogers, 224
Kidd, Daniel, 82–84
King, Tom, 350n91
Klein, Naomi, 8, 212, 371n51
  The Shock Doctrine, 92, 98, 281
  This Changes Everything, 108
Klinenberg, Eric, 341n3
Klose, Christian, 159
Klum, Heidi, 74
Kobe, Japan
  1995 earthquake and recovery, 95–96,
    346n28
Koreman, Kristian, 203
Kousky, Caroline, 221
Kovel, Joel, 8
Koven, Charles, 36
Krall, Lisi, 221
Krieger, Nancy, 123
Kron, Wolfgang, 149
Kropotkin, Peter, 114
Kukavskaya, Elena, 34, 37
Kuksin, Grigory, 333n72
Kumar, P.M. Dilip, 306
Kunreuther, Howard, 220

La Fortune, Jean de, 64, 343n50
landslides, 4, 8, 99, 145, 149, 151–52, 199,
    226, 243, 310–11
  in Baja California, 157
  in China, 248
  damming rivers and, 153, 158, 310

in El Salvador, 154
geothermal energy and, 157
in the Himalaya, 308
in Honduras, 226
human causes of, 145, 151–54
hydroelectric power and, 323
at Kedarnath Temple, 1, 301, 315
in Nepal, 153
in Peru, 117
in Philippines, 42, 45, 48, 57, 59, 61,
    148
physical defenses against, limits of,
    200
at Sainte-Victoire Mountain, 144
in Seattle, 153–54
at Sonprayag, India, 303–4
slums and, 226–27
in Switzerland, 157
at Traverse Mountains, Utah, 153
urban, 226
in Uttarakhand, India, 308–9, 311,
    315
Vaiont dam and, 153, 158
Lapindo gas drilling company, 123, 126,
    128, 132–33, 135–36
  slow to compensate victims, 134–39
L'Aquila, Italy, 246
  old city as ghost town, 163
  2009 earthquake, 159–63, 166, 172, 259,
    322
    probabilities of, 169
    problems of predicting, 161, 165–66,
      168–71
  reconstruction effort, 164, 172–73
  and CASE program, 172
L'Aquila Seven, 160, 165, 167–68,
    173–74
Lea, David, 268–69, 276
Le Breton, Bridie, 21
Le Breton, Christie, 21
Le Breton, Connor, 21
Le Breton, Phil, 21
Lechuga, Lucas, 175
Lester, Janeen, 272
Levine, Philip, 186, 193
Lim-Napoles, Janet, 52–53
Lisbon earthquake (1755), 100–101
Living with Risk, 10
Loesch, Ronnie, 86

Long Term Recovery Organization
    (LTRO), 77
López-Marrero, Tania, 242
Lovato, Grant, 12
Lugas, Jubar, 38
Lumpur Lapindo resistance movement,
    138
Lupi, M., 354n18
Lusi mud volcano, 123–24, 126–27, 129,
    145, 154, 174, 246
    coincidences in, 353n16
    dispute over cause, 123, 128–32
    drilling trigger hypothesis, 128, 131–32,
        355n21
    earthquake trigger hypothesis, 129–32,
        138, 354n18, 355n21
    gases from, 131, 325n1, 355n22
    government recompense to victims and,
        139
    naming of, 352n6
    plans for, 140, 246
    Porong river and, 140–41
    rebuilding efforts and, 356n38
    refugees, protest actions of, 134–35,
        141
Luy, Benhur, 52–53
Lwasa, Shuaib, 235–36

Maeslantkering, 202
Mahler, Jonathan, 81
Make the Road New York (immigrant
    organization), 70
Malanyaon, Corazon, 61
Maldives, 208
Mama Allan, 235–36
Manga, Michael, 129–30
mangroves, 150, 199
    in Bangkok, 150
    in East Java, Indonesia, 141
    ecological benefits of, 243–46
    India and, 244
    Miami Beach and, 179
    Mumbai and, 228, 233
    in Philippines, 62–63
    Sri Lanka and, 244
    tsunamis and, 243–45
Manla, Antonio, 42
Manla, Estanislao, 42
Manuel-Navarrete, David, 320

Marchetti, Victor, 370n41
Marcos, Ferdinand, 335n4
Marston, Richard, 300
Martin, George, 261
Marx, Karl, 93
Marzocchi, Warner, 168–71
Masters, Jeff, 147
Mazzini, Adriano, 130–31, 354n20
McCollum, Lonnie, 279, 281
McDonald, Lynette, 134
McGovern, Gail, 74
McKenna, John, 246
McPhee, John
    The Control of Nature, 199
McQueary, Kristen, 109
Meade, Robin, 268, 272–73
Megadike, 200
Mejia, Lilah, 83
merchants of optimism (on climate
    change), 6–7
Miami, Florida, 178–79
    carbon emissions and, 366n36
    climate change denial and, 186
    climate threats to, 181–84
    founding of, 179
    future of, 194–97
    insurance companies and, 188–89
    1926 hurricane, 180–81, 188
    real estate growth in, 184–86
Miami Beach, Florida
    flooding in, 175, 181, 183
    mangroves and, 179
    pumping water out of, 177–78, 183,
        193
    road raising in, 178
Miami-Dade Climate Change Advisory
    Task Force, 186–87
Miami-Dade County, Florida
    Dutch experts and, 193–94
    flooding and, 177, 182, 184, 189
    income inequality in, 189
Miller, Stephen, 131
Ministry of Emergency Situations
    (EMERCOM), 32–33
Molenaar, Arnoud, 202–4
Montserrat (West Indies), 256–60
    disastourism and, 275–76
    evacuations in, 262, 268–70
    geothermal energy and, 274–75

Hurricane Hugo and, 261
rebuilding, 274
    governmental response to, 259,
        263–71
    recording studio in, 261
    residential tourism in, 260
    resilience and, 273
    volcanic eruptions in, 259, 262
Montserrat Volcano Observatory (MVO),
    264–65
Moore, Oklahoma, 295
Moses, Robert, 81, 83
Moszczynski, Frank, 87
Mowry, Bruce, 175, 177, 183, 193, 195–96
Mumbai, India
    development projects in, 228
    flooding in, 226–30
    mangroves and, 228, 233
    slums in, 227–32

Nandini (co-founder of School for
        Natural Creativity in Guptkashi),
        314–15
Napoles, Janet. See Lim-Napoles, Janet
National Commission on Major Risks
    (Italy), 165
National Flood Insurance Program
    (NFIP), 215–16, 220
natural disasters, 174, 325n6
    anthropogenic climate change and,
        2, 6
    human contribution to, 2–6, 148, 226
    See also disasters
natural-looking hazards, 145
Natural Resources Conservation Service,
    104
Neiman, Susan, 101
Nepal
    earthquakes in, 110, 153, 253, 307, 311,
        315
    landslides in, 153
Netherlands, 201
    floating prisons in, 207
    See also Rotterdam, Netherlands
New Deal programs, 102–3
New Meadowlands (Rebuild by Design
        proposal), 203
New People's Army (NPA), in Philippines,
    54, 62

New York City
    high winds danger in, 341n3
    marathon canceled after Superstorm
        Sandy, 66–68
    waterfront
        climate-proofing plans, 82
        "resilience" and, 81
    See also Superstorm Sandy
New York City Economic Development
        program, 79
New Zealand Earthquake Commission,
    221
Nicoll, Kathleen, 153
Nietzsche, Friedrich, 12
North/South divide
    geoclimatic-hazard insurance and,
        249
    greenhouse emissions and, 251–52
    international disaster relief insurance
        idea and, 252–55
Norris, Ella, 86
Noy, Ilan, 95–96
Nuwer, Rachel, 131–32

Obama, Barack, 62, 82, 106, 109–10, 114,
    287, 296
Occupy Sandy, 68–70, 74, 76–77, 114–15
    mold in houses and, 71
    Staten Island and, 70, 73–74, 76–77
Occupy Wall Street, 68, 70
Ogallala Aquifer, 290–92
Oklahoma Corporation Commission
    (OCC), 156–57
Oklahoma earthquakes, 154–57
    insurance problems for, 156
Oklahoma Geologic Survey (OGS), 156
Oliver-Smith, Anthony, 3, 118–21,
    352n121
Olson, Richard Stuart, 123
Olthius, Koen, 207–8
One57 (skyscraper), 66, 83
One Manhattan Square, 84
Office of Disaster Preparedness and
        Emergency Management
        (ODPEM), 11
Operation Bayanihan (Philippines),
    54–56
operational earthquake forecasting (OEF),
    171

Operation Enduring Freedom, 54
Oreskes, Naomi, 6
Orissa, India
    2013 cyclone, 104
Ormoc, Philippines, 43
Ostrom, Elinor, 14
Ovink, Henk, 194, 201

Pabillo, Broderick (bishop of Manila), 49
Paje, Ramon, 47–49
Palazzo, Steven, 107
Pangloss, Doctor *(Candide),* 100
Paris Agreement (2015), 252–53
Paris Climate Change Conference (2015),
    317
Pascoe, B. Lynn, 133
Patterson, Rolston, 273–74
Pattullo, Polly, 266–67, 269, 273
    Fire from the Mountain, 260
Pelling, Mark, 99, 319–20
Perlas, Nicanor, 50
permafrost melting, 334n76
Peru, 1970 earthquake, 117–18
Peterson, Amy, 75–76
Philippines
    armed rebellion in, 54
    barangays in, 335n4
    calamities in 2013, 57–58
    coping as culture in, 43–44
    earthquakes in, 40, 45, 58–59, 61, 91
    illegal logging in, 48
    mangroves and, 62–63
    political corruption in, 53
    "pork barrel" scandals in, 52–53
    U.S. military in, 55–56, 63
    volcanoes in, 45, 55, 58, 200
    voluntourism in, 47, 50, 63, 246
Philippines Water Code, 42
Phragmites, 343n50
Pielke, Roger, Jr., 6–7
Piñeda, Miguel Ángel, 69
PlaNYC, 79
Plater-Zyberk, Elizabeth, 190–91
Pombal, Marquês de, 100–101, 104
Popp, Aaron, 99, 325n7
Popper, Deborah Epstein, 292
Popper, Frank, 292
Porong mud volcano. *See* Lusi mud
    volcano

Prakash, Gyan
    Mumbai Fables, 373n17
Prasetyo, Dwinanto, 127–28, 132
Prasetyo, Hardi, 127–28, 132, 139–40, 246
Préval, René, 113
prevention spending, elections and,
    105–6
Priority Development Assistance Fund
    (PDAF), in Philippines, 52–53
Putin, Vladimir
    2007 Forest Code Reform, 32–33
Pyne, Stephen, 16–17, 21–24, 35
    Burning Bush, 16
    on fire, 332n50
    Fire on Earth, 29

Qureshi, Aabeda, 229
Qureshi, Mumtz, 229
Qureshi, Zahida, 229

Radovich, Arna, 17–18
    "Mosaic of Loss," 17
Raines, Beth, 24
Rambara, India, 300
Rawat, Harish, 315
Reagan, Ronald, 106
Rebuild by Design competition, 82–83,
    201, 203
Red Cross, 74
Red River Valley flood, 118
Reid, Julian, 319–20
    Resilient Life, 10, 319
reinsurance industry, 216–17
Renojoyo, Indonesia, 136
resilience, 3, 79–80, 320
    as adaptation machine, 11, 13
    critique of, 319
    different forms of, 320
    as ecological concept, 9–10, 18
    economic power and, 324
    fragility in, 14
    in global capitalism, 14
    as managed retreat, 319
    mobility of concept, 20
    questions about nature of, 321–23
    subversive, 320
    sustainability and, distinction between,
        14
    UN definition of, 9–10

resilience doctrine, 318–19, 352n121
  critical analysis of, 10–11
  inversions in, 12–13
  omissions in, 10
  prison crews and, 12–13
  victim blaming and, 352n121
Revi, Aromar, 229
Reyes, Damaris, 83
Ribot, Jesse, 255
"risk society," 325n8
river deltas, shrinking of, 150
Romasko, Victor, 33
Romeo, Donaldson, 259, 263–66, 269–70,
  272, 274–76, 320
Room for the River (Netherlands), 206
Roosevelt, Franklin, 101–3
Rossi, James, 86
Rotterdam, Netherlands, 202–5, 208
Rotterdam Climate Initiative (RTI),
  204
Rousseau, Jean-Jacques, 101
Rundlet, Karen, 185
Russia, prescribed burning in, 35–36
Ruvin, Harvey, 187

Sage, Roza, 115
Sainte-Victoire Mountain
  landslides at, 155
  1989 fire, 143
Sakurajima volcano, 97
Samuels, Richard, 110
Saño, Naderev "Yeb," 45–46, 48, 62, 238,
  339n88
Sati, S.P., 300, 310
Savage, Frank, 265, 271
Schulz, William, 153–54
Scott, Rick, 186
Seattle, Washington
  landslides in, 153
seawalls, problems of, 199–200
Seeber, Leonardo, 159
Sen, Amartya, 344n10
Sendai Framework (UN), 317, 321
Sessa, Frank, 180
Sewall's Point, Florida, 189–90
Sheik, Halema, 229–30
Sheik, Shamshaad, 229
Short, Claire, 268, 271
Shorto, Russell, 202

Shoyajee, Daryoush, 164–65, 173
Sibelius, Kathleen, 279, 286
Siberia, 31–37
  carbon in forests, 31–32
  fire regimes in, 32, 34–36
  fires in, 33–34
  taiga in, 36–37
Sidoarjo Mud Mitigation Agency (BPLS),
  126–27, 132, 139–40
Sigmon, B.M., 102
The Silver Lining: The Benefits of Natural
  Disasters, 13
Simon, Howard, 193
Singh, Chaddar Daar, 304–5
Singh, Harjeet, 254, 317
Sino-Singapore Tianjin Eco-City. See
  Tianjin Eco-City
Sitapur, India, 303
Skidmore, Mark, 92–94
Slumdog Millionaire (film), 229, 232
slums, disaster risks in, 232–33
Sobel, Russell, 106
Soil Conservation Service, 104
Soil Erosion Service, 103
solar radiation management, 211–12,
  371n51
  problems of, 213
Soliman, Dinky, 55
Solnit, Rebecca, 117, 120
  A Paradise Built in Hell, 114, 116
Sonprayag, India, 303–4
Sosa, Rebecca, 187
sōtegai, 96
Sothern, Bill, 71
Soufrière Hills volcano, 257, 259–64, 267,
  272–73
South Asian subcontinent, disasters in, 31
Speth, James Gustav, 8
Spirit of Sandy, 114–15
Sri Lanka, 2004 tsunami, 98
Srinagar, India
  hydroelectric projects and flooding in,
  310
Staten Island, 68, 78
  Occupy Sandy in, 70, 73–74, 76–77
  public buyouts in, 85–86
  Superstorm Sandy and, 85
Steinberg, Ted, 145
  Acts of God, 200

Stern, Nicholas, 7–8
Stern Report, 7
Stewart, Rod, 264–65, 272–73
Stoddard, Philip, 185–86, 188–89, 191–92, 194–97
Stone, Deborah, 122, 198
Strömberg, David, 94
structural functionalism, 386n15
Sukachev Institute of Forest, 32
Sumatra, 2004 tsunami, 244
Superstorm Research Lab, 72, 73, 80
Superstorm Sandy, 64–67, 90, 109, 146, 216, 259
    climate change and, 78–80
    resilience and, 79–80
    Staten Island and, 85
Sweetwater, Florida, 191–93

taiga, climate change and, 36–37
A Tale of Two Sandys, 72, 80
Tao, Rongjia, 199
Tasmania, 25–31
    fire season changes in, 27
Tchebakova, Nadja, 36, 37
"teleconnections," in climate science, 148–49
Third UN World Conference on Disaster Risk Reduction, 317
Thomas, Harry, Jr., 56
Tianjin, China, 209
    explosion, 370n40
    flooding and, 210
Tianjin Binhai New Area, 211
Tianjin Eco-City, 209–11
Tingay, Mark, 128–31, 354n19, 355n21
Tokar, Brian, 9
tornado deaths, mobile homes and, 383n44
tornadoes
    EF scale, 347n50
    warm temperatures and, 146–47
    See also Gainesville, Georgia; "Great Walls of America"; Greensburg, Kansas; Joplin, Missouri; Moore, Oklahoma
Toya, Hideki, 92–94
Trainer, Ted, 8
"transformative adaptation" (Pelling), 320

Traverse Mountains, Utah
    landslides in, 153
Tribal, Adarsh, 313
Tropical Depression Zoraida, 57
Tropical Storm Amang, 63
Tropical Storm Maring, 57
Tropical Storm Sendong (Washi), 47
Tropical Storm Vinta, 57
tsunamis, mangroves and, 243–45
Tuitt, Joseph, 266
Tulfo, Ramon, 49
Tupelo, Mississippi
    1936 tornado, 103
Turnbull, Malcolm, 331n48
Turner, Edith, 116–17
Turner, Victor, 116, 118–19
    The Ritual Process, 117
Typhoon Labuyo, 57
Typhoon Odette, 57
Typhoon Pablo (Bopha), 40, 43, 45–49, 54, 57, 60
    agricultural damage from 51–52
    UN appeals for aid and, 51–51
Typhoon Ruby, 63
typhoons, categorization of, 334–35n3
Typhoon Santi, 57
Typhoon Yolanda (Haiyan), 38–42, 45, 57–58, 60, 115, 146, 317
    aftermath of, 62–63
    rebuilding after, 41–42

Uganda. See Kampala, Uganda; Kamwokya slum
UNISDR (United Nations International Strategy for Disaster Reduction), 4
universal/social disaster insurance for US, idea for, 219–23, 252–53
    arguments against, 221–22
    arguments for, 220–23
UN Office for Disaster Risk Reduction, 10
US Forest Service, firefighting budget, 147–48
Utomo, Paring Waluyo, 138
Uttarakhand state, India, 313–14
    abandoned construction sites in, 312–13
    climate change and, 306–8
    economic losses in, 312–13
    earthquakes in, 307–8, 311, 315
    floods and, 306, 310–11

hydroelectric projects in, 310–11
landslides in, 308–9, 315
post-disaster strategy, 314
roadbuilding and, 308–9
tourism in, 308–9, 312
Uy, Arturo, 48

Vaiont dam disaster, in Italy, 152–53
Vardeman, Johnny, 102
Venice, Italy, 194
Vilsack, Tom, 147–48
Vittorini, Vincenzo, 160, 167, 173
VJ Jingo, 234
volcanoes
    in Iceland, 199–200
    Lusi mud volcano, 123–24, 126–35,
        140–41, 145, 154, 174, 246, 354n18,
        355n21, 355n22, 355n22, 356n28
    in Monserrat (West Indies), 259, 262
    in Philippines, 45, 55, 58, 200
    populations near, 267
Voltaire, 88, 101
    Candide, 100

Wallach, Daniel, 279, 281, 287, 289
Wanless, Harold, 182–83, 192, 194–97, 205
Water Square Benthemplain, 202–3
Weather Underground, 147
Wedel, Ruth Ann, 278, 283, 288–89
Weekes, Yvonne, 269–70
    Volcano, 269

Weiler, Veronica, 86
Wenchuan earthquakes, 158–59, 247
Whight, Sandra, 27
White, Stacey Swearingen, 282, 288
Whitehead, Andrew, 295–86
Widanigrum, Wina, 134
Widodo, Joko (Jokowi), 139
wildland/urban interface (WUI), in
    Australia, 22
Wilhelminakade (Rotterdam), 204–5
Williams, Sharon, 257
Wilson, Frederica, 217
Wine, Bobi, 233
Winmalee, Australia
    fires in, 20
Wisner, Ben, 242
Wolfarth, Robert, 193
Worster, Donald
    Dust Bowl, 292

Yager, Kristy, 361n56
Yaroshenko, Alexey, 35
Yudhoyono, Susilo Bambang, 133–34, 137
Yungay, Peru
    1970 earthquake, 119

Ziegler, Alan, 309
Zipingpu dam, 152, 362n72
Zones Urbaines Sensibles (ZUS), 203
Zotino Tall Tower Observation Facility
    (ZOTTO), 31

ABOUT THE AUTHORS

Stan Cox is research coordinator at The Land Institute in Salina, Kansas, where he lives. His books include *Losing Our Cool* and *Any Way You Slice It* (both published by The New Press).

Paul Cox is an anthropologist and writer based in Copenhagen, Denmark. His work covers development and disaster around the world, with publications strewn all the way from the journal *Disasters* to the *New Inquiry* and *Hyperallergic*.

PUBLISHING IN THE PUBLIC INTEREST

Thank you for reading this book published by The New Press. The New Press is a nonprofit, public interest publisher. New Press books and authors play a crucial role in sparking conversations about the key political and social issues of our day.

We hope you enjoyed this book and that you will stay in touch with The New Press. Here are a few ways to stay up to date with our books, events, and the issues we cover:

- Sign up at www.thenewpress.com/subscribe to receive updates on New Press authors and issues and to be notified about local events
- Like us on Facebook: www.facebook.com/newpressbooks
- Follow us on Twitter: www.twitter.com/thenewpress
- Please consider buying New Press books for yourself; for friends and family; or to donate to schools, libraries, community centers, prison libraries, and other organizations involved with the issues our authors write about.

The New Press is a 501(c)(3) nonprofit organization. You can also support our work with a tax-deductible gift by visiting www.thenewpress.com/donate.